Matrix
Analysis
and
Applied
Linear
Algebra

Matrix Analysis and Applied Linear Algebra

Solutions Manual

Carl Meyer

North Carolina State University
Raleigh, North Carolina

Society for Industrial and Applied Mathematics
Philadelphia

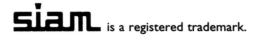 is a registered trademark.

Contents

Solutions for Chapter 1 1

Solutions for Chapter 2 5

Solutions for Chapter 3 11

Solutions for Chapter 4 27

Solutions for Chapter 5 51

Solutions for Chapter 6 115

Solutions for Chapter 7 125

Solutions for Chapter 8 167

Solutions for Chapter 1

Solutions for exercises in section 1. 2

1.2.1. $(1, 0, 0)$

1.2.2. $(1, 2, 3)$

1.2.3. $(1, 0, -1)$

1.2.4. $(-1/2, 1/2, 0, 1)$

1.2.5. $\begin{pmatrix} 2 & -4 & 3 \\ 4 & -7 & 4 \\ 5 & -8 & 4 \end{pmatrix}$

1.2.6. Every row operation is reversible. In particular the "inverse" of any row operation is again a row operation of the same type.

1.2.7. $\frac{\pi}{2}, \pi, 0$

1.2.8. The third equation in the triangularized form is $0x_3 = 1$, which is impossible to solve.

1.2.9. The third equation in the triangularized form is $0x_3 = 0$, and all numbers are solutions. This means that you can start the back substitution with any value whatsoever and consequently produce infinitely many solutions for the system.

1.2.10. $\alpha = -3$, $\beta = \frac{11}{2}$, and $\gamma = -\frac{3}{2}$

1.2.11. (a) If $x_i = $ the number initially in chamber $\#i$, then

$$.4x_1 + 0x_2 + 0x_3 + .2x_4 = 12$$
$$0x_1 + .4x_2 + .3x_3 + .2x_4 = 25$$
$$0x_1 + .3x_2 + .4x_3 + .2x_4 = 26$$
$$.6x_1 + .3x_2 + .3x_3 + .4x_4 = 37$$

and the solution is $x_1 = 10$, $x_2 = 20$, $x_3 = 30$, and $x_4 = 40$.
(b) 16, 22, 22, 40

1.2.12. To interchange rows i and j, perform the following sequence of Type II and Type III operations.

$$R_j \leftarrow R_j + R_i \quad \text{(replace row } j \text{ by the sum of row } j \text{ and } i)$$
$$R_i \leftarrow R_i - R_j \quad \text{(replace row } i \text{ by the difference of row } i \text{ and } j)$$
$$R_j \leftarrow R_j + R_i \quad \text{(replace row } j \text{ by the sum of row } j \text{ and } i)$$
$$R_i \leftarrow -R_i \quad \text{(replace row } i \text{ by its negative)}$$

1.2.13. (a) This has the effect of interchanging the order of the unknowns— x_j and x_k are permuted. (b) The solution to the new system is the same as the solution to the old system except that the solution for the jth unknown of the new system is $\hat{x}_j = \frac{1}{\alpha}x_j$. This has the effect of "changing the units" of the jth unknown.   (c) The solution to the new system is the same as the solution for the old system except that the solution for the kth unknown in the new system is $\hat{x}_k = x_k - \alpha x_j$.

1.2.14. $h_{ij} = \frac{1}{i+j-1}$

1.2.16. If $\mathbf{x} = \begin{pmatrix} x_1 \\ x_2 \\ \vdots \\ x_m \end{pmatrix}$ and $\mathbf{y} = \begin{pmatrix} y_1 \\ y_2 \\ \vdots \\ y_m \end{pmatrix}$ are two different solutions, then

$$\mathbf{z} = \frac{\mathbf{x} + \mathbf{y}}{2} = \begin{pmatrix} \frac{x_1+y_1}{2} \\ \frac{x_2+y_2}{2} \\ \vdots \\ \frac{x_m+y_m}{2} \end{pmatrix}$$

is a third solution different from both $\mathbf{x}$ and $\mathbf{y}$.

Solutions for exercises in section 1. 3

1.3.1. $(1, 0, -1)$

1.3.2. $(2, -1, 0, 0)$

1.3.3. $\begin{pmatrix} 1 & 1 & 1 \\ 1 & 2 & 2 \\ 1 & 2 & 3 \end{pmatrix}$

Solutions for exercises in section 1. 4

1.4.2. Use $y'(t_k) = y'_k \approx \dfrac{y_{k+1} - y_{k-1}}{2h}$ and $y''(t_k) = y''_k \approx \dfrac{y_{k-1} - 2y_k + y_{k+1}}{h^2}$ to write

$$f(t_k) = f_k = y''_k - y'_k \approx \frac{2y_{k-1} - 4y_k + 2y_{k+1}}{2h^2} - \frac{hy_{k+1} - hy_{k-1}}{2h^2}, \quad k = 1, 2, \ldots, n,$$

with $y_0 = y_{n+1} = 0$. These discrete approximations form the tridiagonal system

$$\begin{pmatrix} -4 & 2-h & & & \\ 2+h & -4 & 2-h & & \\ & \ddots & \ddots & \ddots & \\ & & 2+h & -4 & 2-h \\ & & & 2+h & -4 \end{pmatrix} \begin{pmatrix} y_1 \\ y_2 \\ \vdots \\ y_{n-1} \\ y_n \end{pmatrix} = 2h^2 \begin{pmatrix} f_1 \\ f_2 \\ \vdots \\ f_{n-1} \\ f_n \end{pmatrix}.$$

Solutions for exercises in section 1. 5

1.5.1. (a) $(0, -1)$ (c) $(1, -1)$ (e) $\left(\frac{1}{1.001}, \frac{-1}{1.001}\right)$

1.5.2. (a) $(0, 1)$ (b) $(2, 1)$ (c) $(2, 1)$ (d) $\left(\frac{2}{1.0001}, \frac{1.0003}{1.0001}\right)$

1.5.3. Without PP: $(1.01, 1.03)$ With PP: $(1, 1)$ Exact: $(1, 1)$

1.5.4. (a)
$$\left(\begin{array}{ccc|c} 1 & .500 & .333 & .333 \\ .500 & .333 & .250 & .333 \\ .333 & .250 & .200 & .200 \end{array}\right) \longrightarrow \left(\begin{array}{ccc|c} 1 & .500 & .333 & .333 \\ 0 & .083 & .083 & .166 \\ 0 & .083 & .089 & .089 \end{array}\right)$$

$$\longrightarrow \left(\begin{array}{ccc|c} 1 & .500 & .333 & .333 \\ 0 & .083 & .083 & .166 \\ 0 & 0 & .006 & -.077 \end{array}\right) \quad z = -.077/.006 = -12.8,$$

$$y = (.166 - .083z)/.083 = 14.8, \quad x = .333 - (.5y + .333z) = -2.81$$

(b)
$$\left(\begin{array}{ccc|c} 1 & .500 & .333 & .333 \\ .500 & .333 & .250 & .333 \\ .333 & .250 & .200 & .200 \end{array}\right) \longrightarrow \left(\begin{array}{ccc|c} 1 & .500 & .333 & .333 \\ 1 & .666 & .500 & .666 \\ 1 & .751 & .601 & .601 \end{array}\right)$$

$$\longrightarrow \left(\begin{array}{ccc|c} 1 & .500 & .333 & .333 \\ 0 & .166 & .167 & .333 \\ 0 & .251 & .268 & .268 \end{array}\right) \longrightarrow \left(\begin{array}{ccc|c} 1 & .500 & .333 & .333 \\ 0 & .251 & .268 & .268 \\ 0 & .166 & .167 & .333 \end{array}\right)$$

$$\longrightarrow \left(\begin{array}{ccc|c} 1 & .500 & .333 & .333 \\ 0 & .251 & .268 & .268 \\ 0 & 0 & -.01 & .156 \end{array}\right) \quad z = -.156/.01 = -15.6,$$

$$y = (.268 - .268z)/.251 = 17.7, \quad x = .333 - (.5y + .333z) = -3.33$$

(c)
$$\left(\begin{array}{ccc|c} 1 & .500 & .333 & .333 \\ .500 & .333 & .250 & .333 \\ .333 & .250 & .200 & .200 \end{array}\right) \longrightarrow \left(\begin{array}{ccc|c} 1 & .500 & .333 & .333 \\ 1 & .666 & .500 & .666 \\ 1 & .751 & .601 & .601 \end{array}\right)$$

$$\longrightarrow \left(\begin{array}{ccc|c} 1 & .500 & .333 & .333 \\ 0 & .166 & .167 & .333 \\ 0 & .251 & .268 & .268 \end{array}\right) \longrightarrow \left(\begin{array}{ccc|c} 1 & .500 & .333 & .333 \\ 0 & .994 & 1 & 1.99 \\ 0 & .937 & 1 & 1 \end{array}\right)$$

$$\longrightarrow \left(\begin{array}{ccc|c} 1 & .500 & .333 & .333 \\ 0 & .994 & 1 & 1.99 \\ 0 & 0 & .057 & -.880 \end{array}\right) \quad z = -.88/.057 = -15.4,$$

$$y = (1.99 - z)/.994 = 17.5, \quad x = .333 - (.5y + .333z) = -3.29$$

(d) $\quad x = -3, \quad y = 16, \quad z = -14$

1.5.5. (a)

$$.0055x + .095y + 960z = 5000$$
$$.0011x + .\ 01y + 112z = 600$$
$$.0093x + .025y + 560z = 3000$$

(b) 3-digit solution = (55, 900 lbs. silica, 8, 600 lbs. iron, 4.04 lbs. gold).
Exact solution (to 10 digits) = (56, 753.68899, 8, 626.560726, 4.029511918). The
relative error (rounded to 3 digits) is $e_r = 1.49 \times 10^{-2}$.

(c) Let $u = x/2000$, $v = y/1000$, and $w = 12z$ to obtain the system

$$11u + 95v + \quad 80w = 5000$$
$$2.2u + 10v + 9.33w = 600$$
$$18.6u + 25v + 46.7w = 3000.$$

(d) 3-digit solution = (28.5 tons silica, 8.85 half-tons iron, 48.1 troy oz. gold).
Exact solution (to 10 digits) = (28.82648317, 8.859282804, 48.01596023). The
relative error (rounded to 3 digits) is $e_r = 5.95 \times 10^{-3}$. So, partial pivoting
applied to the column-scaled system yields higher relative accuracy than partial
pivoting applied to the unscaled system.

1.5.6. (a) $(-8.1, -6.09)$ = 3-digit solution with partial pivoting but no scaling.
(b) No! Scaled partial pivoting produces the exact solution—the same as with
complete pivoting.

1.5.7. (a) 2^{n-1} (b) 2

(c) This is a famous example that shows that there are indeed cases where partial pivoting will fail due to the large growth of some elements during elimination,
but complete pivoting will be successful because all elements remain relatively
small and of the same order of magnitude.

1.5.8. Use the fact that with partial pivoting no multiplier can exceed 1 together with
the triangle inequality $|\alpha + \beta| \leq |\alpha| + |\beta|$ and proceed inductively.

Solutions for exercises in section 1. 6

1.6.1. (a) There are no 5-digit solutions. (b) This doesn't help—there are now infinitely
many 5-digit solutions. (c) 6-digit solution = $(1.23964, -1.3)$ and exact solution
= $(1, -1)$ (d) $r_1 = r_2 = 0$ (e) $r_1 = -10^{-6}$ and $r_2 = 10^{-7}$ (f) Even if computed
residuals are 0, you can't be sure you have the exact solution.

1.6.2. (a) $(1, -1.0015)$ (b) Ill-conditioning guarantees that the solution will be very
sensitive to *some* small perturbation but not necessarily to *every* small perturbation. It is usually difficult to determine beforehand those perturbations for which
an ill-conditioned system will not be sensitive, so one is forced to be pessimistic
whenever ill-conditioning is suspected.

1.6.3. (a) $m_1(5) = m_2(5) = -1.2519$, $m_1(6) = -1.25187$, and $m_2(6) = -1.25188$
(c) An optimally well-conditioned system represents orthogonal (i.e., perpendicular) lines, planes, etc.

1.6.4. They rank as (b) = Almost optimally well-conditioned. (a) = Moderately well-
conditioned. (c) = Badly ill-conditioned.

1.6.5. Original solution = $(1, 1, 1)$. Perturbed solution = $(-238, 490, -266)$. System
is ill-conditioned.

Solutions for Chapter 2

Solutions for exercises in section 2. 1

2.1.1. (a) $\begin{pmatrix} 1 & 2 & 3 & 3 \\ 0 & 2 & 1 & 0 \\ 0 & 0 & 0 & 3 \end{pmatrix}$ is one possible answer. Rank = 3 and the basic columns

are $\{\mathbf{A}_{*1}, \mathbf{A}_{*2}, \mathbf{A}_{*4}\}$. (b) $\begin{pmatrix} 1 & 2 & 3 \\ 0 & 2 & 2 \\ 0 & 0 & -8 \\ 0 & 0 & 0 \\ 0 & 0 & 0 \end{pmatrix}$ is one possible answer. Rank = 3 and

every column in $\mathbf{A}$ is basic.

(c) $\begin{pmatrix} 2 & 1 & 1 & 3 & 0 & 4 & 1 \\ 0 & 0 & 2 & -2 & 1 & -3 & 3 \\ 0 & 0 & 0 & 0 & -1 & 3 & -1 \\ 0 & 0 & 0 & 0 & 0 & 0 & 0 \\ 0 & 0 & 0 & 0 & 0 & 0 & 0 \\ 0 & 0 & 0 & 0 & 0 & 0 & 0 \end{pmatrix}$ is one possible answer. The rank is 3, and

the basic columns are $\{\mathbf{A}_{*1}, \mathbf{A}_{*3}, \mathbf{A}_{*5}\}$.

2.1.2. (c) and (d) are in row echelon form.

2.1.3. (a) Since any row or column can contain at most one pivot, the number of pivots cannot exceed the number of rows nor the number of columns. (b) A zero row cannot contain a pivot. (c) If one row is a multiple of another, then one of them can be annihilated by the other to produce a zero row. Now the result of the previous part applies. (d) One row can be annihilated by the associated combination of row operations. (e) If a column is zero, then there are fewer than n basic columns because each basic column must contain a pivot.

2.1.4. (a) $rank\,(\mathbf{A}) = 3$ (b) 3-digit $rank\,(\mathbf{A}) = 2$ (c) With PP, 3-digit $rank\,(\mathbf{A}) = 3$

2.1.5. 15

2.1.6. (a) No, consider the form $\left(\begin{array}{ccc|c} * & * & * & * \\ 0 & 0 & 0 & 0 \\ 0 & 0 & 0 & * \end{array} \right)$ (b) Yes—in fact, $\mathbf{E}$ is a row echelon form obtainable from $\mathbf{A}$.

Solutions for exercises in section 2. 2

2.2.1. (a) $\begin{pmatrix} 1 & 0 & 2 & 0 \\ 0 & 1 & \frac{1}{2} & 0 \\ 0 & 0 & 0 & 1 \end{pmatrix}$ and $\mathbf{A}_{*3} = 2\mathbf{A}_{*1} + \frac{1}{2}\mathbf{A}_{*2}$

$$\text{(b)} \quad \begin{pmatrix} 1 & \frac{1}{2} & 0 & 2 & 0 & 2 & 0 \\ 0 & 0 & 1 & -1 & 0 & 0 & 1 \\ 0 & 0 & 0 & 0 & 1 & -3 & 1 \\ 0 & 0 & 0 & 0 & 0 & 0 & 0 \\ 0 & 0 & 0 & 0 & 0 & 0 & 0 \\ 0 & 0 & 0 & 0 & 0 & 0 & 0 \end{pmatrix} \quad \text{and}$$

$$\mathbf{A}_{*2} = \tfrac{1}{2}\mathbf{A}_{*1}, \quad \mathbf{A}_{*4} = 2\mathbf{A}_{*1} - \mathbf{A}_{*3}, \quad \mathbf{A}_{*6} = 2\mathbf{A}_{*1} - 3\mathbf{A}_{*5}, \quad \mathbf{A}_{*7} = \mathbf{A}_{*3} + \mathbf{A}_{*5}$$

2.2.2. No.

2.2.3. The same would have to hold in $\mathbf{E_A}$, and there you can see that this means not all columns can be basic. Remember, $rank\,(\mathbf{A}) =$ number of basic columns.

2.2.4. (a) $\begin{pmatrix} 1 & 0 & 0 \\ 0 & 1 & 0 \\ 0 & 0 & 1 \end{pmatrix}$ (b) $\begin{pmatrix} 1 & 0 & -1 \\ 0 & 1 & 2 \\ 0 & 0 & 0 \end{pmatrix}$ $\mathbf{A}_{*3}$ is almost a combination of $\mathbf{A}_{*1}$ and $\mathbf{A}_{*2}$. In particular, $\mathbf{A}_{*3} \approx -\mathbf{A}_{*1} + 2\mathbf{A}_{*2}$.

2.2.5. $\mathbf{E}_{*1} = 2\mathbf{E}_{*2} - \mathbf{E}_{*3}$ and $\mathbf{E}_{*2} = \tfrac{1}{2}\mathbf{E}_{*1} + \tfrac{1}{2}\mathbf{E}_{*3}$

Solutions for exercises in section 2. 3

2.3.1. (a), (b)—There is no need to do any arithmetic for this one because the right-hand side is entirely zero so that you know $(0,0,0)$ is automatically one solution. (d), (f)

2.3.3. It is always true that $rank\,(\mathbf{A}) \le rank[\mathbf{A}|\mathbf{b}] \le m$. Since $rank\,(\mathbf{A}) = m$, it follows that $rank[\mathbf{A}|\mathbf{b}] = rank\,(\mathbf{A})$.

2.3.4. Yes—Consistency implies that $\mathbf{b}$ and $\mathbf{c}$ are each combinations of the basic columns in $\mathbf{A}$. If $\mathbf{b} = \sum \beta_i \mathbf{A}_{*b_i}$ and $\mathbf{c} = \sum \gamma_i \mathbf{A}_{*b_i}$ where the $\mathbf{A}_{*b_i}$'s are the basic columns, then $\mathbf{b} + \mathbf{c} = \sum(\beta_i + \gamma_i)\mathbf{A}_{*b_i} = \sum \xi_i \mathbf{A}_{*b_i}$, where $\xi_i = \beta_i + \gamma_i$ so that $\mathbf{b} + \mathbf{c}$ is also a combination of the basic columns in $\mathbf{A}$.

2.3.5. Yes—because the 4×3 system $\alpha + \beta x_i + \gamma x_i^2 = y_i$ obtained by using the four given points (x_i, y_i) is consistent.

2.3.6. The system is inconsistent using 5-digits but consistent when 6-digits are used.

2.3.7. If x, y, and z denote the number of pounds of the respective brands applied, then the following constraints must be met.

$$\text{total \# units of phosphorous} = 2x + \ y + z = 10$$
$$\text{total \# units of potassium} = 3x + 3y \qquad = 9$$
$$\text{total \# units of nitrogen} = 5x + 4y + z = 19$$

Since this is a consistent system, the recommendation can be satisfied exactly. Of course, the solution tells how much of each brand to apply.

2.3.8. No—if one or more such rows were ever present, how could you possibly eliminate all of them with row operations? You could eliminate all but one, but then there is no way to eliminate the last remaining one, and hence it would have to appear in the final form.

Solutions for exercises in section 2. 4

2.4.1. (a) $x_2 \begin{pmatrix} -2 \\ 1 \\ 0 \\ 0 \end{pmatrix} + x_4 \begin{pmatrix} -1 \\ 0 \\ -1 \\ 1 \end{pmatrix}$ (b) $y \begin{pmatrix} -\frac{1}{2} \\ 1 \\ 0 \end{pmatrix}$ (c) $x_3 \begin{pmatrix} -1 \\ -1 \\ 1 \\ 0 \end{pmatrix} + x_4 \begin{pmatrix} -1 \\ 1 \\ 0 \\ 1 \end{pmatrix}$

(d) The trivial solution is the only solution.

2.4.2. $\begin{pmatrix} 0 \\ 0 \\ 0 \end{pmatrix}$ and $\begin{pmatrix} 1 \\ -\frac{1}{2} \\ 0 \end{pmatrix}$

2.4.3. $x_2 \begin{pmatrix} -2 \\ 1 \\ 0 \\ 0 \\ 0 \end{pmatrix} + x_4 \begin{pmatrix} -2 \\ 0 \\ -1 \\ 1 \\ 0 \end{pmatrix}$

2.4.4. $rank\,(\mathbf{A}) = 3$

2.4.5. (a) 2—because the maximum rank is 4.   (b) 5—because the minimum rank is 1.

2.4.6. Because $r = rank\,(\mathbf{A}) \leq m < n \implies n - r > 0$.

2.4.7. There are many different correct answers. One approach is to answer the question "What must $\mathbf{E_A}$ look like?" The form of the general solution tells you that $rank\,(\mathbf{A}) = 2$ and that the first and third columns are basic. Consequently,

$$\mathbf{E_A} = \begin{pmatrix} 1 & \alpha & 0 & \beta \\ 0 & 0 & 1 & \gamma \\ 0 & 0 & 0 & 0 \end{pmatrix}$$ so that $x_1 = -\alpha x_2 - \beta x_4$ and $x_3 = -\gamma x_4$ gives rise

to the general solution $x_2 \begin{pmatrix} -\alpha \\ 1 \\ 0 \\ 0 \end{pmatrix} + x_4 \begin{pmatrix} -\beta \\ 0 \\ -\gamma \\ 1 \end{pmatrix}$. Therefore, $\alpha = 2$, $\beta = 3$,

and $\gamma = -2$. Any matrix $\mathbf{A}$ obtained by performing row operations to $\mathbf{E_A}$ will be the coefficient matrix for a homogeneous system with the desired general solution.

2.4.8. If $\sum_i x_{f_i} \mathbf{h}_i$ is the general solution, then there must exist scalars α_i and β_i such that $\mathbf{c}_1 = \sum_i \alpha_i \mathbf{h}_i$ and $\mathbf{c}_2 = \sum_i \beta_i \mathbf{h}_i$. Therefore, $\mathbf{c}_1 + \mathbf{c}_2 = \sum_i (\alpha_i + \beta_i)\mathbf{h}_i$, and this shows that $\mathbf{c}_1 + \mathbf{c}_2$ is the solution obtained when the free variables x_{f_i} assume the values $x_{f_i} = \alpha_i + \beta_i$.

Solutions for exercises in section 2. 5

2.5.1. (a) $\begin{pmatrix} 1 \\ 0 \\ 2 \\ 0 \end{pmatrix} + x_2 \begin{pmatrix} -2 \\ 1 \\ 0 \\ 0 \end{pmatrix} + x_4 \begin{pmatrix} -1 \\ 0 \\ -1 \\ 1 \end{pmatrix}$ (b) $\begin{pmatrix} 1 \\ 0 \\ 2 \end{pmatrix} + y \begin{pmatrix} -\frac{1}{2} \\ 1 \\ 0 \end{pmatrix}$

(c) $\begin{pmatrix} 2 \\ -1 \\ 0 \\ 0 \end{pmatrix} + x_3 \begin{pmatrix} -1 \\ -1 \\ 1 \\ 0 \end{pmatrix} + x_4 \begin{pmatrix} -1 \\ 1 \\ 0 \\ 1 \end{pmatrix}$ (d) $\begin{pmatrix} 3 \\ -3 \\ -1 \end{pmatrix}$

2.5.2. From Example 2.5.1, the solutions of the linear equations are:

$$x_1 = 1 - x_3 - 2x_4$$
$$x_2 = 1 - x_3$$
$$x_3 \text{ is free}$$
$$x_4 \text{ is free}$$
$$x_5 = -1$$

Substitute these into the two constraints to get $x_3 = \pm 1$ and $x_4 = \pm 1$. Thus there are exactly four solutions:

$$\left\{ \begin{pmatrix} -2 \\ 0 \\ 1 \\ 1 \\ -1 \end{pmatrix}, \begin{pmatrix} 2 \\ 0 \\ 1 \\ -1 \\ -1 \end{pmatrix}, \begin{pmatrix} 0 \\ 2 \\ -1 \\ 1 \\ -1 \end{pmatrix}, \begin{pmatrix} 4 \\ 2 \\ -1 \\ -1 \\ -1 \end{pmatrix} \right\}$$

2.5.3. (a) $\{(3,0,4),\ (2,1,5),\ (1,2,6),\ (0,3,7)\}$ See the solution to Exercise 2.3.7 for the underlying system.   (b) $(3,0,4)$ costs \$15 and is least expensive.

2.5.4. (a) Consistent for all α.   (b) $\alpha \neq 3$, in which case the solution is $(1,-1,0)$.

(c) $\alpha = 3$, in which case the general solution is $\begin{pmatrix} 1 \\ -1 \\ 0 \end{pmatrix} + z \begin{pmatrix} 0 \\ -\frac{3}{2} \\ 1 \end{pmatrix}$.

2.5.5. No

2.5.6.

$$\mathbf{E_A} = \begin{pmatrix} 1 & 0 & \cdots & 0 \\ 0 & 1 & \cdots & 0 \\ \vdots & \vdots & \ddots & \vdots \\ 0 & 0 & \cdots & 1 \\ 0 & 0 & \cdots & 0 \\ \vdots & \vdots & \cdots & \vdots \\ 0 & 0 & \cdots & 0 \end{pmatrix}_{m \times n}$$

2.5.7. See the solution to Exercise 2.4.7.

2.5.8. (a) $\begin{pmatrix} -.3976 \\ 0 \\ 1 \end{pmatrix} + y \begin{pmatrix} -.7988 \\ 1 \\ 0 \end{pmatrix}$ (b) There are no solutions in this case.

(c) $\begin{pmatrix} 1.43964 \\ -2.3 \\ 1 \end{pmatrix}$

Solutions for exercises in section 2. 6

2.6.1. (a) $(1/575)(383, 533, 261, 644, -150, -111)$

2.6.2. $(1/211)(179, 452, 36)$

2.6.3. $(18, 10)$

2.6.4. (a) 4   (b) 6   (c) 7 loops but only 3 simple loops.   (d) Show that $rank\,([\mathbf{A}|\mathbf{b}]) = 3$ (g) $5/6$

I fear explanations explanatory of things explained.
— *Abraham Lincoln* (1809–1865)

Solutions for Chapter 3

Solutions for exercises in section 3. 2

3.2.1. (a) $\mathbf{X} = \begin{pmatrix} 0 & 1 \\ 2 & 3 \end{pmatrix}$ (b) $x = -\frac{1}{2}$, $y = -6$, and $z = 0$

3.2.2. (a) Neither   (b) Skew symmetric   (c) Symmetric   (d) Neither

3.2.3. The 3×3 zero matrix trivially satisfies all conditions, and it is the only possible answer for part (a). The only possible answers for (b) are real symmetric matrices. There are many nontrivial possibilities for (c).

3.2.4. $\mathbf{A} = \mathbf{A}^T$ and $\mathbf{B} = \mathbf{B}^T \implies (\mathbf{A} + \mathbf{B})^T = \mathbf{A}^T + \mathbf{B}^T = \mathbf{A} + \mathbf{B}$. Yes—the skew-symmetric matrices are also closed under matrix addition.

3.2.5. (a) $\mathbf{A} = -\mathbf{A}^T \implies a_{ij} = -a_{ji}$. If $i = j$, then $a_{jj} = -a_{jj} \implies a_{jj} = 0$.

(b) $\mathbf{A} = -\mathbf{A}^* \implies a_{ij} = -\overline{a_{ji}}$. If $i = j$, then $a_{jj} = -\overline{a_{jj}}$. Write $a_{jj} = x + iy$ to see that $a_{jj} = -\overline{a_{jj}} \implies x + iy = -x + iy \implies x = 0 \implies a_{jj}$ is pure imaginary.

(c) $\mathbf{B}^* = (i\mathbf{A})^* = -i\mathbf{A}^* = -i\overline{\mathbf{A}}^T = -i\mathbf{A}^T = -i\mathbf{A} = -\mathbf{B}$.

3.2.6. (a) Let $\mathbf{S} = \mathbf{A} + \mathbf{A}^T$ and $\mathbf{K} = \mathbf{A} - \mathbf{A}^T$. Then $\mathbf{S}^T = \mathbf{A}^T + \mathbf{A}^{TT} = \mathbf{A}^T + \mathbf{A} = \mathbf{S}$. Likewise, $\mathbf{K}^T = \mathbf{A}^T - \mathbf{A}^{TT} = \mathbf{A}^T - \mathbf{A} = -\mathbf{K}$.

(b) $\mathbf{A} = \frac{\mathbf{S}}{2} + \frac{\mathbf{K}}{2}$ is one such decomposition. To see it is unique, suppose $\mathbf{A} = \mathbf{X} + \mathbf{Y}$, where $\mathbf{X} = \mathbf{X}^T$ and $\mathbf{Y} = -\mathbf{Y}^T$. Thus, $\mathbf{A}^T = \mathbf{X}^T + \mathbf{Y}^T = \mathbf{X} - \mathbf{Y} \implies \mathbf{A} + \mathbf{A}^T = 2\mathbf{X}$, so that $\mathbf{X} = \frac{\mathbf{A} + \mathbf{A}^T}{2} = \frac{\mathbf{S}}{2}$. A similar argument shows that $\mathbf{Y} = \frac{\mathbf{A} - \mathbf{A}^T}{2} = \frac{\mathbf{K}}{2}$.

3.2.7. (a) $[(\mathbf{A} + \mathbf{B})^*]_{ij} = \overline{[\mathbf{A} + \mathbf{B}]}_{ji} = \overline{[\mathbf{A} + \mathbf{B}]}_{ji} = \overline{[\mathbf{A}]}_{ji} + \overline{[\mathbf{B}]}_{ji} = [\mathbf{A}^*]_{ij} + [\mathbf{B}^*]_{ij} = [\mathbf{A}^* + \mathbf{B}^*]_{ij}$

(b) $[(\alpha\mathbf{A})^*]_{ij} = \overline{[\alpha\mathbf{A}]}_{ji} = \overline{[\bar{\alpha}\mathbf{A}]}_{ji} = \bar{\alpha}\overline{[\mathbf{A}]}_{ji} = \bar{\alpha}[\mathbf{A}^*]_{ij}$

3.2.8. $k \begin{pmatrix} 1 & -1 & 0 & \cdots & 0 & 0 \\ -1 & 2 & -1 & \cdots & 0 & 0 \\ 0 & -1 & 2 & \cdots & 0 & 0 \\ \vdots & \vdots & \vdots & \ddots & \vdots & \vdots \\ 0 & 0 & 0 & \cdots & 2 & -1 \\ 0 & 0 & 0 & \cdots & -1 & 1 \end{pmatrix}$

Solutions for exercises in section 3. 3

3.3.1. Functions (b) and (f) are linear. For example, to check if (b) is linear, let
$$\mathbf{A} = \begin{pmatrix} a_1 \\ a_2 \end{pmatrix} \text{ and } \mathbf{B} = \begin{pmatrix} b_1 \\ b_2 \end{pmatrix}, \text{ and check if } f(\mathbf{A} + \mathbf{B}) = f(\mathbf{A}) + f(\mathbf{B}) \text{ and}$$

$f(\alpha \mathbf{A}) = \alpha f(\mathbf{A})$. Do so by writing

$$f(\mathbf{A} + \mathbf{B}) = f\left(\begin{array}{c} a_1 + b_1 \\ a_2 + b_2 \end{array}\right) = \left(\begin{array}{c} a_2 + b_2 \\ a_1 + b_1 \end{array}\right) = \left(\begin{array}{c} a_2 \\ a_1 \end{array}\right) + \left(\begin{array}{c} b_2 \\ b_1 \end{array}\right) = f(\mathbf{A}) + f(\mathbf{B}),$$

$$f(\alpha \mathbf{A}) = f\left(\begin{array}{c} \alpha a_1 \\ \alpha a_2 \end{array}\right) = \left(\begin{array}{c} \alpha a_2 \\ \alpha a_1 \end{array}\right) = \alpha \left(\begin{array}{c} a_2 \\ a_1 \end{array}\right) = \alpha f(\mathbf{A}).$$

3.3.2. Write $f(\mathbf{x}) = \sum_{i=1}^{n} \xi_i x_i$. For all points $\mathbf{x} = \left(\begin{array}{c} x_1 \\ x_2 \\ \vdots \\ x_n \end{array}\right)$ and $\mathbf{y} = \left(\begin{array}{c} y_1 \\ y_2 \\ \vdots \\ y_n \end{array}\right)$, and for all scalars α, it is true that

$$f(\alpha \mathbf{x} + \mathbf{y}) = \sum_{i=1}^{n} \xi_i(\alpha x_i + y_i) = \sum_{i=1}^{n} \xi_i \alpha x_i + \sum_{i=1}^{n} \xi_i y_i$$

$$= \alpha \sum_{i=1}^{n} \xi_i x_i + \sum_{i=1}^{n} \xi_i y_i = \alpha f(\mathbf{x}) + f(\mathbf{y}).$$

3.3.3. There are many possibilities. Two of the simplest and most common are Hooke's law for springs that says that $F = kx$ (see Example 3.2.1) and Newton's second law that says that $F = ma$ (i.e., force = mass $\times$ acceleration).

3.3.4. They are all linear. To see that rotation is linear, use trigonometry to deduce that if $\mathbf{p} = \left(\begin{array}{c} x_1 \\ x_2 \end{array}\right)$, then $f(\mathbf{p}) = \mathbf{u} = \left(\begin{array}{c} u_1 \\ u_2 \end{array}\right)$, where

$$u_1 = (\cos\theta)x_1 - (\sin\theta)x_2$$
$$u_2 = (\sin\theta)x_1 + (\cos\theta)x_2.$$

f is linear because this is a special case of Example 3.3.2. To see that reflection is linear, write $\mathbf{p} = \left(\begin{array}{c} x_1 \\ x_2 \end{array}\right)$ and $f(\mathbf{p}) = \left(\begin{array}{c} x_1 \\ -x_2 \end{array}\right)$. Verification of linearity is straightforward. For the projection function, use the Pythagorean theorem to conclude that if $\mathbf{p} = \left(\begin{array}{c} x_1 \\ x_2 \end{array}\right)$, then $f(\mathbf{p}) = \frac{x_1 + x_2}{2}\left(\begin{array}{c} 1 \\ 1 \end{array}\right)$. Linearity is now easily verified.

Solutions for exercises in section 3. 4

3.4.1. Refer to the solution for Exercise 3.3.4. If $\mathbf{Q}$, $\mathbf{R}$, and $\mathbf{P}$ denote the matrices associated with the rotation, reflection, and projection, respectively, then

$$\mathbf{Q} = \begin{pmatrix} \cos\theta & -\sin\theta \\ \sin\theta & \cos\theta \end{pmatrix}, \quad \mathbf{R} = \begin{pmatrix} 1 & 0 \\ 0 & -1 \end{pmatrix}, \quad \text{and} \quad \mathbf{P} = \begin{pmatrix} \frac{1}{2} & \frac{1}{2} \\ \frac{1}{2} & \frac{1}{2} \end{pmatrix}.$$

3.4.2. Refer to the solution for Exercise 3.4.1 and write

$$\mathbf{RQ} = \begin{pmatrix} 1 & 0 \\ 0 & -1 \end{pmatrix} \begin{pmatrix} \cos\theta & -\sin\theta \\ \sin\theta & \cos\theta \end{pmatrix} = \begin{pmatrix} \cos\theta & -\sin\theta \\ -\sin\theta & -\cos\theta \end{pmatrix}.$$

If $Q(\mathbf{x})$ is the rotation function and $R(\mathbf{x})$ is the reflection function, then the composition is

$$R\big(Q(\mathbf{x})\big) = \begin{pmatrix} (\cos\theta)x_1 - (\sin\theta)x_2 \\ -(\sin\theta)x_1 - (\cos\theta)x_2 \end{pmatrix}.$$

3.4.3. Refer to the solution for Exercise 3.4.1 and write

$$\mathbf{PQR} = \begin{pmatrix} a_{11}x_1 + a_{12}x_2 \\ a_{21}x_1 + a_{22}x_2 \end{pmatrix} \begin{pmatrix} \cos\theta & -\sin\theta \\ \sin\theta & \cos\theta \end{pmatrix} \begin{pmatrix} 1 & 0 \\ 0 & -1 \end{pmatrix}$$

$$= \frac{1}{2} \begin{pmatrix} \cos\theta + \sin\theta & \sin\theta - \cos\theta \\ \cos\theta + \sin\theta & \sin\theta - \cos\theta \end{pmatrix}.$$

Therefore, the composition of the three functions in the order asked for is

$$P\Big(Q\big(R(\mathbf{x})\big)\Big) = \frac{1}{2} \begin{pmatrix} (\cos\theta + \sin\theta)x_1 + (\sin\theta - \cos\theta)x_2 \\ (\cos\theta + \sin\theta)x_1 + (\sin\theta - \cos\theta)x_2 \end{pmatrix}.$$

Solutions for exercises in section 3. 5

3.5.1. (a) $\mathbf{AB} = \begin{pmatrix} 10 & 15 \\ 12 & 8 \\ 28 & 52 \end{pmatrix}$ (b) $\mathbf{BA}$ does not exist   (c) $\mathbf{CB}$ does not exist

(d) $\mathbf{C}^T\mathbf{B} = (\, 10 \quad 31 \,)$ (e) $\mathbf{A}^2 = \begin{pmatrix} 13 & -1 & 19 \\ 16 & 13 & 12 \\ 36 & -17 & 64 \end{pmatrix}$ (f) $\mathbf{B}^2$ does not exist

(g) $\mathbf{C}^T\mathbf{C} = 14$ (h) $\mathbf{CC}^T = \begin{pmatrix} 1 & 2 & 3 \\ 2 & 4 & 6 \\ 3 & 6 & 9 \end{pmatrix}$ (i) $\mathbf{BB}^T = \begin{pmatrix} 5 & 8 & 17 \\ 8 & 16 & 28 \\ 17 & 28 & 58 \end{pmatrix}$

(j) $\mathbf{B}^T\mathbf{B} = \begin{pmatrix} 10 & 23 \\ 23 & 69 \end{pmatrix}$ (k) $\mathbf{C}^T\mathbf{AC} = 76$

3.5.2. (a) $\mathbf{A} = \begin{pmatrix} 2 & 1 & 1 \\ 4 & 0 & 2 \\ 2 & 2 & 0 \end{pmatrix}$, $\mathbf{x} = \begin{pmatrix} x_1 \\ x_2 \\ x_3 \end{pmatrix}$, $\mathbf{b} = \begin{pmatrix} 3 \\ 10 \\ -2 \end{pmatrix}$ (b) $\mathbf{s} = \begin{pmatrix} 1 \\ -2 \\ 3 \end{pmatrix}$

(c) $\mathbf{b} = \mathbf{A}_{*1} - 2\mathbf{A}_{*2} + 3\mathbf{A}_{*3}$

3.5.3. (a) $\mathbf{EA} = \begin{pmatrix} \mathbf{A}_{1*} \\ \mathbf{A}_{2*} \\ 3\mathbf{A}_{1*} + \mathbf{A}_{3*} \end{pmatrix}$ (b) $\mathbf{AE} = \begin{pmatrix} \mathbf{A}_{*1} + 3\mathbf{A}_{*3} & \mathbf{A}_{*2} & \mathbf{A}_{*3} \end{pmatrix}$

3.5.4. (a) $\mathbf{A}_{*j}$ (b) $\mathbf{A}_{i*}$ (c) a_{ij}

3.5.5. $\mathbf{Ax} = \mathbf{Bx} \; \forall \; \mathbf{x} \implies \mathbf{Ae}_j = \mathbf{Be}_j \; \forall \; \mathbf{e}_j \implies \mathbf{A}_{*j} = \mathbf{B}_{*j} \; \forall \; j \implies \mathbf{A} = \mathbf{B}$.

(The symbol $\forall$ is mathematical shorthand for the phrase "for all.")

3.5.6. The limit is the zero matrix.

3.5.7. If $\mathbf{A}$ is $m \times p$ and $\mathbf{B}$ is $p \times n$, write the product as

$$\mathbf{AB} = \begin{pmatrix} \mathbf{A}_{*1} & \mathbf{A}_{*2} & \cdots & \mathbf{A}_{*p} \end{pmatrix} \begin{pmatrix} \mathbf{B}_{1*} \\ \mathbf{B}_{2*} \\ \vdots \\ \mathbf{B}_{p*} \end{pmatrix} = \mathbf{A}_{*1}\mathbf{B}_{1*} + \mathbf{A}_{*2}\mathbf{B}_{2*} + \cdots + \mathbf{A}_{*p}\mathbf{B}_{p*}$$

$$= \sum_{k=1}^{p} \mathbf{A}_{*k}\mathbf{B}_{k*}.$$

3.5.8. (a) $[\mathbf{AB}]_{ij} = \mathbf{A}_{i*}\mathbf{B}_{*j} = \begin{pmatrix} 0 & \cdots & 0 & a_{ii} & \cdots & a_{in} \end{pmatrix} \begin{pmatrix} b_{1j} \\ \vdots \\ b_{jj} \\ 0 \\ \vdots \\ 0 \end{pmatrix}$ is 0 when $i > j$.

(b) When $i = j$, the only nonzero term in the product $\mathbf{A}_{i*}\mathbf{B}_{*i}$ is $a_{ii}b_{ii}$.

(c) Yes.

3.5.9. Use $[\mathbf{AB}]_{ij} = \sum_k a_{ik}b_{kj}$ along with the rules of differentiation to write

$$\frac{d[\mathbf{AB}]_{ij}}{dt} = \frac{d\left(\sum_k a_{ik}b_{kj}\right)}{dt} = \sum_k \frac{d(a_{ik}b_{kj})}{dt}$$

$$= \sum_k \left(\frac{da_{ik}}{dt}b_{kj} + a_{ik}\frac{db_{kj}}{dt}\right) = \sum_k \frac{da_{ik}}{dt}b_{kj} + \sum_k a_{ik}\frac{db_{kj}}{dt}$$

$$= \left[\frac{d\mathbf{A}}{dt}\mathbf{B}\right]_{ij} + \left[\mathbf{A}\frac{d\mathbf{B}}{dt}\right]_{ij} = \left[\frac{d\mathbf{A}}{dt}\mathbf{B} + \mathbf{A}\frac{d\mathbf{B}}{dt}\right]_{ij}.$$

3.5.10. (a) $[\mathbf{Ce}]_i =$ the total number of paths *leaving* node i.

(b) $[\mathbf{e}^T\mathbf{C}]_i =$ the total number of paths *entering* node i.

3.5.11. At time t, the concentration of salt in tank i is $\frac{x_i(t)}{V}$ lbs/gal. For tank 1,

$$\frac{dx_1}{dt} = \frac{\text{lbs}}{\text{sec}} \text{ coming in} - \frac{\text{lbs}}{\text{sec}} \text{ going out} = 0\frac{\text{lbs}}{\text{sec}} - \left(r\frac{\text{gal}}{\text{sec}} \times \frac{x_1(t)}{V}\frac{\text{lbs}}{\text{gal}} \right)$$

$$= -\frac{r}{V}x_1(t)\frac{\text{lbs}}{\text{sec}}.$$

For tank 2,

$$\frac{dx_2}{dt} = \frac{\text{lbs}}{\text{sec}} \text{ coming in} - \frac{\text{lbs}}{\text{sec}} \text{ going out} = \frac{r}{V}x_1(t)\frac{\text{lbs}}{\text{sec}} - \left(r\frac{\text{gal}}{\text{sec}} \times \frac{x_2(t)}{V}\frac{\text{lbs}}{\text{gal}} \right)$$

$$= \frac{r}{V}x_1(t)\frac{\text{lbs}}{\text{sec}} - \frac{r}{V}x_2(t)\frac{\text{lbs}}{\text{sec}} = \frac{r}{V}\left(x_1(t) - x_2(t) \right),$$

and for tank 3,

$$\frac{dx_3}{dt} = \frac{\text{lbs}}{\text{sec}} \text{ coming in} - \frac{\text{lbs}}{\text{sec}} \text{ going out} = \frac{r}{V}x_2(t)\frac{\text{lbs}}{\text{sec}} - \left(r\frac{\text{gal}}{\text{sec}} \times \frac{x_3(t)}{V}\frac{\text{lbs}}{\text{gal}} \right)$$

$$= \frac{r}{V}x_2(t)\frac{\text{lbs}}{\text{sec}} - \frac{r}{V}x_3(t)\frac{\text{lbs}}{\text{sec}} = \frac{r}{V}\left(x_2(t) - x_3(t) \right).$$

This is a system of three linear first-order differential equations

$$\frac{dx_1}{dt} = \frac{r}{V}\left(-x_1(t) \right)$$
$$\frac{dx_2}{dt} = \frac{r}{V}\left(x_1(t) - x_2(t) \right)$$
$$\frac{dx_3}{dt} = \frac{r}{V}\left(x_2(t) - x_3(t) \right)$$

that can be written as a single matrix differential equation

$$\begin{pmatrix} dx_1/dt \\ dx_2/dt \\ dx_3/dt \end{pmatrix} = \frac{r}{V}\begin{pmatrix} -1 & 0 & 0 \\ 1 & -1 & 0 \\ 0 & 1 & -1 \end{pmatrix}\begin{pmatrix} x_1(t) \\ x_2(t) \\ x_3(t) \end{pmatrix}.$$

Solutions for exercises in section 3. 6

3.6.1.

$$\mathbf{AB} = \begin{pmatrix} \mathbf{A}_{11} & \mathbf{A}_{12} & \mathbf{A}_{13} \\ \mathbf{A}_{21} & \mathbf{A}_{22} & \mathbf{A}_{23} \end{pmatrix} \begin{pmatrix} \mathbf{B}_1 \\ \mathbf{B}_2 \\ \mathbf{B}_3 \end{pmatrix} = \begin{pmatrix} \mathbf{A}_{11}\mathbf{B}_1 + \mathbf{A}_{12}\mathbf{B}_2 + \mathbf{A}_{13}\mathbf{B}_3 \\ \mathbf{A}_{21}\mathbf{B}_1 + \mathbf{A}_{22}\mathbf{B}_2 + \mathbf{A}_{23}\mathbf{B}_3 \end{pmatrix}$$

$$= \left(\begin{array}{cc} -10 & -19 \\ -10 & -19 \\ \hline -1 & -1 \end{array} \right)$$

3.6.2. Use block multiplication to verify $\mathbf{L}^2 = \mathbf{I}$—be careful not to commute any of the terms when forming the various products.

3.6.3. Partition the matrix as $\mathbf{A} = \begin{pmatrix} \mathbf{I} & \mathbf{C} \\ \mathbf{0} & \mathbf{C} \end{pmatrix}$, where $\mathbf{C} = \frac{1}{3} \begin{pmatrix} 1 & 1 & 1 \\ 1 & 1 & 1 \\ 1 & 1 & 1 \end{pmatrix}$ and observe

that $\mathbf{C}^2 = \mathbf{C}$. Use this together with block multiplication to conclude that

$$\mathbf{A}^k = \begin{pmatrix} \mathbf{I} & \mathbf{C} + \mathbf{C}^2 + \mathbf{C}^3 + \cdots + \mathbf{C}^k \\ \mathbf{0} & \mathbf{C}^k \end{pmatrix} = \begin{pmatrix} \mathbf{I} & k\mathbf{C} \\ \mathbf{0} & \mathbf{C} \end{pmatrix}.$$

Therefore, $\mathbf{A}^{300} = \begin{pmatrix} 1 & 0 & 0 & 100 & 100 & 100 \\ 0 & 1 & 0 & 100 & 100 & 100 \\ 0 & 0 & 1 & 100 & 100 & 100 \\ 0 & 0 & 0 & 1/3 & 1/3 & 1/3 \\ 0 & 0 & 0 & 1/3 & 1/3 & 1/3 \\ 0 & 0 & 0 & 1/3 & 1/3 & 1/3 \end{pmatrix}.$

3.6.4. $(\mathbf{A}^*\mathbf{A})^* = \mathbf{A}^*\mathbf{A}^{**} = \mathbf{A}^*\mathbf{A}$ and $(\mathbf{AA}^*)^* = \mathbf{A}^{**}\mathbf{A}^* = \mathbf{AA}^*.$

3.6.5. $(\mathbf{AB})^T = \mathbf{B}^T\mathbf{A}^T = \mathbf{BA} = \mathbf{AB}.$ It is easy to construct a 2×2 example to show that this need not be true when $\mathbf{AB} \neq \mathbf{BA}.$

3.6.6.

$$[(\mathbf{D} + \mathbf{E})\mathbf{F}]_{ij} = (\mathbf{D} + \mathbf{E})_{i*}\mathbf{F}_{*j} = \sum_k [\mathbf{D} + \mathbf{E}]_{ik}[\mathbf{F}]_{kj} = \sum_k ([\mathbf{D}]_{ik} + [\mathbf{E}]_{ik}) [\mathbf{F}]_{kj}$$

$$= \sum_k ([\mathbf{D}]_{ik}[\mathbf{F}]_{kj} + [\mathbf{E}]_{ik}[\mathbf{F}]_{kj}) = \sum_k [\mathbf{D}]_{ik}[\mathbf{F}]_{kj} + \sum_k [\mathbf{E}]_{ik}[\mathbf{F}]_{kj}$$

$$= \mathbf{D}_{i*}\mathbf{F}_{*j} + \mathbf{E}_{i*}\mathbf{F}_{*j} = [\mathbf{DF}]_{ij} + [\mathbf{EF}]_{ij}$$

$$= [\mathbf{DF} + \mathbf{EF}]_{ij}.$$

3.6.7. If a matrix $\mathbf{X}$ did indeed exist, then

$$\mathbf{I} = \mathbf{AX} - \mathbf{XA} \implies trace\,(\mathbf{I}) = trace\,(\mathbf{AX} - \mathbf{XA})$$
$$\implies n = trace\,(\mathbf{AX}) - trace\,(\mathbf{XA}) = 0,$$

which is impossible.

3.6.8. (a) $\mathbf{y}^T\mathbf{A} = \mathbf{b}^T \implies (\mathbf{y}^T\mathbf{A})^T = \mathbf{b}^{T\,T} \implies \mathbf{A}^T\mathbf{y} = \mathbf{b}$. This is an $n \times m$ system of equations whose coefficient matrix is $\mathbf{A}^T$. (b) They are the same.

3.6.9. Draw a transition diagram similar to that in Example 3.6.3 with North and South replaced by ON and OFF, respectively. Let x_k be the proportion of switches in the ON state, and let y_k be the proportion of switches in the OFF state after k clock cycles have elapsed. According to the given information,

$$x_k = x_{k-1}(.1) + y_{k-1}(.3)$$
$$y_k = x_{k-1}(.9) + y_{k-1}(.7)$$

so that $\mathbf{p}_k = \mathbf{p}_{k-1}\mathbf{P}$, where

$$\mathbf{p}_k = (\,x_k \quad y_k\,) \quad \text{and} \quad \mathbf{P} = \begin{pmatrix} .1 & .9 \\ .3 & .7 \end{pmatrix}.$$

Just as in Example 3.6.3, $\mathbf{p}_k = \mathbf{p}_0\mathbf{P}^k$. Compute a few powers of $\mathbf{P}$ to find

$$\mathbf{P}^2 = \begin{pmatrix} .280 & .720 \\ .240 & .760 \end{pmatrix}, \quad \mathbf{P}^3 = \begin{pmatrix} .244 & .756 \\ .252 & .748 \end{pmatrix}$$

$$\mathbf{P}^4 = \begin{pmatrix} .251 & .749 \\ .250 & .750 \end{pmatrix}, \quad \mathbf{P}^5 = \begin{pmatrix} .250 & .750 \\ .250 & .750 \end{pmatrix}$$

and deduce that $\mathbf{P}^\infty = \lim_{k \to \infty} \mathbf{P}^k = \begin{pmatrix} 1/4 & 3/4 \\ 1/4 & 3/4 \end{pmatrix}$. Thus

$$\mathbf{p}_k \to \mathbf{p}_0\mathbf{P}^\infty = (\,\tfrac{1}{4}(x_0 + y_0) \quad \tfrac{3}{4}(x_0 + y_0)\,) = (\,\tfrac{1}{4} \quad \tfrac{3}{4}\,).$$

For practical purposes, the device can be considered to be in equilibrium after about 5 clock cycles—regardless of the initial proportions.

3.6.10. $(-4 \quad 1 \quad -6 \quad 5)$

3.6.11. (a) $trace\,(\mathbf{ABC}) = trace\,(\mathbf{A}\{\mathbf{BC}\}) = trace\,(\{\mathbf{BC}\}\mathbf{A}) = trace\,(\mathbf{BCA})$. The other equality is similar. (b) Use almost any set of 2×2 matrices to construct an example that shows equality need not hold. (c) Use the fact that $trace\,(\mathbf{C}^T) = trace\,(\mathbf{C})$ for all square matrices to conclude that

$$trace\,(\mathbf{A}^T\mathbf{B}) = trace\,\left((\mathbf{A}^T\mathbf{B})^T\right) = trace\,\left(\mathbf{B}^T\mathbf{A}^{T\,T}\right)$$

$$= trace\,(\mathbf{B}^T\mathbf{A}) = trace\,(\mathbf{AB}^T).$$

3.6.12. (a) $\mathbf{x}^T\mathbf{x} = 0 \iff \sum_{k=1}^n x_i^2 = 0 \iff x_i = 0$ for each $i \iff \mathbf{x} = \mathbf{0}$.

(b) $trace\,(\mathbf{A}^T\mathbf{A}) = 0 \iff \sum_i [\mathbf{A}^T\mathbf{A}]_{ii} = 0 \iff \sum_i (\mathbf{A}^T)_{i*}\mathbf{A}_{*i} = 0$

$$\iff \sum_i \sum_k [\mathbf{A}^T]_{ik}[\mathbf{A}]_{ki} = 0 \iff \sum_i \sum_k [\mathbf{A}]_{ki}[\mathbf{A}]_{ki} = 0$$

$$\iff \sum_i \sum_k [\mathbf{A}]_{ki}^2 = 0$$

$$\iff [\mathbf{A}]_{ki} = 0 \text{ for each } k \text{ and } i \iff \mathbf{A} = \mathbf{0}$$

Solutions for exercises in section 3. 7

3.7.1. (a) $\begin{pmatrix} 3 & -2 \\ -1 & 1 \end{pmatrix}$ (b) Singular (c) $\begin{pmatrix} 2 & -4 & 3 \\ 4 & -7 & 4 \\ 5 & -8 & 4 \end{pmatrix}$ (d) Singular

(e) $\begin{pmatrix} 2 & -1 & 0 & 0 \\ -1 & 2 & -1 & 0 \\ 0 & -1 & 2 & -1 \\ 0 & 0 & -1 & 1 \end{pmatrix}$

3.7.2. Write the equation as $(\mathbf{I} - \mathbf{A})\mathbf{X} = \mathbf{B}$ and compute

$$\mathbf{X} = (\mathbf{I} - \mathbf{A})^{-1}\mathbf{B} = \begin{pmatrix} 1 & -1 & 1 \\ 0 & 1 & -1 \\ 0 & 0 & 1 \end{pmatrix} \begin{pmatrix} 1 & 2 \\ 2 & 1 \\ 3 & 3 \end{pmatrix} = \begin{pmatrix} 2 & 4 \\ -1 & -2 \\ 3 & 3 \end{pmatrix}.$$

3.7.3. In each case, the given information implies that $rank\,(\mathbf{A}) < n$ —see the solution for Exercise 2.1.3.

3.7.4. (a) If $\mathbf{D}$ is diagonal, then $\mathbf{D}^{-1}$ exists if and only if each $d_{ii} \neq 0$, in which case

$$\begin{pmatrix} d_{11} & 0 & \cdots & 0 \\ 0 & d_{22} & \cdots & 0 \\ \vdots & \vdots & \ddots & \vdots \\ 0 & 0 & \cdots & d_{nn} \end{pmatrix}^{-1} = \begin{pmatrix} 1/d_{11} & 0 & \cdots & 0 \\ 0 & 1/d_{22} & \cdots & 0 \\ \vdots & \vdots & \ddots & \vdots \\ 0 & 0 & \cdots & 1/d_{nn} \end{pmatrix}.$$

(b) If $\mathbf{T}$ is triangular, then $\mathbf{T}^{-1}$ exists if and only if each $t_{ii} \neq 0$. If $\mathbf{T}$ is upper (lower) triangular, then $\mathbf{T}^{-1}$ is also upper (lower) triangular with $[\mathbf{T}^{-1}]_{ii} = 1/t_{ii}$.

3.7.5. $\left(\mathbf{A}^{-1}\right)^{T} = \left(\mathbf{A}^{T}\right)^{-1} = \mathbf{A}^{-1}$.

3.7.6. Start with $\mathbf{A}(\mathbf{I} - \mathbf{A}) = (\mathbf{I} - \mathbf{A})\mathbf{A}$ and apply $(\mathbf{I} - \mathbf{A})^{-1}$ to both sides, first on one side and then on the other.

3.7.7. Use the result of Example 3.6.5 that says that $trace\,(\mathbf{AB}) = trace\,(\mathbf{BA})$ to write

$$m = trace\,(\mathbf{I}_m) = trace\,(\mathbf{AB}) = trace\,(\mathbf{BA}) = trace\,(\mathbf{I}_n) = n.$$

3.7.8. Use the reverse order law for inversion to write

$$\left[\mathbf{A}(\mathbf{A} + \mathbf{B})^{-1}\mathbf{B}\right]^{-1} = \mathbf{B}^{-1}(\mathbf{A} + \mathbf{B})\mathbf{A}^{-1} = \mathbf{B}^{-1} + \mathbf{A}^{-1}$$

and

$$\left[\mathbf{B}(\mathbf{A} + \mathbf{B})^{-1}\mathbf{A}\right]^{-1} = \mathbf{A}^{-1}(\mathbf{A} + \mathbf{B})\mathbf{B}^{-1} = \mathbf{B}^{-1} + \mathbf{A}^{-1}.$$

3.7.9. (a) $(\mathbf{I} - \mathbf{S})\mathbf{x} = \mathbf{0} \implies \mathbf{x}^{T}(\mathbf{I} - \mathbf{S})\mathbf{x} = 0 \implies \mathbf{x}^{T}\mathbf{x} = \mathbf{x}^{T}\mathbf{S}\mathbf{x}$. Taking transposes on both sides yields $\mathbf{x}^{T}\mathbf{x} = -\mathbf{x}^{T}\mathbf{S}\mathbf{x}$, so that $\mathbf{x}^{T}\mathbf{x} = 0$, and thus $\mathbf{x} = \mathbf{0}$

(recall Exercise 3.6.12). The conclusion follows from property (3.7.8).

(b) First notice that Exercise 3.7.6 implies that $\mathbf{A} = (\mathbf{I} + \mathbf{S})(\mathbf{I} - \mathbf{S})^{-1} = (\mathbf{I} - \mathbf{S})^{-1}(\mathbf{I} + \mathbf{S})$. By using the reverse order laws, transposing both sides yields exactly the same thing as inverting both sides.

3.7.10. Use block multiplication to verify that the product of the matrix with its inverse is the identity matrix.

3.7.11. Use block multiplication to verify that the product of the matrix with its inverse is the identity matrix.

3.7.12. Let $\mathbf{M} = \begin{pmatrix} \mathbf{A} & \mathbf{B} \\ \mathbf{C} & \mathbf{D} \end{pmatrix}$ and $\mathbf{X} = \begin{pmatrix} \mathbf{D}^T & -\mathbf{B}^T \\ -\mathbf{C}^T & \mathbf{A}^T \end{pmatrix}$. The hypothesis implies that $\mathbf{MX} = \mathbf{I}$, and hence (from the discussion in Example 3.7.2) it must also be true that $\mathbf{XM} = \mathbf{I}$, from which the conclusion follows. **Note:** This problem appeared on a past Putnam Exam—a national mathematics competition for undergraduate students that is considered to be quite challenging. This means that you can be proud of yourself if you solved it before looking at this solution.

Solutions for exercises in section 3. 8

3.8.1. (a) $\mathbf{B}^{-1} = \begin{pmatrix} 1 & 2 & -1 \\ 0 & -1 & 1 \\ 1 & 4 & -2 \end{pmatrix}$

(b) Let $\mathbf{c} = \begin{pmatrix} 0 \\ 0 \\ 1 \end{pmatrix}$ and $\mathbf{d}^T = \begin{pmatrix} 0 & 2 & 1 \end{pmatrix}$ to obtain $\mathbf{C}^{-1} = \begin{pmatrix} 0 & -2 & 1 \\ 1 & 3 & -1 \\ -1 & -4 & 2 \end{pmatrix}$

3.8.2. $\mathbf{A}_{*j}$ needs to be removed, and $\mathbf{b}$ needs to be inserted in its place. This is accomplished by writing $\mathbf{B} = \mathbf{A} + (\mathbf{b} - \mathbf{A}_{*j})\mathbf{e}_j^T$. Applying the Sherman–Morrison formula with $\mathbf{c} = \mathbf{b} - \mathbf{A}_{*j}$ and $\mathbf{d}^T = \mathbf{e}_j^T$ yields

$$\mathbf{B}^{-1} = \mathbf{A}^{-1} - \frac{\mathbf{A}^{-1}(\mathbf{b} - \mathbf{A}_{*j})\mathbf{e}_j^T \mathbf{A}^{-1}}{1 + \mathbf{e}_j^T \mathbf{A}^{-1}(\mathbf{b} - \mathbf{A}_{*j})} = \mathbf{A}^{-1} - \frac{\mathbf{A}^{-1}\mathbf{b}\mathbf{e}_j^T \mathbf{A}^{-1} - \mathbf{e}_j \mathbf{e}_j^T \mathbf{A}^{-1}}{1 + \mathbf{e}_j^T \mathbf{A}^{-1}\mathbf{b} - \mathbf{e}_j^T \mathbf{e}_j}$$

$$= \mathbf{A}^{-1} - \frac{\mathbf{A}^{-1}\mathbf{b}[\mathbf{A}^{-1}]_{j*} - \mathbf{e}_j[\mathbf{A}^{-1}]_{j*}}{[\mathbf{A}^{-1}]_{j*}\mathbf{b}} = \mathbf{A}^{-1} - \frac{(\mathbf{A}^{-1}\mathbf{b} - \mathbf{e}_j)[\mathbf{A}^{-1}]_{j*}}{[\mathbf{A}^{-1}]_{j*}\mathbf{b}}.$$

3.8.3. Use the Sherman–Morrison formula to write

$$\mathbf{z} = (\mathbf{A} + \mathbf{c}\mathbf{d}^T)^{-1}\mathbf{b} = \left(\mathbf{A}^{-1} - \frac{\mathbf{A}^{-1}\mathbf{c}\mathbf{d}^T \mathbf{A}^{-1}}{1 + \mathbf{d}^T \mathbf{A}^{-1}\mathbf{c}}\right)\mathbf{b} = \mathbf{A}^{-1}\mathbf{b} - \frac{\mathbf{A}^{-1}\mathbf{c}\mathbf{d}^T \mathbf{A}^{-1}\mathbf{b}}{1 + \mathbf{d}^T \mathbf{A}^{-1}\mathbf{c}}$$

$$= \mathbf{x} - \frac{\mathbf{y}\mathbf{d}^T \mathbf{x}}{1 + \mathbf{d}^T \mathbf{y}}.$$

3.8.4. (a) For a nonsingular matrix $\mathbf{A}$, the Sherman–Morrison formula guarantees that $\mathbf{A} + \alpha\mathbf{e}_i\mathbf{e}_j^T$ is also nonsingular when $1 + \alpha[\mathbf{A}^{-1}]_{ji} \neq 0$, and this certainly will be true if α is sufficiently small.

(b)   Write $\mathbf{E}_{m \times m} = [\epsilon_{ij}] = \sum_{i,j=1}^{m} \epsilon_{ij} \mathbf{e}_i \mathbf{e}_j^T$ and successively apply part (a) to

$$\mathbf{I} + \mathbf{E} = \left(\left(\left(\mathbf{I} + \epsilon_{11} \mathbf{e}_1 \mathbf{e}_1^T \right) + \epsilon_{12} \mathbf{e}_1 \mathbf{e}_2^T \right) + \cdots + \epsilon_{mm} \mathbf{e}_m \mathbf{e}_m^T \right)$$

to conclude that when the ϵ_{ij}'s are sufficiently small,

$$\mathbf{I} + \epsilon_{11} \mathbf{e}_1 \mathbf{e}_1^T, \quad \left(\left(\mathbf{I} + \epsilon_{11} \mathbf{e}_1 \mathbf{e}_1^T \right) + \epsilon_{12} \mathbf{e}_1 \mathbf{e}_2^T \right), \quad \ldots, \quad \mathbf{I} + \mathbf{E}$$

are each nonsingular.

3.8.5. Write $\mathbf{A} + \epsilon \mathbf{B} = \mathbf{A}(\mathbf{I} + \mathbf{A}^{-1} \epsilon \mathbf{B})$. You can either use the Neumann series result (3.8.5) or Exercise 3.8.4 to conclude that $(\mathbf{I} + \mathbf{A}^{-1} \epsilon \mathbf{B})$ is nonsingular whenever the entries of $\mathbf{A}^{-1} \epsilon \mathbf{B}$ are sufficiently small in magnitude, and this can be insured by restricting ϵ to a small enough interval about the origin. Since the product of two nonsingular matrices is again nonsingular—see (3.7.14)—it follows that $\mathbf{A} + \epsilon \mathbf{B} = \mathbf{A}(\mathbf{I} + \mathbf{A}^{-1} \epsilon \mathbf{B})$ must be nonsingular.

3.8.6. Since

$$\begin{pmatrix} \mathbf{I} & \mathbf{C} \\ \mathbf{0} & \mathbf{I} \end{pmatrix} \begin{pmatrix} \mathbf{A} & \mathbf{C} \\ \mathbf{D}^T & -\mathbf{I} \end{pmatrix} \begin{pmatrix} \mathbf{I} & \mathbf{0} \\ \mathbf{D}^T & \mathbf{I} \end{pmatrix} = \begin{pmatrix} \mathbf{A} + \mathbf{C}\mathbf{D}^T & \mathbf{0} \\ \mathbf{0} & -\mathbf{I} \end{pmatrix},$$

we can use $\mathbf{R} = \mathbf{D}^T$ and $\mathbf{B} = -\mathbf{I}$ in part (a) of Exercise 3.7.11 to obtain

$$\begin{pmatrix} \mathbf{I} & \mathbf{0} \\ -\mathbf{D}^T & \mathbf{I} \end{pmatrix} \begin{pmatrix} \mathbf{A}^{-1} + \mathbf{A}^{-1}\mathbf{C}\mathbf{S}^{-1}\mathbf{D}^T\mathbf{A}^{-1} & -\mathbf{A}^{-1}\mathbf{C}\mathbf{S}^{-1} \\ -\mathbf{S}^{-1}\mathbf{D}^T\mathbf{A}^{-1} & \mathbf{S}^{-1} \end{pmatrix} \begin{pmatrix} \mathbf{I} & -\mathbf{C} \\ \mathbf{0} & \mathbf{I} \end{pmatrix} =$$
$$\begin{pmatrix} (\mathbf{A} + \mathbf{C}\mathbf{D}^T)^{-1} & \mathbf{0} \\ \mathbf{0} & -\mathbf{I} \end{pmatrix},$$

where $\mathbf{S} = -\left(\mathbf{I} + \mathbf{D}^T \mathbf{A}^{-1} \mathbf{C}\right)$. Comparing the upper-left-hand blocks produces

$$\left(\mathbf{A} + \mathbf{C}\mathbf{D}^T\right)^{-1} = \mathbf{A}^{-1} - \mathbf{A}^{-1}\mathbf{C}\left(\mathbf{I} + \mathbf{D}^T\mathbf{A}^{-1}\mathbf{C}\right)^{-1}\mathbf{D}^T\mathbf{A}^{-1}.$$

3.8.7. The ranking from best to worst condition is $\mathbf{A}$, $\mathbf{B}$, $\mathbf{C}$, because

$$\mathbf{A}^{-1} = \frac{1}{100} \begin{pmatrix} 2 & 1 & 1 \\ 1 & 2 & 1 \\ 1 & 1 & 1 \end{pmatrix} \implies \kappa(\mathbf{A}) = 20 = 2 \times 10^1$$

$$\mathbf{B}^{-1} = \begin{pmatrix} -1465 & -161 & 17 \\ 173 & 19 & -2 \\ -82 & -9 & 1 \end{pmatrix} \implies \kappa(\mathbf{B}) = 149,513 \approx 1.5 \times 10^5$$

$$\mathbf{C}^{-1} = \begin{pmatrix} -42659 & 39794 & -948 \\ 2025 & -1889 & 45 \\ 45 & -42 & 1 \end{pmatrix} \implies \kappa(\mathbf{C}) = 82,900,594 \approx 8.2 \times 10^7.$$

3.8.8. (a)   Differentiate $\mathbf{A}(t)\mathbf{A}(t)^{-1} = \mathbf{I}$ with the product rule for differentiation (recall Exercise 3.5.9).

(b)   Use the product rule for differentiation together with part (a) to differentiate $\mathbf{A}(t)\mathbf{x}(t) = \mathbf{b}(t)$.

Solutions for exercises in section 3. 9

3.9.1. (a) If $\mathbf{G}_1, \mathbf{G}_2, \ldots, \mathbf{G}_k$ is the sequence of elementary matrices that corresponds to the elementary row operations used in the reduction $[\mathbf{A}|\mathbf{I}] \longrightarrow [\mathbf{B}|\mathbf{P}]$, then

$$\mathbf{G}_k \cdots \mathbf{G}_2\mathbf{G}_1[\mathbf{A}|\mathbf{I}] = [\mathbf{B}|\mathbf{P}] \implies [\mathbf{G}_k \cdots \mathbf{G}_2\mathbf{G}_1\mathbf{A} \mid \mathbf{G}_k \cdots \mathbf{G}_2\mathbf{G}_1\mathbf{I}] = [\mathbf{B}|\mathbf{P}]$$
$$\implies \mathbf{G}_k \cdots \mathbf{G}_2\mathbf{G}_1\mathbf{A} = \mathbf{B} \quad \text{and} \quad \mathbf{G}_k \cdots \mathbf{G}_2\mathbf{G}_1 = \mathbf{P}.$$

(b) Use the same argument given above, but apply it on the right-hand side.

(c) $[\mathbf{A}|\mathbf{I}] \xrightarrow{\text{Gauss–Jordan}} [\mathbf{E_A}|\mathbf{P}]$ yields

$$\left(\begin{array}{cccc|ccc} 1 & 2 & 3 & 4 & 1 & 0 & 0 \\ 2 & 4 & 6 & 7 & 0 & 1 & 0 \\ 1 & 2 & 3 & 6 & 0 & 0 & 1 \end{array}\right) \longrightarrow \left(\begin{array}{cccc|ccc} 1 & 2 & 3 & 0 & -7 & 4 & 0 \\ 0 & 0 & 0 & 1 & 2 & -1 & 0 \\ 0 & 0 & 0 & 0 & -5 & 2 & 1 \end{array}\right).$$

Thus $\mathbf{P} = \begin{pmatrix} -7 & 4 & 0 \\ 2 & -1 & 0 \\ -5 & 2 & 1 \end{pmatrix}$ is the product of the elementary matrices corresponding to the operations used in the reduction, and $\mathbf{PA} = \mathbf{E_A}$.

(d) You already have $\mathbf{P}$ such that $\mathbf{PA} = \mathbf{E_A}$. Now find $\mathbf{Q}$ such that $\mathbf{E_A}\mathbf{Q} = \mathbf{N}_r$ by column reducing $\mathbf{E_A}$. Proceed using part (b) to accumulate $\mathbf{Q}$.

$$\left[\frac{\mathbf{E_A}}{\mathbf{I}_4}\right] \longrightarrow \left(\begin{array}{cccc} 1 & 2 & 3 & 0 \\ 0 & 0 & 0 & 1 \\ 0 & 0 & 0 & 0 \\ \hline 1 & 0 & 0 & 0 \\ 0 & 1 & 0 & 0 \\ 0 & 0 & 1 & 0 \\ 0 & 0 & 0 & 1 \end{array}\right) \longrightarrow \left(\begin{array}{cccc} 1 & 0 & 2 & 3 \\ 0 & 1 & 0 & 0 \\ 0 & 0 & 0 & 0 \\ \hline 1 & 0 & 0 & 0 \\ 0 & 0 & 1 & 0 \\ 0 & 0 & 0 & 1 \\ 0 & 1 & 0 & 0 \end{array}\right) \longrightarrow \left(\begin{array}{cccc} 1 & 0 & 0 & 0 \\ 0 & 1 & 0 & 0 \\ 0 & 0 & 0 & 0 \\ \hline 1 & 0 & -2 & -3 \\ 0 & 0 & 1 & 0 \\ 0 & 0 & 0 & 1 \\ 0 & 1 & 0 & 0 \end{array}\right)$$

3.9.2. (a) Yes—because $rank(\mathbf{A}) = rank(\mathbf{B})$. (b) Yes—because $\mathbf{E_A} = \mathbf{E_B}$.
(c) No—because $\mathbf{E}_{\mathbf{A}^T} \neq \mathbf{E}_{\mathbf{B}^T}$.

3.9.3. The positions of the basic columns in $\mathbf{A}$ correspond to those in $\mathbf{E_A}$. Because $\mathbf{A} \overset{\text{row}}{\sim} \mathbf{B} \Longleftrightarrow \mathbf{E_A} = \mathbf{E_B}$, it follows that the basic columns in $\mathbf{A}$ and $\mathbf{B}$ must be in the same positions.

3.9.4. An elementary interchange matrix (a Type I matrix) has the form $\mathbf{E} = \mathbf{I} - \mathbf{u}\mathbf{u}^T$, where $\mathbf{u} = \mathbf{e}_i - \mathbf{e}_j$, and it follows from (3.9.1) that $\mathbf{E} = \mathbf{E}^T = \mathbf{E}^{-1}$. If $\mathbf{P} = \mathbf{E}_1\mathbf{E}_2 \cdots \mathbf{E}_k$ is a product of elementary interchange matrices, then the reverse order laws yield

$$\mathbf{P}^{-1} = (\mathbf{E}_1\mathbf{E}_2 \cdots \mathbf{E}_k)^{-1} = \mathbf{E}_k^{-1} \cdots \mathbf{E}_2^{-1}\mathbf{E}_1^{-1}$$
$$= \mathbf{E}_k^T \cdots \mathbf{E}_2^T\mathbf{E}_1^T = (\mathbf{E}_1\mathbf{E}_2 \cdots \mathbf{E}_k)^T = \mathbf{P}^T.$$

3.9.5. They are all true! $\mathbf{A} \sim \mathbf{I} \sim \mathbf{A}^{-1}$ because $rank\,(\mathbf{A}) = n = rank\,(\mathbf{A}^{-1})$, $\mathbf{A} \overset{\text{row}}{\sim}$ $\mathbf{A}^{-1}$ because $\mathbf{PA} = \mathbf{A}^{-1}$ with $\mathbf{P} = (\mathbf{A}^{-1})^2 = \mathbf{A}^{-2}$, and $\mathbf{A} \overset{\text{col}}{\sim} \mathbf{A}^{-1}$ because $\mathbf{AQ} = \mathbf{A}^{-1}$ with $\mathbf{Q} = \mathbf{A}^{-2}$. The fact that $\mathbf{A} \overset{\text{row}}{\sim} \mathbf{I}$ and $\mathbf{A} \overset{\text{col}}{\sim} \mathbf{I}$ follows since $\mathbf{A}^{-1}\mathbf{A} = \mathbf{A}\mathbf{A}^{-1} = \mathbf{I}$.

3.9.6. (a), (c), (d), and (e) are true.

3.9.7. Rows i and j can be interchanged with the following sequence of Type II and Type III operations—this is Exercise 1.2.12 on p. 14.

$$R_j \leftarrow R_j + R_i \quad \text{(replace row } j \text{ by the sum of row } j \text{ and } i)$$
$$R_i \leftarrow R_i - R_j \quad \text{(replace row } i \text{ by the difference of row } i \text{ and } j)$$
$$R_j \leftarrow R_j + R_i \quad \text{(replace row } j \text{ by the sum of row } j \text{ and } i)$$
$$R_i \leftarrow -R_i \quad \text{(replace row } i \text{ by its negative)}$$

Translating these to elementary matrices (remembering to build from the right to the left) produces

$$(\mathbf{I} - 2\mathbf{e}_i\mathbf{e}_i^T)(\mathbf{I} + \mathbf{e}_j\mathbf{e}_i^T)(\mathbf{I} - \mathbf{e}_i\mathbf{e}_j^T)(\mathbf{I} + \mathbf{e}_j\mathbf{e}_i^T) = \mathbf{I} - \mathbf{uu}^T, \quad \text{where} \quad \mathbf{u} = \mathbf{e}_i - \mathbf{e}_j.$$

3.9.8. Let $\mathbf{B}_{m \times r} = [\mathbf{A}_{*b_1}\mathbf{A}_{*b_2} \cdots \mathbf{A}_{*b_r}]$ contain the basic columns of $\mathbf{A}$, and let $\mathbf{C}_{r \times n}$ contain the nonzero rows of $\mathbf{E_A}$. If $\mathbf{A}_{*k}$ is basic—say $\mathbf{A}_{*k} = \mathbf{A}_{*b_j}$—then $\mathbf{C}_{*k} = \mathbf{e}_j$, and

$$(\mathbf{BC})_{*k} = \mathbf{BC}_{*k} = \mathbf{Be}_j = \mathbf{B}_{*j} = \mathbf{A}_{*b_j} = \mathbf{A}_{*k}.$$

If $\mathbf{A}_{*k}$ is nonbasic, then $\mathbf{C}_{*k}$ is nonbasic and has the form

$$\mathbf{C}_{*k} = \begin{pmatrix} \mu_1 \\ \mu_2 \\ \vdots \\ \mu_j \\ \vdots \\ 0 \end{pmatrix} = \mu_1 \begin{pmatrix} 1 \\ 0 \\ \vdots \\ 0 \\ \vdots \\ 0 \end{pmatrix} + \mu_2 \begin{pmatrix} 0 \\ 1 \\ \vdots \\ 0 \\ \vdots \\ 0 \end{pmatrix} + \cdots + \mu_j \begin{pmatrix} 0 \\ 0 \\ \vdots \\ 1 \\ \vdots \\ 0 \end{pmatrix}$$
$$= \mu_1\mathbf{e}_1 + \mu_2\mathbf{e}_2 + \cdots + \mu_j\mathbf{e}_j,$$

where the $\mathbf{e}_i$'s are the basic columns to the left of $\mathbf{C}_{*k}$. Because $\mathbf{A} \overset{\text{row}}{\sim} \mathbf{E_A}$, the relationships that exist among the columns of $\mathbf{A}$ are exactly the same as the relationships that exist among the columns of $\mathbf{E_A}$. In particular,

$$\mathbf{A}_{*k} = \mu_1\mathbf{A}_{*b_1} + \mu_2\mathbf{A}_{*b_2} + \cdots + \mu_j\mathbf{A}_{*b_j},$$

where the $\mathbf{A}_{*b_i}$'s are the basic columns to the left of $\mathbf{A}_{*k}$. Therefore,

$$(\mathbf{BC})_{*k} = \mathbf{BC}_{*k} = \mathbf{B}(\mu_1\mathbf{e}_1 + \mu_2\mathbf{e}_2 + \cdots + \mu_j\mathbf{e}_j)$$
$$= \mu_1\mathbf{B}_{*1} + \mu_2\mathbf{B}_{*2} + \cdots + \mu_j\mathbf{B}_{*j}$$
$$= \mu_1\mathbf{A}_{*b_1} + \mu_2\mathbf{A}_{*b_2} + \cdots + \mu_j\mathbf{A}_{*b_j}$$
$$= \mathbf{A}_{*k}.$$

3.9.9. If $\mathbf{A} = \mathbf{uv}^T$, where $\mathbf{u}_{m \times 1}$ and $\mathbf{v}_{n \times 1}$ are nonzero columns, then

$$\mathbf{u} \overset{row}{\sim} \mathbf{e}_1 \quad \text{and} \quad \mathbf{v}^T \overset{col}{\sim} \mathbf{e}_1^T \quad \Longrightarrow \quad \mathbf{A} = \mathbf{uv}^T \sim \mathbf{e}_1\mathbf{e}_1^T = \mathbf{N}_1 \quad \Longrightarrow \quad rank\,(\mathbf{A}) = 1.$$

Conversely, if $rank\,(\mathbf{A}) = 1$, then the existence of $\mathbf{u}$ and $\mathbf{v}$ follows from Exercise 3.9.8. If you do not wish to rely on Exercise 3.9.8, write $\mathbf{PAQ} = \mathbf{N}_1 = \mathbf{e}_1\mathbf{e}_1^T$, where $\mathbf{e}_1$ is $m \times 1$ and $\mathbf{e}_1^T$ is $1 \times n$ so that

$$\mathbf{A} = \mathbf{P}^{-1}\mathbf{e}_1\mathbf{e}_1^T\mathbf{Q}^{-1} = \left(\mathbf{P}^{-1}\right)_{*1}\left(\mathbf{Q}^{-1}\right)_{1*} = \mathbf{uv}^T.$$

3.9.10. Use Exercise 3.9.9 and write

$$\mathbf{A} = \mathbf{uv}^T \quad \Longrightarrow \quad \mathbf{A}^2 = \left(\mathbf{uv}^T\right)\left(\mathbf{uv}^T\right) = \mathbf{u}\left(\mathbf{v}^T\mathbf{u}\right)\mathbf{v}^T = \tau\mathbf{uv}^T = \tau\mathbf{A},$$

where $\tau = \mathbf{v}^T\mathbf{u}$. Recall from Example 3.6.5 that $trace\,(\mathbf{AB}) = trace\,(\mathbf{BA})$, and write

$$\tau = trace(\tau) = trace\left(\mathbf{v}^T\mathbf{u}\right) = trace\left(\mathbf{uv}^T\right) = trace\,(\mathbf{A}).$$

Solutions for exercises in section 3. 10

3.10.1. (a) $\mathbf{L} = \begin{pmatrix} 1 & 0 & 0 \\ 4 & 1 & 0 \\ 3 & 2 & 1 \end{pmatrix}$ and $\mathbf{U} = \begin{pmatrix} 1 & 4 & 5 \\ 0 & 2 & 6 \\ 0 & 0 & 3 \end{pmatrix}$ (b) $\mathbf{x}_1 = \begin{pmatrix} 110 \\ -36 \\ 8 \end{pmatrix}$ and

$\mathbf{x}_2 = \begin{pmatrix} 112 \\ -39 \\ 10 \end{pmatrix}$

(c) $\mathbf{A}^{-1} = \frac{1}{6}\begin{pmatrix} 124 & -40 & 14 \\ -42 & 15 & -6 \\ 10 & -4 & 2 \end{pmatrix}$

3.10.2. (a) The second pivot is zero. (b) $\mathbf{P}$ is the permutation matrix associated with the permutation $\mathbf{p} = (2 \quad 4 \quad 1 \quad 3)$. $\mathbf{P}$ is constructed by permuting the rows of $\mathbf{I}$ in this manner.

$$\mathbf{L} = \begin{pmatrix} 1 & 0 & 0 & 0 \\ 0 & 1 & 0 & 0 \\ 1/3 & 0 & 1 & 0 \\ 2/3 & -1/2 & 1/2 & 1 \end{pmatrix} \quad \text{and} \quad \mathbf{U} = \begin{pmatrix} 3 & 6 & -12 & 3 \\ 0 & 2 & -2 & 6 \\ 0 & 0 & 8 & 16 \\ 0 & 0 & 0 & -5 \end{pmatrix}$$

(c) $\mathbf{x} = \begin{pmatrix} 2 \\ -1 \\ 0 \\ 1 \end{pmatrix}$

3.10.3. $\xi = 0, \ \pm\sqrt{2}, \ \pm\sqrt{3}$

3.10.4. **A** possesses an LU factorization if and only if all leading principal submatrices are nonsingular. The argument associated with equation (3.10.13) proves that

$$\begin{pmatrix} \mathbf{L}_k & \mathbf{0} \\ \mathbf{c}^T\mathbf{U}_k^{-1} & 1 \end{pmatrix} \begin{pmatrix} \mathbf{U}_k & \mathbf{L}_k^{-1}\mathbf{b} \\ \mathbf{0} & a_{k+1,k+1} - \mathbf{c}^T\mathbf{A}_k^{-1}\mathbf{b} \end{pmatrix} = \mathbf{L}_{k+1}\mathbf{U}_{k+1}$$

is the LU factorization for $\mathbf{A}_{k+1}$. The desired conclusion follows from the fact that the $k+1^{th}$ pivot is the $(k+1, k+1)$-entry in $\mathbf{U}_{k+1}$. This pivot must be nonzero because $\mathbf{U}_{k+1}$ is nonsingular.

3.10.5. If **L** and **U** are both triangular with 1's on the diagonal, then $\mathbf{L}^{-1}$ and $\mathbf{U}^{-1}$ contain only integer entries, and consequently $\mathbf{A}^{-1} = \mathbf{U}^{-1}\mathbf{L}^{-1}$ is an integer matrix.

3.10.6. (b) $\mathbf{L} = \begin{pmatrix} 1 & 0 & 0 & 0 \\ -1/2 & 1 & 0 & 0 \\ 0 & -2/3 & 1 & 0 \\ 0 & 0 & -3/4 & 1 \end{pmatrix}$ and $\mathbf{U} = \begin{pmatrix} 2 & -1 & 0 & 0 \\ 0 & 3/2 & -1 & 0 \\ 0 & 0 & 4/3 & -1 \\ 0 & 0 & 0 & 1/4 \end{pmatrix}$

3.10.7. Observe how the LU factors evolve from Gaussian elimination. Following the procedure described in Example 3.10.1 where multipliers ℓ_{ij} are stored in the positions they annihilate (i.e., in the (i, j)-position), and where $\star$'s are put in positions that can be nonzero, the reduction of a 5×5 band matrix with bandwidth $w = 2$ proceeds as shown below.

$$\begin{pmatrix} \star & \star & \star & 0 & 0 \\ \star & \star & \star & \star & 0 \\ \star & \star & \star & \star & \star \\ 0 & \star & \star & \star & \star \\ 0 & 0 & \star & \star & \star \end{pmatrix} \longrightarrow \begin{pmatrix} \star & \star & \star & 0 & 0 \\ l_{21} & \star & \star & \star & 0 \\ l_{31} & \star & \star & \star & \star \\ 0 & \star & \star & \star & \star \\ 0 & 0 & \star & \star & \star \end{pmatrix} \longrightarrow \begin{pmatrix} \star & \star & \star & 0 & 0 \\ l_{21} & \star & \star & \star & 0 \\ l_{31} & l_{32} & \star & \star & \star \\ 0 & l_{42} & \star & \star & \star \\ 0 & 0 & \star & \star & \star \end{pmatrix}$$

$$\longrightarrow \begin{pmatrix} \star & \star & \star & 0 & 0 \\ l_{21} & \star & \star & \star & 0 \\ l_{31} & l_{32} & \star & \star & \star \\ 0 & l_{42} & l_{43} & \star & \star \\ 0 & 0 & l_{53} & \star & \star \end{pmatrix} \longrightarrow \begin{pmatrix} \star & \star & \star & 0 & 0 \\ l_{21} & \star & \star & \star & 0 \\ l_{31} & l_{32} & \star & \star & \star \\ 0 & l_{42} & l_{43} & \star & \star \\ 0 & 0 & l_{53} & l_{54} & \star \end{pmatrix}$$

Thus $\mathbf{L} = \begin{pmatrix} 1 & 0 & 0 & 0 & 0 \\ l_{21} & 1 & 0 & 0 & 0 \\ l_{31} & l_{32} & 1 & 0 & 0 \\ 0 & l_{42} & l_{43} & 1 & 0 \\ 0 & 0 & l_{53} & l_{54} & 1 \end{pmatrix}$ and $\mathbf{U} = \begin{pmatrix} \star & \star & \star & 0 & 0 \\ 0 & \star & \star & \star & 0 \\ 0 & 0 & \star & \star & \star \\ 0 & 0 & 0 & \star & \star \\ 0 & 0 & 0 & 0 & \star \end{pmatrix}$.

3.10.8. (a) $\mathbf{A} = \begin{pmatrix} 0 & 1 \\ 1 & 0 \end{pmatrix}$ (b) $\mathbf{A} = \begin{pmatrix} 1 & 0 \\ 0 & -1 \end{pmatrix}$

3.10.9. (a) $\mathbf{L} = \begin{pmatrix} 1 & 0 & 0 \\ 4 & 1 & 0 \\ 3 & 2 & 1 \end{pmatrix}$, $\mathbf{D} = \begin{pmatrix} 1 & 0 & 0 \\ 0 & 2 & 0 \\ 0 & 0 & 3 \end{pmatrix}$, and $\mathbf{U} = \begin{pmatrix} 1 & 4 & 5 \\ 0 & 1 & 3 \\ 0 & 0 & 1 \end{pmatrix}$

(b) Use the same argument given for the uniqueness of the LU factorization with minor modifications.

(c) $\mathbf{A} = \mathbf{A}^T \implies \mathbf{LDU} = \mathbf{U}^T\mathbf{D}^T\mathbf{L}^T = \mathbf{U}^T\mathbf{DL}^T$. These are each LDU factorizations for $\mathbf{A}$, and consequently the uniqueness of the LDU factorization means that $\mathbf{U} = \mathbf{L}^T$.

3.10.10. $\mathbf{A}$ is symmetric with pivots 1, 4, 9. The Cholesky factor is $\mathbf{R} = \begin{pmatrix} 1 & 0 & 0 \\ 2 & 2 & 0 \\ 3 & 3 & 3 \end{pmatrix}$.

It is unworthy of excellent men to lose hours
like slaves in the labor of calculations.
— *Baron Gottfried Wilhelm von Leibnitz (1646–1716)*

Solutions for Chapter 4

Solutions for exercises in section 4. 1

4.1.1. Only (b) and (d) are subspaces.

4.1.2. (a), (b), (f), (g), and (i) are subspaces.

4.1.3. All of $\Re^3$.

4.1.4. If $\mathbf{v} \in \mathcal{V}$ is a nonzero vector in a space $\mathcal{V}$, then all scalar multiples $\alpha\mathbf{v}$ must also be in $\mathcal{V}$.

4.1.5. (a) A line.   (b) The (x,y)-plane.   (c) $\Re^3$

4.1.6. Only (c) and (e) span $\Re^3$. To see that (d) does not span $\Re^3$, ask whether or not every vector $(x, y, z) \in \Re^3$ can be written as a linear combination of the vectors in (d). It's convenient to think in terms columns, so rephrase the question by asking if every $\mathbf{b} = \begin{pmatrix} x \\ y \\ z \end{pmatrix}$ can be written as a linear combination of $\left\{ \mathbf{v}_1 = \begin{pmatrix} 1 \\ 2 \\ 1 \end{pmatrix}, \ \mathbf{v}_2 = \begin{pmatrix} 2 \\ 0 \\ -1 \end{pmatrix}, \ \mathbf{v}_3 = \begin{pmatrix} 4 \\ 4 \\ 1 \end{pmatrix} \right\}$. That is, for each $\mathbf{b} \in \Re^3$, are there scalars $\alpha_1, \alpha_2, \alpha_3$ such that $\alpha_1\mathbf{v}_1 + \alpha_2\mathbf{v}_2 + \alpha_3\mathbf{v}_3 = \mathbf{b}$ or, equivalently, is

$$\begin{pmatrix} 1 & 2 & 4 \\ 2 & 0 & 4 \\ 1 & -1 & 1 \end{pmatrix} \begin{pmatrix} \alpha_1 \\ \alpha_2 \\ \alpha_3 \end{pmatrix} = \begin{pmatrix} x \\ y \\ z \end{pmatrix} \quad \text{consistent for all} \quad \begin{pmatrix} x \\ y \\ z \end{pmatrix}?$$

This is a system of the form $\mathbf{Ax} = \mathbf{b}$, and it is consistent for all $\mathbf{b}$ if and only if $rank\,([\mathbf{A}|\mathbf{b}]) = rank\,(\mathbf{A})$ for all $\mathbf{b}$. Since

$$\left(\begin{array}{ccc|c} 1 & 2 & 4 & x \\ 2 & 0 & 4 & y \\ 1 & -1 & 1 & z \end{array} \right) \rightarrow \left(\begin{array}{ccc|c} 1 & 2 & 4 & x \\ 0 & -4 & -4 & y - 2x \\ 0 & -3 & -3 & z - x \end{array} \right)$$

$$\rightarrow \left(\begin{array}{ccc|c} 1 & 2 & 4 & x \\ 0 & -4 & -4 & y - 2x \\ 0 & 0 & 0 & (x/2) - (3y/4) + z \end{array} \right),$$

it's clear that there exist $\mathbf{b}$'s (e.g., $\mathbf{b} = (1, 0, 0)^T$) for which $\mathbf{Ax} = \mathbf{b}$ is not consistent, and hence not all $\mathbf{b}$'s are a combination of the $\mathbf{v}_i$'s. Therefore, the $\mathbf{v}_i$'s don't span $\Re^3$.

4.1.7. This follows from (4.1.2).

4.1.8. (a) $\mathbf{u}, \mathbf{v} \in \mathcal{X} \cap \mathcal{Y} \implies \mathbf{u}, \mathbf{v} \in \mathcal{X}$ and $\mathbf{u}, \mathbf{v} \in \mathcal{Y}$. Because $\mathcal{X}$ and $\mathcal{Y}$ are closed with respect to addition, it follows that $\mathbf{u} + \mathbf{v} \in \mathcal{X}$ and $\mathbf{u} + \mathbf{v} \in \mathcal{Y}$, and therefore $\mathbf{u} + \mathbf{v} \in \mathcal{X} \cap \mathcal{Y}$. Because $\mathcal{X}$ and $\mathcal{Y}$ are both closed with respect to scalar multiplication, we have that $\alpha \mathbf{u} \in \mathcal{X}$ and $\alpha \mathbf{u} \in \mathcal{Y}$ for all α, and consequently $\alpha \mathbf{u} \in \mathcal{X} \cap \mathcal{Y}$ for all α.

(b) The union of two different lines through the origin in $\Re^2$ is not a subspace.

4.1.9. (a) **(A1)** holds because $\mathbf{x}_1, \mathbf{x}_2 \in \mathbf{A}(\mathcal{S}) \implies \mathbf{x}_1 = \mathbf{A}\mathbf{s}_1$ and $\mathbf{x}_2 = \mathbf{A}\mathbf{s}_2$ for some $\mathbf{s}_1, \mathbf{s}_2 \in \mathcal{S} \implies \mathbf{x}_1 + \mathbf{x}_2 = \mathbf{A}(\mathbf{s}_1 + \mathbf{s}_2)$. Since $\mathcal{S}$ is a subspace, it is closed under vector addition, so $\mathbf{s}_1 + \mathbf{s}_2 \in \mathcal{S}$. Therefore, $\mathbf{x}_1 + \mathbf{x}_2$ is the image of something in $\mathcal{S}$ —namely, $\mathbf{s}_1 + \mathbf{s}_2$ —and this means that $\mathbf{x}_1 + \mathbf{x}_2 \in \mathbf{A}(\mathcal{S})$. To see that **(M1)** holds, consider $\alpha \mathbf{x}$, where α is an arbitrary scalar and $\mathbf{x} \in \mathbf{A}(\mathcal{S})$. Now, $\mathbf{x} \in \mathbf{A}(\mathcal{S}) \implies \mathbf{x} = \mathbf{A}\mathbf{s}$ for some $\mathbf{s} \in \mathcal{S} \implies \alpha \mathbf{x} = \alpha \mathbf{A}\mathbf{s} = \mathbf{A}(\alpha \mathbf{s})$. Since $\mathcal{S}$ is a subspace, we are guaranteed that $\alpha \mathbf{s} \in \mathcal{S}$, and therefore $\alpha \mathbf{x}$ is the image of something in $\mathcal{S}$. This is what it means to say $\alpha \mathbf{x} \in \mathbf{A}(\mathcal{S})$.

(b) Prove equality by demonstrating that $span \{\mathbf{A}\mathbf{s}_1, \mathbf{A}\mathbf{s}_2, \ldots, \mathbf{A}\mathbf{s}_k\} \subseteq \mathbf{A}(\mathcal{S})$ and $\mathbf{A}(\mathcal{S}) \subseteq span \{\mathbf{A}\mathbf{s}_1, \mathbf{A}\mathbf{s}_2, \ldots, \mathbf{A}\mathbf{s}_k\}$. To show $span \{\mathbf{A}\mathbf{s}_1, \mathbf{A}\mathbf{s}_2, \ldots, \mathbf{A}\mathbf{s}_k\} \subseteq \mathbf{A}(\mathcal{S})$, write

$$\mathbf{x} \in span \{\mathbf{A}\mathbf{s}_1, \mathbf{A}\mathbf{s}_2, \ldots, \mathbf{A}\mathbf{s}_k\} \implies \mathbf{x} = \sum_{i=1}^{k} \alpha_i (\mathbf{A}\mathbf{s}_i) = \mathbf{A}\left(\sum_{i=1}^{k} \alpha_i \mathbf{s}_i\right) \in \mathbf{A}(\mathcal{S}).$$

Inclusion in the reverse direction is established by saying

$$\mathbf{x} \in \mathbf{A}(\mathcal{S}) \implies \mathbf{x} = \mathbf{A}\mathbf{s} \text{ for some } \mathbf{s} \in \mathcal{S} \implies \mathbf{s} = \sum_{i=1}^{k} \beta_i \mathbf{s}_i$$

$$\implies \mathbf{x} = \mathbf{A}\left(\sum_{i=1}^{k} \beta_i \mathbf{s}_i\right) = \sum_{i=1}^{k} \beta_i (\mathbf{A}\mathbf{s}_i) \in span \{\mathbf{A}\mathbf{s}_1, \mathbf{A}\mathbf{s}_2, \ldots, \mathbf{A}\mathbf{s}_k\}.$$

4.1.10. (a) Yes, all of the defining properties are satisfied.

(b) Yes, this is essentially $\Re^2$.

(c) No, it is not closed with respect to scalar multiplication.

4.1.11. If $span (\mathcal{M}) = span (\mathcal{N})$, then every vector in $\mathcal{N}$ must be a linear combination of vectors from $\mathcal{M}$. In particular, $\mathbf{v}$ must be a linear combination of the $\mathbf{m}_i$'s, and hence $\mathbf{v} \in span (\mathcal{M})$. To prove the converse, first notice that $span (\mathcal{M}) \subseteq span (\mathcal{N})$. The desired conclusion will follow if it can be demonstrated that $span (\mathcal{M}) \supseteq span (\mathcal{N})$. The hypothesis that $\mathbf{v} \in span (\mathcal{M})$ guarantees that $\mathbf{v} = \sum_{i=1}^{r} \beta_i \mathbf{m}_i$. If $\mathbf{z} \in span (\mathcal{N})$, then

$$\mathbf{z} = \sum_{i=1}^{r} \alpha_i \mathbf{m}_i + \alpha_{r+1} \mathbf{v} = \sum_{i=1}^{r} \alpha_i \mathbf{m}_i + \alpha_{r+1} \sum_{i=1}^{r} \beta_i \mathbf{m}_i$$

$$= \sum_{i=1}^{r} \left(\alpha_i + \alpha_{r+1}\beta_i\right) \mathbf{m}_i,$$

which shows $\mathbf{z} \in span\,(\mathcal{M})$, and therefore $span\,(\mathcal{M}) \supseteq span\,(\mathcal{N})$.

4.1.12. To show $span\,(\mathcal{S}) \subseteq \mathcal{M}$, observe that $\mathbf{x} \in span\,(\mathcal{S}) \implies \mathbf{x} = \sum_i \alpha_i \mathbf{v}_i$. If $\mathcal{V}$ is any subspace containing $\mathcal{S}$, then $\sum_i \alpha_i \mathbf{v}_i \in \mathcal{V}$ because $\mathcal{V}$ is closed under addition and scalar multiplication, and therefore $\mathbf{x} \in \mathcal{M}$. The fact that $\mathcal{M} \subseteq span\,(\mathcal{S})$ follows because if $\mathbf{x} \in \mathcal{M}$, then $\mathbf{x} \in span\,(\mathcal{S})$ because $span\,(\mathcal{S})$ is one particular subspace that contains $\mathcal{S}$.

Solutions for exercises in section 4. 2

4.2.1. $R\,(\mathbf{A}) = span \left\{ \begin{pmatrix} 1 \\ -2 \\ 1 \end{pmatrix}, \begin{pmatrix} 1 \\ 0 \\ 2 \end{pmatrix} \right\}, \quad N\,(\mathbf{A}^T) = span \left\{ \begin{pmatrix} 4 \\ 1 \\ -2 \end{pmatrix} \right\},$

$$N\,(\mathbf{A}) = span \left\{ \begin{pmatrix} -2 \\ 1 \\ 0 \\ 0 \\ 0 \end{pmatrix}, \begin{pmatrix} 2 \\ 0 \\ -3 \\ 1 \\ 0 \end{pmatrix}, \begin{pmatrix} -1 \\ 0 \\ -4 \\ 0 \\ 1 \end{pmatrix} \right\},$$

$$R\,(\mathbf{A}^T) = span \left\{ \begin{pmatrix} 1 \\ 2 \\ 0 \\ -2 \\ 1 \end{pmatrix}, \begin{pmatrix} 0 \\ 0 \\ 1 \\ 3 \\ 4 \end{pmatrix} \right\}.$$

4.2.2. (a) This is simply a restatement of equation (4.2.3).

(b) $\mathbf{Ax} = \mathbf{b}$ has a unique solution if and only if $rank\,(\mathbf{A}) = n$ (i.e., there are no free variables—see §2.5), and (4.2.10) says $rank\,(\mathbf{A}) = n \iff N\,(\mathbf{A}) = \{\mathbf{0}\}$.

4.2.3. (a) It is consistent because $\mathbf{b} \in R\,(\mathbf{A})$.

(b) It is nonunique because $N\,(\mathbf{A}) \neq \{\mathbf{0}\}$—see Exercise 4.2.2.

4.2.4. Yes, because $rank[\mathbf{A}|\mathbf{b}] = rank\,(\mathbf{A}) = 3 \implies \exists\,\mathbf{x}$ such that $\mathbf{Ax} = \mathbf{b}$—i.e., $\mathbf{Ax} = \mathbf{b}$ is consistent.

4.2.5. (a) If $R\,(\mathbf{A}) = \Re^n$, then

$$R\,(\mathbf{A}) = R\,(\mathbf{I}_n) \implies \mathbf{A} \overset{\text{col}}{\sim} \mathbf{I}_n \implies rank\,(\mathbf{A}) = rank\,(\mathbf{I}_n) = n.$$

(b) $R\,(\mathbf{A}) = R\,(\mathbf{A}^T) = \Re^n$ and $N\,(\mathbf{A}) = N\,(\mathbf{A}^T) = \{\mathbf{0}\}$.

4.2.6. $\mathbf{E_A} \neq \mathbf{E_B}$ means that $R\,(\mathbf{A}^T) \neq R\,(\mathbf{B}^T)$ and $N\,(\mathbf{A}) \neq N\,(\mathbf{B})$. However, $\mathbf{E}_{\mathbf{A}^T} = \mathbf{E}_{\mathbf{B}^T}$ implies that $R\,(\mathbf{A}) = R\,(\mathbf{B})$ and $N\,(\mathbf{A}^T) = N\,(\mathbf{B}^T)$.

4.2.7. Demonstrate that $rank\,(\mathbf{A}_{n\times n}) = n$ by using (4.2.10). If $\mathbf{x} \in N\,(\mathbf{A})$, then

$$\mathbf{Ax} = \mathbf{0} \implies \mathbf{A}_1\mathbf{x} = \mathbf{0} \quad \text{and} \quad \mathbf{A}_2\mathbf{x} = \mathbf{0}$$
$$\implies \mathbf{x} \in N\,(\mathbf{A}_1) = R\,(\mathbf{A}_2^T) \implies \exists\,\mathbf{y}^T \text{ such that } \mathbf{x}^T = \mathbf{y}^T\mathbf{A}_2$$
$$\implies \mathbf{x}^T\mathbf{x} = \mathbf{y}^T\mathbf{A}_2\mathbf{x} = \mathbf{0} \implies \sum_i x_i^2 = 0 \implies \mathbf{x} = \mathbf{0}.$$

4.2.8. $\mathbf{y}^T\mathbf{b} = 0 \; \forall \; \mathbf{y} \in N\left(\mathbf{A}^T\right) = R\left(\mathbf{P}_2^T\right) \implies \mathbf{P}_2\mathbf{b} = 0 \implies \mathbf{b} \in N\left(\mathbf{P}_2\right) = R\left(\mathbf{A}\right)$

4.2.9. $\mathbf{x} \in R\left(\mathbf{A} \mid \mathbf{B}\right) \iff \exists \; \mathbf{y} \text{ such that } \mathbf{x} = \left(\mathbf{A} \mid \mathbf{B}\right)\mathbf{y} = \left(\mathbf{A} \mid \mathbf{B}\right)\begin{pmatrix} \mathbf{y}_1 \\ \mathbf{y}_2 \end{pmatrix} = \mathbf{A}\mathbf{y}_1 +$ $\mathbf{B}\mathbf{y}_2 \iff \mathbf{x} \in R\left(\mathbf{A}\right) + R\left(\mathbf{B}\right)$

4.2.10. (a) $\mathbf{p} + N\left(\mathbf{A}\right)$ is the set of all possible solutions to $\mathbf{A}\mathbf{x} = \mathbf{b}$. Recall from (2.5.7) that the general solution of a nonhomogeneous equation is a particular solution plus the general solution of the homogeneous equation $\mathbf{A}\mathbf{x} = \mathbf{0}$. The general solution of the homogeneous equation is simply a way of describing all possible solutions of $\mathbf{A}\mathbf{x} = \mathbf{0}$, which is $N\left(\mathbf{A}\right)$.
(b) $rank\left(\mathbf{A}_{3\times 3}\right) = 1$ means that $N\left(\mathbf{A}\right)$ is spanned by two vectors, and hence $N\left(\mathbf{A}\right)$ is a plane through the origin. From the parallelogram law, $\mathbf{p} + N\left(\mathbf{A}\right)$ is a plane parallel to $N\left(\mathbf{A}\right)$ passing through the point defined by $\mathbf{p}$.
(c) This time $N\left(\mathbf{A}\right)$ is spanned by a single vector, and $\mathbf{p} + N\left(\mathbf{A}\right)$ is a line parallel to $N\left(\mathbf{A}\right)$ passing through the point defined by $\mathbf{p}$.

4.2.11. $\mathbf{a} \in R\left(\mathbf{A}^T\right) \iff \exists \; \mathbf{y} \text{ such that } \mathbf{a}^T = \mathbf{y}^T\mathbf{A}$. If $\mathbf{A}\mathbf{x} = \mathbf{b}$, then

$$\mathbf{a}^T\mathbf{x} = \mathbf{y}^T\mathbf{A}\mathbf{x} = \mathbf{y}^T\mathbf{b},$$

which is independent of $\mathbf{x}$.

4.2.12. (a) $\mathbf{b} \in R\left(\mathbf{A}\mathbf{B}\right) \implies \exists \; \mathbf{x} \text{ such that } \mathbf{b} = \mathbf{A}\mathbf{B}\mathbf{x} = \mathbf{A}(\mathbf{B}\mathbf{x}) \implies \mathbf{b} \in R\left(\mathbf{A}\right)$ because $\mathbf{b}$ is the image of $\mathbf{B}\mathbf{x}$.
(b) $\mathbf{x} \in N\left(\mathbf{B}\right) \implies \mathbf{B}\mathbf{x} = \mathbf{0} \implies \mathbf{A}\mathbf{B}\mathbf{x} = \mathbf{0} \implies \mathbf{x} \in N\left(\mathbf{A}\mathbf{B}\right)$.

4.2.13. Given any $\mathbf{z} \in R\left(\mathbf{A}\mathbf{B}\right)$, the object is to show that $\mathbf{z}$ can be written as some linear combination of the $\mathbf{A}\mathbf{b}_i$'s. Argue as follows. $\mathbf{z} \in R\left(\mathbf{A}\mathbf{B}\right) \implies \mathbf{z} = \mathbf{A}\mathbf{B}\mathbf{y}$ for some $\mathbf{y}$. But it is always true that $\mathbf{B}\mathbf{y} \in R\left(\mathbf{B}\right)$, so

$$\mathbf{B}\mathbf{y} = \alpha_1\mathbf{b}_1 + \alpha_2\mathbf{b}_2 + \cdots + \alpha_n\mathbf{b}_n,$$

and therefore $\mathbf{z} = \mathbf{A}\mathbf{B}\mathbf{y} = \alpha_1\mathbf{A}\mathbf{b}_1 + \alpha_2\mathbf{A}\mathbf{b}_2 + \cdots + \alpha_n\mathbf{A}\mathbf{b}_n$.

Solutions for exercises in section 4. 3

4.3.1. (a) and (b) are linearly dependent—all others are linearly independent. To write one vector as a combination of others in a dependent set, place the vectors as columns in $\mathbf{A}$ and find $\mathbf{E_A}$. This reveals the dependence relationships among columns of $\mathbf{A}$.

4.3.2. (a) According to (4.3.12), the basic columns in $\mathbf{A}$ always constitute one maximal linearly independent subset.
(b) Ten—5 sets using two vectors, 4 sets using one vector, and the empty set.

4.3.3. $rank\left(\mathbf{H}\right) \leq 3$, and according to (4.3.11), $rank\left(\mathbf{H}\right)$ is the maximal number of independent rows in $\mathbf{H}$.

4.3.4. The question is really whether or not the columns in

$$\hat{\mathbf{A}} = \begin{array}{c} \\ \#1 \\ \#2 \\ \#3 \\ \#4 \end{array} \begin{pmatrix} \overset{S}{1} & \overset{L}{1} & \overset{F}{1} & 10 \\ 1 & 2 & 1 & 12 \\ 1 & 2 & 2 & 15 \\ 1 & 3 & 2 & 17 \end{pmatrix}$$

are linearly independent. Reducing $\hat{\mathbf{A}}$ to $\mathbf{E}_{\hat{\mathbf{A}}}$ shows that $5 + 2S + 3L - F = 0$.

4.3.5. (a) This follows directly from the definition of linear dependence because there are nonzero values of α such that $\alpha\mathbf{0} = \mathbf{0}$.
(b) This is a consequence of (4.3.13).

4.3.6. If each $t_{ii} \neq 0$, then $\mathbf{T}$ is nonsingular, and the result follows from (4.3.6) and (4.3.7).

4.3.7. It is linearly independent because

$$\alpha_1 \begin{pmatrix} 1 & 0 \\ 0 & 0 \end{pmatrix} + \alpha_2 \begin{pmatrix} 1 & 1 \\ 0 & 0 \end{pmatrix} + \alpha_3 \begin{pmatrix} 1 & 1 \\ 1 & 0 \end{pmatrix} + \alpha_4 \begin{pmatrix} 1 & 1 \\ 1 & 1 \end{pmatrix} = \begin{pmatrix} 0 & 0 \\ 0 & 0 \end{pmatrix}$$

$$\iff \alpha_1 \begin{pmatrix} 1 \\ 0 \\ 0 \\ 0 \end{pmatrix} + \alpha_2 \begin{pmatrix} 1 \\ 1 \\ 0 \\ 0 \end{pmatrix} + \alpha_3 \begin{pmatrix} 1 \\ 1 \\ 1 \\ 0 \end{pmatrix} + \alpha_4 \begin{pmatrix} 1 \\ 1 \\ 1 \\ 1 \end{pmatrix} = \begin{pmatrix} 0 \\ 0 \\ 0 \\ 0 \end{pmatrix}$$

$$\iff \begin{pmatrix} 1 & 1 & 1 & 1 \\ 0 & 1 & 1 & 1 \\ 0 & 0 & 1 & 1 \\ 0 & 0 & 0 & 1 \end{pmatrix} \begin{pmatrix} \alpha_1 \\ \alpha_2 \\ \alpha_3 \\ \alpha_4 \end{pmatrix} = \begin{pmatrix} 0 \\ 0 \\ 0 \\ 0 \end{pmatrix} \iff \begin{pmatrix} \alpha_1 \\ \alpha_2 \\ \alpha_3 \\ \alpha_4 \end{pmatrix} = \begin{pmatrix} 0 \\ 0 \\ 0 \\ 0 \end{pmatrix}.$$

4.3.8. $\mathbf{A}$ is nonsingular because it is diagonally dominant.

4.3.9. $\mathcal{S}$ is linearly independent using exact arithmetic, but using 3-digit arithmetic yields the conclusion that $\mathcal{S}$ is dependent.

4.3.10. If $\mathbf{e}$ is the column vector of all 1's, then $\mathbf{Ae} = \mathbf{0}$, so that $N(\mathbf{A}) \neq \{\mathbf{0}\}$.

4.3.11. (Solution 1.) $\sum_i \alpha_i \mathbf{Pu}_i = \mathbf{0} \implies \mathbf{P} \sum_i \alpha_i \mathbf{u}_i = \mathbf{0} \implies \sum_i \alpha_i \mathbf{u}_i = \mathbf{0} \implies$ each $\alpha_i = 0$ because the $\mathbf{u}_i$'s are linearly independent.

(Solution 2.) If $\mathbf{A}_{m \times n}$ is the matrix containing the $\mathbf{u}_i$'s as columns, then $\mathbf{PA} = \mathbf{B}$ is the matrix containing the vectors in $\mathbf{P}(\mathcal{S})$ as its columns. Now,

$$\mathbf{A} \overset{\text{row}}{\sim} \mathbf{B} \implies rank(\mathbf{B}) = rank(\mathbf{A}) = n,$$

and hence (4.3.3) insures that the columns of $\mathbf{B}$ are linearly independent. The result need not be true if $\mathbf{P}$ is singular—take $\mathbf{P} = \mathbf{0}$ for example.

4.3.12. If $\mathbf{A}_{m \times n}$ is the matrix containing the $\mathbf{u}_i$'s as columns, and if

$$\mathbf{Q}_{n \times n} = \begin{pmatrix} 1 & 1 & \cdots & 1 \\ 0 & 1 & \cdots & 1 \\ \vdots & \vdots & \ddots & \vdots \\ 0 & 0 & \cdots & 1 \end{pmatrix},$$

then the columns of $\mathbf{B} = \mathbf{AQ}$ are the vectors in $\mathcal{S}'$. Clearly, $\mathbf{Q}$ is nonsingular so that $\mathbf{A} \overset{\text{col}}{\sim} \mathbf{B}$, and thus $rank(\mathbf{A}) = rank(\mathbf{B})$. The desired result now follows from (4.3.3).

4.3.13. (a) and (b) are linearly independent because the Wronski matrix $\mathbf{W}(0)$ is nonsingular in each case. (c) is dependent because $\sin^2 x - \cos^2 x + \cos 2x = 0$.

4.3.14. If $\mathcal{S}$ were dependent, then there would exist a constant α such that $x^3 = \alpha|x|^3$ for all values of x. But this would mean that

$$\alpha = \frac{x^3}{|x|^3} = \begin{cases} 1 & \text{if } x > 0, \\ -1 & \text{if } x < 0, \end{cases}$$

which is clearly impossible since α must be constant. The associated Wronski matrix is

$$\mathbf{W}(x) = \begin{cases} \begin{pmatrix} x^3 & x^3 \\ 3x^2 & 3x^2 \end{pmatrix} & \text{when } x \geq 0, \\ \begin{pmatrix} x^3 & -x^3 \\ 3x^2 & -3x^2 \end{pmatrix} & \text{when } x < 0, \end{cases}$$

which is singular for all values of x.

4.3.15. Start with the fact that

$$\mathbf{A}^T \text{ diag. dom.} \implies |b_{ii}| > |d_i| + \sum_{j \neq i} |b_{ji}| \quad \text{and} \quad |\alpha| > \sum_j |c_j|$$

$$\implies \sum_{j \neq i} |b_{ji}| < |b_{ii}| - |d_i| \quad \text{and} \quad \frac{1}{|\alpha|} \sum_{j \neq i} |c_j| < 1 - \frac{|c_i|}{|\alpha|},$$

and then use the forward and backward triangle inequality to write

$$\sum_{j \neq i} |x_{ij}| = \sum_{j \neq i} \left| b_{ji} - \frac{d_i c_j}{\alpha} \right| \leq \sum_{j \neq i} |b_{ji}| + \frac{|d_i|}{|\alpha|} \sum_{j \neq i} |c_j|$$

$$< (|b_{ii}| - |d_i|) + |d_i| \left(1 - \frac{|c_i|}{|\alpha|} \right) = |b_{ii}| - \frac{|d_i| \, |c_i|}{|\alpha|}$$

$$\leq \left| b_{ii} - \frac{d_i c_i}{\alpha} \right| = |x_{ii}|.$$

Now, diagonal dominance of $\mathbf{A}^T$ insures that α is the entry of largest magnitude in the first column of $\mathbf{A}$, so no row interchange is needed at the first step of Gaussian elimination. After one step, the diagonal dominance of $\mathbf{X}$ guarantees that the magnitude of the second pivot is maximal with respect to row interchanges. Proceeding by induction establishes that no step requires partial pivoting.

Solutions for exercises in section 4. 4

4.4.1. $\dim R(\mathbf{A}) = \dim R(\mathbf{A}^T) = rank(\mathbf{A}) = 2,\quad \dim N(\mathbf{A}) = n - r = 4 - 2 = 2,$
and $\dim N(\mathbf{A}^T) = m - r = 3 - 2 = 1.$

4.4.2. $\mathcal{B}_{R(\mathbf{A})} = \left\{ \begin{pmatrix} 1 \\ 3 \\ 2 \end{pmatrix}, \begin{pmatrix} 0 \\ 1 \\ 1 \end{pmatrix} \right\}, \quad \mathcal{B}_{N(\mathbf{A}^T)} = \left\{ \begin{pmatrix} 1 \\ -1 \\ 1 \end{pmatrix} \right\}$

$\mathcal{B}_{R(\mathbf{A}^T)} = \left\{ \begin{pmatrix} 1 \\ 2 \\ 0 \\ 2 \\ 1 \end{pmatrix}, \begin{pmatrix} 0 \\ 0 \\ 1 \\ 3 \\ 3 \end{pmatrix} \right\}, \quad \mathcal{B}_{N(\mathbf{A})} = \left\{ \begin{pmatrix} -2 \\ 1 \\ 0 \\ 0 \\ 0 \end{pmatrix}, \begin{pmatrix} -2 \\ 0 \\ -3 \\ 1 \\ 0 \end{pmatrix}, \begin{pmatrix} -1 \\ 0 \\ -3 \\ 0 \\ 1 \end{pmatrix} \right\}$

4.4.3. $\dim\big(span\,(\mathcal{S})\big) = 3$

4.4.4. (a) $n + 1$ (See Example 4.4.1) (b) mn (c) $\frac{n^2 + n}{2}$

4.4.5. Use the technique of Example 4.4.5. Find $\mathbf{E_A}$ to determine

$$\left\{ \mathbf{h}_1 = \begin{pmatrix} -2 \\ 1 \\ 0 \\ 0 \\ 0 \end{pmatrix}, \ \mathbf{h}_2 = \begin{pmatrix} -2 \\ 0 \\ 1 \\ 1 \\ 0 \end{pmatrix}, \ \mathbf{h}_3 = \begin{pmatrix} -1 \\ 0 \\ -2 \\ 0 \\ 1 \end{pmatrix} \right\}$$

is a basis for $N(\mathbf{A})$. Reducing the matrix $\big(\mathbf{v},\ \mathbf{h}_1,\ \mathbf{h}_2,\ \mathbf{h}_3\big)$ to row echelon form reveals that its first, second, and fourth columns are basic, and hence $\{\mathbf{v},\ \mathbf{h}_1,\ \mathbf{h}_3\}$ is a basis for $N(\mathbf{A})$ that contains $\mathbf{v}$.

4.4.6. Placing the vectors from $\mathcal{A}$ and $\mathcal{B}$ as rows in matrices and reducing

$$\mathbf{A} = \begin{pmatrix} 1 & 2 & 3 \\ 5 & 8 & 7 \\ 3 & 4 & 1 \end{pmatrix} \longrightarrow \mathbf{E_A} = \begin{pmatrix} 1 & 0 & -5 \\ 0 & 1 & 4 \\ 0 & 0 & 0 \end{pmatrix}$$

and

$$\mathbf{B} = \begin{pmatrix} 2 & 3 & 2 \\ 1 & 1 & -1 \end{pmatrix} \longrightarrow \mathbf{E_B} = \begin{pmatrix} 1 & 0 & -5 \\ 0 & 1 & 4 \end{pmatrix}$$

shows $\mathbf{A}$ and $\mathbf{B}$ have the same row space (recall Example 4.2.2), and hence $\mathcal{A}$ and $\mathcal{B}$ span the same space. Because $\mathcal{B}$ is linearly independent, it follows that $\mathcal{B}$ is a basis for $span\,(\mathcal{A})$.

4.4.7. $3 = \dim N(\mathbf{A}) = n - r = 4 - rank(\mathbf{A}) \implies rank(\mathbf{A}) = 1.$ Therefore, any rank-one matrix with no zero entries will do the job.

4.4.8. If $\mathbf{v} = \alpha_1 \mathbf{b}_1 + \alpha_2 \mathbf{b}_2 + \cdots + \alpha_n \mathbf{b}_n$ and $\mathbf{v} = \beta_1 \mathbf{b}_1 + \beta_2 \mathbf{b}_2 + \cdots + \beta_n \mathbf{b}_n$, then subtraction produces

$$\mathbf{0} = (\alpha_1 - \beta_1)\mathbf{b}_1 + (\alpha_2 - \beta_2)\mathbf{b}_2 + \cdots + (\alpha_n - \beta_n)\mathbf{b}_n.$$

But $\mathcal{B}$ is a linearly independent set, so this equality can hold only if $(\alpha_i - \beta_i) = 0$ for each $i = 1, 2, \ldots, n$, and hence the α_i's are unique.

4.4.9. Prove that if $\{\mathbf{s}_1, \mathbf{s}_2, \ldots, \mathbf{s}_k\}$ is a basis for $\mathcal{S}$, then $\{\mathbf{A}\mathbf{s}_1, \mathbf{A}\mathbf{s}_2, \ldots, \mathbf{A}\mathbf{s}_k\}$ is a basis for $\mathbf{A}(\mathcal{S})$. The result of Exercise 4.1.9 insures that

$$span \{\mathbf{A}\mathbf{s}_1, \mathbf{A}\mathbf{s}_2, \ldots, \mathbf{A}\mathbf{s}_k\} = \mathbf{A}(\mathcal{S}),$$

so we need only establish the independence of $\{\mathbf{A}\mathbf{s}_1, \mathbf{A}\mathbf{s}_2, \ldots, \mathbf{A}\mathbf{s}_k\}$. To do this, write

$$\sum_{i=1}^{k} \alpha_i (\mathbf{A}\mathbf{s}_i) = \mathbf{0} \implies \mathbf{A}\left(\sum_{i=1}^{k} \alpha_i \mathbf{s}_i\right) = \mathbf{0} \implies \sum_{i=1}^{k} \alpha_i \mathbf{s}_i \in N(\mathbf{A})$$

$$\implies \sum_{i=1}^{k} \alpha_i \mathbf{s}_i = \mathbf{0} \quad \text{because} \quad \mathcal{S} \cap N(\mathbf{A}) = \mathbf{0}$$

$$\implies \alpha_1 = \alpha_2 = \cdots = \alpha_k = 0$$

because $\{\mathbf{s}_1, \mathbf{s}_2, \ldots, \mathbf{s}_k\}$ is linearly independent. Since $\{\mathbf{A}\mathbf{s}_1, \mathbf{A}\mathbf{s}_2, \ldots, \mathbf{A}\mathbf{s}_k\}$ is a basis for $\mathbf{A}(\mathcal{S})$, it follows that $\dim \mathbf{A}(\mathcal{S}) = k = \dim(\mathcal{S})$.

4.4.10. $rank(\mathbf{A}) = rank(\mathbf{A} - \mathbf{B} + \mathbf{B}) \leq rank(\mathbf{A} - \mathbf{B}) + rank(\mathbf{B})$ implies that

$$rank(\mathbf{A}) - rank(\mathbf{B}) \leq rank(\mathbf{A} - \mathbf{B}).$$

Furthermore, $rank(\mathbf{B}) = rank(\mathbf{B} - \mathbf{A} + \mathbf{A}) \leq rank(\mathbf{B} - \mathbf{A}) + rank(\mathbf{A}) = rank(\mathbf{A} - \mathbf{B}) + rank(\mathbf{A})$ implies that

$$-\left(rank(\mathbf{A}) - rank(\mathbf{B})\right) \leq rank(\mathbf{A} - \mathbf{B}).$$

4.4.11. Example 4.4.8 guarantees that $rank(\mathbf{A} + \mathbf{E}) \leq rank(\mathbf{A}) + rank(\mathbf{E}) = r + k$. Use Exercise 4.4.10 to write

$$rank(\mathbf{A} + \mathbf{E}) = rank(\mathbf{A} - (-\mathbf{E})) \geq rank(\mathbf{A}) - rank(-\mathbf{E}) = r - k.$$

4.4.12. Let $\mathbf{v}_1 \in \mathcal{V}$ such that $\mathbf{v}_1 \neq \mathbf{0}$. If $span\{\mathbf{v}_1\} = \mathcal{V}$, then $\mathcal{S}_1 = \{\mathbf{v}_1\}$ is an independent spanning set for $\mathcal{V}$, and we are finished. If $span\{\mathbf{v}_1\} \neq \mathcal{V}$, then there is a vector $\mathbf{v}_2 \in \mathcal{V}$ such that $\mathbf{v}_2 \notin span\{\mathbf{v}_1\}$, and hence the extension set $\mathcal{S}_2 = \{\mathbf{v}_1, \mathbf{v}_2\}$ is independent. If $span(\mathcal{S}_2) = \mathcal{V}$, then we are finished. Otherwise, we can proceed as described in Example 4.4.5 and continue to build independent extension sets $\mathcal{S}_3, \mathcal{S}_4, \ldots$. Statement (4.3.16) guarantees that the process must eventually yield a linearly independent spanning set $\mathcal{S}_k$ with $k \leq n$.

4.4.13. Since $\mathbf{0} = \mathbf{e}^T \mathbf{E} = \mathbf{E}_{1*} + \mathbf{E}_{2*} + \cdots + \mathbf{E}_{m*}$, any row can be written as a combination of the other $m - 1$ rows, so any set of $m - 1$ rows from $\mathbf{E}$ spans $N(\mathbf{E}^T)$. Furthermore, $rank(\mathbf{E}) = m - 1$ insures that no fewer than $m - 1$ vectors can span $N\left(\mathbf{E}^{T}\right)$, and therefore any set of $m-1$ rows from $\mathbf{E}$ is a minimal spanning set, and hence a basis.

4.4.14. $\left[\mathbf{EE}^{T}\right]_{ij} = \mathbf{E}_{i*}\left(\mathbf{E}^{T}\right)_{*j} = \mathbf{E}_{i*}\left(\mathbf{E}_{j*}\right)^{T} = \sum_{k} e_{ik}e_{jk}$. Observe that edge E_{k} touches node N_{i} if and only if $e_{ik} = \pm 1$ or, equivalently, $e_{ik}^{2} = 1$. Thus $\left[\mathbf{EE}^{T}\right]_{ii} = \sum_{k} e_{ik}^{2} =$ the number of edges touching N_{i}. If $i \neq j$, then

$$e_{ik}e_{jk} = \begin{cases} -1 & \text{if } E_{k} \text{ is between } N_{i} \text{ and } N_{j} \\ 0 & \text{if } E_{k} \text{ is not between } N_{i} \text{ and } N_{j} \end{cases}$$

so that $\left[\mathbf{EE}^{T}\right]_{ij} = \sum_{k} e_{ik}e_{jk} = -$ (the number of edges between N_{i} and N_{j}).

4.4.15. Apply (4.4.19) to $span\left(\mathcal{M}\cup\mathcal{N}\right) = span\left(\mathcal{M}\right) + span\left(\mathcal{N}\right)$ (see Exercise 4.1.7).

4.4.16. (a) Exercise 4.2.9 says $R\left(\mathbf{A} \mid \mathbf{B}\right) = R\left(\mathbf{A}\right) + R\left(\mathbf{B}\right)$. Since rank is the same as dimension of the range, (4.4.19) yields

$$rank\left(\mathbf{A} \mid \mathbf{B}\right) = \dim R\left(\mathbf{A} \mid \mathbf{B}\right) = \dim\left(R\left(\mathbf{A}\right) + R\left(\mathbf{B}\right)\right)$$

$$= \dim R\left(\mathbf{A}\right) + \dim R\left(\mathbf{B}\right) - \dim\left(R\left(\mathbf{A}\right) \cap R\left(\mathbf{B}\right)\right)$$

$$= rank\left(\mathbf{A}\right) + rank\left(\mathbf{B}\right) - \dim\left(R\left(\mathbf{A}\right) \cap R\left(\mathbf{B}\right)\right).$$

(b) Use the results of part (a) to write

$$\dim N\left(\mathbf{A} \mid \mathbf{B}\right) = n + k - rank\left(\mathbf{A} \mid \mathbf{B}\right)$$

$$= \left(n - rank\left(\mathbf{A}\right)\right) + \left(k - rank\left(\mathbf{B}\right)\right) + \dim\left(R\left(\mathbf{A}\right) \cap R\left(\mathbf{B}\right)\right)$$

$$= \dim N\left(\mathbf{A}\right) + \dim N\left(\mathbf{B}\right) + \dim\left(R\left(\mathbf{A}\right) \cap R\left(\mathbf{B}\right)\right).$$

(c) Let $\mathbf{A} = \begin{pmatrix} -1 & 1 & -2 \\ -1 & 0 & -4 \\ -1 & 0 & -5 \\ -1 & 0 & -6 \\ -1 & 0 & -6 \end{pmatrix}$ and $\mathbf{B} = \begin{pmatrix} 3 & -2 \\ 2 & -1 \\ 1 & 0 \\ 0 & 1 \\ 0 & 1 \end{pmatrix}$ contain bases for $R\left(\mathbf{C}\right)$ and $N\left(\mathbf{C}\right)$, respectively, so that $R\left(\mathbf{A}\right) = R\left(\mathbf{C}\right)$ and $R\left(\mathbf{B}\right) = N\left(\mathbf{C}\right)$. Use either part (a) or part (b) to obtain

$$\dim\left(R\left(\mathbf{C}\right) \cap N\left(\mathbf{C}\right)\right) = \dim\left(R\left(\mathbf{A}\right) \cap R\left(\mathbf{B}\right)\right) = 2.$$

Using $R\left(\mathbf{A} \mid \mathbf{B}\right) = R\left(\mathbf{A}\right) + R\left(\mathbf{B}\right)$ produces

$$\dim\left(R\left(\mathbf{C}\right) + N\left(\mathbf{C}\right)\right) = \dim\left(R\left(\mathbf{A}\right) + R\left(\mathbf{B}\right)\right) = rank\left(\mathbf{A} \mid \mathbf{B}\right) = 3.$$

4.4.17. Suppose $\mathbf{A}$ is $m \times n$. Existence of a solution for every $\mathbf{b}$ implies $R\left(\mathbf{A}\right) = \Re^{m}$. Recall from §2.5 that uniqueness of the solution implies $rank\left(\mathbf{A}\right) = n$. Thus $m = \dim R\left(\mathbf{A}\right) = rank\left(\mathbf{A}\right) = n$ so that $\mathbf{A}$ is $m \times m$ of rank m.

4.4.18. (a) $\mathbf{x} \in \mathcal{S} \implies \mathbf{x} \in span\,(\mathcal{S}_{max})$ —otherwise, the extension set $\mathcal{E} = \mathcal{S}_{max} \cup \{\mathbf{x}\}$ would be linearly independent—which is impossible because $\mathcal{E}$ would contain more independent solutions than $\mathcal{S}_{max}$. Now show $span\,(\mathcal{S}_{max}) \subseteq span\,\{\mathbf{p}\} + N\,(\mathbf{A})$. Since $\mathcal{S} = \mathbf{p} + N\,(\mathbf{A})$ (see Exercise 4.2.10), $\mathbf{s}_i \in \mathcal{S}$ means there must exist a corresponding vector $\mathbf{n}_i \in N\,(\mathbf{A})$ such that $\mathbf{s}_i = \mathbf{p} + \mathbf{n}_i$, and hence

$$\mathbf{x} \in span\,(\mathcal{S}_{max}) \implies \mathbf{x} = \sum_{i=1}^{t} \alpha_i \mathbf{s}_i = \sum_{i=1}^{t} \alpha_i\,(\mathbf{p} + \mathbf{n}_i) = \sum_{i=1}^{t} \alpha_i \mathbf{p} + \sum_{i=1}^{t} \alpha_i \mathbf{n}_i$$

$$\implies \mathbf{x} \in span\,\{\mathbf{p}\} + N\,(\mathbf{A})$$

$$\implies span\,(\mathcal{S}_{max}) \subseteq span\,\{\mathbf{p}\} + N\,(\mathbf{A}).$$

To prove the reverse inclusion, observe that if $\mathbf{x} \in span\,\{\mathbf{p}\} + N\,(\mathbf{A})$, then there exists a scalar α and a vector $\mathbf{n} \in N\,(\mathbf{A})$ such that

$$\mathbf{x} = \alpha \mathbf{p} + \mathbf{n} = (\alpha - 1)\mathbf{p} + (\mathbf{p} + \mathbf{n}).$$

Because $\mathbf{p}$ and $(\mathbf{p} + \mathbf{n})$ are both solutions, $\mathcal{S} \subseteq span(\mathcal{S}_{max})$ guarantees that $\mathbf{p}$ and $(\mathbf{p} + \mathbf{n})$ each belong to $span\,(\mathcal{S}_{max})$, and the closure properties of a subspace insure that $\mathbf{x} \in span\,(\mathcal{S}_{max})$. Thus $span\,\{\mathbf{p}\} + N\,(\mathbf{A}) \subseteq span\,(\mathcal{S}_{max})$.
(b) The problem is really to determine the value of t in $\mathcal{S}_{max}$. The fact that $\mathcal{S}_{max}$ is a basis for $span\,(\mathcal{S}_{max})$ together with (4.4.19) produces

$$t = \dim\,\big(span\,(\mathcal{S}_{max})\big) = \dim\,\Big(span\,\{\mathbf{p}\} + N\,(\mathbf{A})\Big)$$

$$= \dim\,\big(span\,\{\mathbf{p}\}\big) + \dim N\,(\mathbf{A}) - \dim\,\Big(span\,\{\mathbf{p}\} \cap N\,(\mathbf{A})\Big)$$

$$= 1 + (n - r) - 0.$$

4.4.19. To show $\mathcal{S}_{max}$ is linearly independent, suppose

$$\mathbf{0} = \alpha_0 \mathbf{p} + \sum_{i=1}^{n-r} \alpha_i\,(\mathbf{p} + \mathbf{h}_i) = \left(\sum_{i=0}^{n-r} \alpha_i\right) \mathbf{p} + \sum_{i=1}^{n-r} \alpha_i \mathbf{h}_i.$$

Multiplication by $\mathbf{A}$ yields $\mathbf{0} = \left(\sum_{i=0}^{n-r} \alpha_i\right) \mathbf{b}$, which implies $\sum_{i=0}^{n-r} \alpha_i = 0$, and hence $\sum_{i=1}^{n-r} \alpha_i \mathbf{h}_i = \mathbf{0}$. Because $\mathcal{H}$ is independent, we may conclude that $\alpha_1 = \alpha_2 = \cdots = \alpha_{n-r} = 0$. Consequently, $\alpha_0 \mathbf{p} = \mathbf{0}$, and therefore $\alpha_0 = 0$ (because $\mathbf{p} \neq \mathbf{0}$), so that $\mathcal{S}_{max}$ is an independent set. By Exercise 4.4.18, it must also be maximal because it contains $n - r + 1$ vectors.

4.4.20. The proof depends on the observation that if $\mathbf{B} = \mathbf{P}^T \mathbf{A} \mathbf{P}$, where $\mathbf{P}$ is a permutation matrix, then the graph $\mathcal{G}(\mathbf{B})$ is the same as $\mathcal{G}(\mathbf{A})$ except that the nodes in $\mathcal{G}(\mathbf{B})$ have been renumbered according to the permutation defining $\mathbf{P}$. This follows because $\mathbf{P}^T = \mathbf{P}^{-1}$ implies $\mathbf{A} = \mathbf{P}\mathbf{B}\mathbf{P}^T$, so if the rows (and columns) in $\mathbf{P}$ are the unit vectors that appear according to the permutation $\boldsymbol{\pi} = \begin{pmatrix} 1 & 2 & \cdots & n \\ \pi_1 & \pi_2 & \cdots & \pi_n \end{pmatrix}$, then

$$a_{ij} = \left[\mathbf{P}\mathbf{B}\mathbf{P}^T\right]_{ij} = \left[\begin{pmatrix} \mathbf{e}_{\pi_1}^T \\ \vdots \\ \mathbf{e}_{\pi_n}^T \end{pmatrix} \mathbf{B}\begin{pmatrix} \mathbf{e}_{\pi_1} & \cdots & \mathbf{e}_{\pi_n} \end{pmatrix}\right]_{ij} = \mathbf{e}_{\pi_i}^T \mathbf{B}\mathbf{e}_{\pi_j} = b_{\pi_i \pi_j}.$$

Consequently, $a_{ij} \neq 0$ if and only if $b_{\pi_i \pi_j} \neq 0$, and thus $\mathcal{G}(\mathbf{A})$ and $\mathcal{G}(\mathbf{B})$ are the same except for the fact that node N_k in $\mathcal{G}(\mathbf{A})$ is node N_{π_k} in $\mathcal{G}(\mathbf{B})$ for each $k = 1, 2, \ldots, n$. Now we can prove $\mathcal{G}(\mathbf{A})$ is *not* strongly connected $\iff \mathbf{A}$ is reducible. If $\mathbf{A}$ is reducible, then there is a permutation matrix such that $\mathbf{P}^T \mathbf{A} \mathbf{P} = \mathbf{B} = \begin{pmatrix} \mathbf{X} & \mathbf{Y} \\ \mathbf{0} & \mathbf{Z} \end{pmatrix}$, where $\mathbf{X}$ is $r \times r$ and $\mathbf{Z}$ is $n - r \times n - r$. The zero pattern in $\mathbf{B}$ indicates that the nodes $\{N_1, N_2, \ldots, N_r\}$ in $\mathcal{G}(\mathbf{B})$ are inaccessible from nodes $\{N_{r+1}, N_{r+2}, \ldots, N_n\}$, and hence $\mathcal{G}(\mathbf{B})$ is not strongly connected—e.g., there is no sequence of edges leading from N_{r+1} to N_1. Since $\mathcal{G}(\mathbf{B})$ is the same as $\mathcal{G}(\mathbf{A})$ except that the nodes have different numbers, we may conclude that $\mathcal{G}(\mathbf{A})$ is also not strongly connected. Conversely, if $\mathcal{G}(\mathbf{A})$ is not strongly connected, then there are two nodes in $\mathcal{G}(\mathbf{A})$ such that one is inaccessible from the other by any sequence of directed edges. Relabel the nodes in $\mathcal{G}(\mathbf{A})$ so that this pair is N_1 and N_n, where N_1 is inaccessible from N_n. If there are additional nodes—excluding N_n itself—which are also inaccessible from N_n, label them $N_2, N_3, \ldots, N_r$ so that the set of all nodes that are inaccessible from N_n—with the possible exception of N_n itself—is $\overline{N_n} = \{N_1, N_2, \ldots, N_r\}$ (inaccessible nodes). Label the remaining nodes—which are all accessible from N_n—as $\underline{N_n} = \{N_{r+1}, N_{r+2}, \ldots, N_{n-1}\}$ (accessible nodes). It follows that no node in $\overline{N_n}$ can be accessible from any node in $\underline{N_n}$, for otherwise nodes in $\overline{N_n}$ would be accessible from N_n through nodes in $\underline{N_n}$. In other words, if $N_{r+k} \in \underline{N_n}$ and $N_{r+k} \to N_i \in \overline{N_n}$, then $N_n \to N_{r+k} \to N_i$, which is impossible. This means that if $\boldsymbol{\pi} = \begin{pmatrix} 1 & 2 & \cdots & n \\ \pi_1 & \pi_2 & \cdots & \pi_n \end{pmatrix}$ is the permutation generated by the relabeling process, then $a_{\pi_i \pi_j} = 0$ for each $i = r+1, r+2, \ldots, n-1$ and $j = 1, 2, \ldots, r$. Therefore, if $\mathbf{B} = \mathbf{P}^T \mathbf{A} \mathbf{P}$, where $\mathbf{P}$ is the permutation matrix corresponding to the permutation $\boldsymbol{\pi}$, then $b_{ij} = a_{\pi_i \pi_j}$, so $\mathbf{P}^T \mathbf{A} \mathbf{P} = \mathbf{B} = \begin{pmatrix} \mathbf{X} & \mathbf{Y} \\ \mathbf{0} & \mathbf{Z} \end{pmatrix}$, where $\mathbf{X}$ is $r \times r$ and $\mathbf{Z}$ is $n - r \times n - r$, and thus $\mathbf{A}$ is reducible.

Solutions for exercises in section 4. 5

4.5.1. $rank\left(\mathbf{A}^T \mathbf{A}\right) = rank\left(\mathbf{A}\right) = rank\left(\mathbf{A}\mathbf{A}^T\right) = 2$

4.5.2. $\dim N\left(\mathbf{A}\right) \cap R\left(\mathbf{B}\right) = rank\left(\mathbf{B}\right) - rank\left(\mathbf{A}\mathbf{B}\right) = 2 - 1 = 1.$

4.5.3. Gaussian elimination yields $\mathbf{X} = \begin{pmatrix} 1 & 1 \\ -1 & 1 \\ 2 & 2 \end{pmatrix}$, $\mathbf{V} = \begin{pmatrix} 1 \\ 1 \end{pmatrix}$, and $\mathbf{XV} = \begin{pmatrix} 2 \\ 0 \\ 4 \end{pmatrix}$.

4.5.4. Statement (4.5.2) says that the rank of a product cannot exceed the rank of any factor.

4.5.5. $rank(\mathbf{A}) = rank(\mathbf{A}^T\mathbf{A}) = 0 \implies \mathbf{A} = \mathbf{0}$.

4.5.6. $rank(\mathbf{A}) = 2$, and there are *six* 2×2 nonsingular submatrices in $\mathbf{A}$.

4.5.7. Yes. $\mathbf{A} = \begin{pmatrix} 1 & 1 \\ 1 & 1 \end{pmatrix}$ and $\mathbf{B} = \begin{pmatrix} 1 & 1 \\ -1 & -1 \end{pmatrix}$ is one of many examples.

4.5.8. No—it is not difficult to construct a counterexample using two singular matrices. If either matrix is nonsingular, then the statement is true.

4.5.9. Transposition does not alter rank, so (4.5.1) says

$$
\begin{aligned}
rank(\mathbf{AB}) = rank(\mathbf{AB})^T &= rank(\mathbf{B}^T\mathbf{A}^T) \\
&= rank(\mathbf{A}^T) - \dim N(\mathbf{B}^T) \cap R(\mathbf{A}^T) \\
&= rank(\mathbf{A}) - \dim N(\mathbf{B}^T) \cap R(\mathbf{A}^T).
\end{aligned}
$$

4.5.10. This follows immediately from (4.5.1) because $\dim N(\mathbf{AB}) = p - rank(\mathbf{AB})$ and $\dim N(\mathbf{B}) = p - rank(\mathbf{B})$.

4.5.11. (a)   First notice that $N(\mathbf{B}) \subseteq N(\mathbf{AB})$ (Exercise 4.2.12) for all conformable $\mathbf{A}$ and $\mathbf{B}$, so, by (4.4.5), $\dim N(\mathbf{B}) \le \dim N(\mathbf{AB})$, or $\nu(\mathbf{B}) \le \nu(\mathbf{AB})$, is always true—this also answers the second half of part (b). If $\mathbf{A}$ and $\mathbf{B}$ are both $n \times n$, then the rank-plus-nullity theorem together with (4.5.2) produces

$$\nu(\mathbf{A}) = \dim N(\mathbf{A}) = n - rank(\mathbf{A}) \le n - rank(\mathbf{AB}) = \dim N(\mathbf{AB}) = \nu(\mathbf{AB}),$$

so, together with the first observation, we have $\max\{\nu(\mathbf{A}), \nu(\mathbf{B})\} \le \nu(\mathbf{AB})$. The rank-plus-nullity theorem applied to (4.5.3) yields $\nu(\mathbf{AB}) \le \nu(\mathbf{A}) + \nu(\mathbf{B})$.

(b)   To see that $\nu(\mathbf{A}) > \nu(\mathbf{AB})$ is possible for rectangular matrices, consider $\mathbf{A} = \begin{pmatrix} 1 & 1 \end{pmatrix}$ and $\mathbf{B} = \begin{pmatrix} 1 \\ 1 \end{pmatrix}$.

4.5.12. (a) $rank(\mathbf{B}_{n \times p}) = n \implies R(\mathbf{B}) = \Re^n \implies N(\mathbf{A}) \cap R(\mathbf{B}) = N(\mathbf{A}) \implies$

$$
\begin{aligned}
rank(\mathbf{AB}) &= rank(\mathbf{B}) - \dim N(\mathbf{A}) \cap R(\mathbf{B}) = n - \dim N(\mathbf{A}) \\
&= n - (n - rank(\mathbf{A})) = rank(\mathbf{A})
\end{aligned}
$$

It's always true that $R(\mathbf{AB}) \subseteq R(\mathbf{A})$. When $\dim R(\mathbf{AB}) = \dim R(\mathbf{A})$ (i.e., when $rank(\mathbf{AB}) = rank(\mathbf{A})$), (4.4.6) implies $R(\mathbf{AB}) = R(\mathbf{A})$.

(b) $rank(\mathbf{A}_{m \times n}) = n \implies N(\mathbf{A}) = \{\mathbf{0}\} \implies N(\mathbf{A}) \cap R(\mathbf{B}) = \{\mathbf{0}\} \implies$

$$rank(\mathbf{AB}) = rank(\mathbf{B}) - \dim N(\mathbf{A}) \cap R(\mathbf{B}) = rank(\mathbf{B})$$

Assuming the product exists, it is always the case that $N(\mathbf{B}) \subseteq N(\mathbf{AB})$. Use $rank(\mathbf{B}) = rank(\mathbf{AB}) \implies p - rank(\mathbf{B}) = p - rank(\mathbf{AB}) \implies \dim N(\mathbf{B}) = \dim N(\mathbf{AB})$ together with (4.4.6) to conclude that $N(\mathbf{B}) = N(\mathbf{AB})$.

4.5.13. (a) $rank(\mathbf{A}) = 2$, and the unique exact solution is $(-1,\, 1)$.

(b) Same as part (a).

(c) The 3-digit rank is 2, and the unique 3-digit solution is $(-1,\, 1)$.

(d) The 3-digit normal equations $\begin{pmatrix} 6 & 12 \\ 12 & 24 \end{pmatrix}\begin{pmatrix} x_1 \\ x_2 \end{pmatrix} = \begin{pmatrix} 6.01 \\ 12 \end{pmatrix}$ have infinitely many 3-digit solutions.

4.5.14. Use an indirect argument. Suppose $\mathbf{x} \in N(\mathbf{I}+\mathbf{F})$ in which $x_i \neq 0$ is a component of maximal magnitude. Use the triangle inequality together with $\mathbf{x} = -\mathbf{F}\mathbf{x}$ to conclude

$$|x_i| = \left| \sum_{j=1}^{r} f_{ij}x_j \right| \leq \sum_{j=1}^{r} |f_{ij}x_j| = \sum_{j=1}^{r} |f_{ij}|\,|x_j| \leq \left(\sum_{j=1}^{r} |f_{ij}| \right)|x_i| < |x_i|,$$

which is impossible. Therefore, $N(\mathbf{I}+\mathbf{F}) = \mathbf{0}$, and hence $\mathbf{I}+\mathbf{F}$ is nonsingular.

4.5.15. Follow the approach used in (4.5.8) to write

$$\mathbf{A} \sim \begin{pmatrix} \mathbf{W} & \mathbf{0} \\ \mathbf{0} & \mathbf{S} \end{pmatrix}, \quad \text{where} \quad \mathbf{S} = \mathbf{Z} - \mathbf{Y}\mathbf{W}^{-1}\mathbf{X}.$$

$rank(\mathbf{A}) = rank(\mathbf{W}) \implies rank(\mathbf{S}) = 0 \implies \mathbf{S} = \mathbf{0}$, so $\mathbf{Z} = \mathbf{Y}\mathbf{W}^{-1}\mathbf{X}$. The desired conclusion now follows by taking $\mathbf{B} = \mathbf{Y}\mathbf{W}^{-1}$ and $\mathbf{C} = \mathbf{W}^{-1}\mathbf{X}$.

4.5.16. (a) Suppose that $\mathbf{A}$ is nonsingular, and let $\mathbf{E}_k = \mathbf{A}_k - \mathbf{A}$ so that $\lim_{k \to \infty} \mathbf{E}_k = \mathbf{0}$. This together with (4.5.9) implies there exists a sufficiently large value of k such that

$$rank(\mathbf{A}_k) = rank(\mathbf{A} + \mathbf{E}_k) \geq rank(\mathbf{A}) = n,$$

which is impossible because each $\mathbf{A}_k$ is singular. Therefore, the supposition that $\mathbf{A}$ is nonsingular must be false.

(b) No!—consider $\left[\frac{1}{k}\right]_{1 \times 1} \to [0]$.

4.5.17. $\mathcal{M} \subseteq \mathcal{N}$ because $R(\mathbf{BC}) \subseteq R(\mathbf{B})$, and therefore $\dim\mathcal{M} \leq \dim\mathcal{N}$. Formula (4.5.1) guarantees $\dim\mathcal{M} = rank(\mathbf{BC}) - rank(\mathbf{ABC})$ and $\dim\mathcal{N} = rank(\mathbf{B}) - rank(\mathbf{AB})$, so the desired conclusion now follows.

4.5.18. $N(\mathbf{A}) \subseteq N(\mathbf{A}^2)$ and $R(\mathbf{A}^2) \subseteq R(\mathbf{A})$ always hold, so (4.4.6) insures

$$N(\mathbf{A}) = N(\mathbf{A}^2) \iff \dim N(\mathbf{A}) = \dim N(\mathbf{A}^2)$$
$$\iff n - rank(\mathbf{A}) = n - rank(\mathbf{A}^2)$$
$$\iff rank(\mathbf{A}) = rank(\mathbf{A}^2)$$
$$\iff R(\mathbf{A}) = R(\mathbf{A}^2).$$

Formula (4.5.1) says $rank(\mathbf{A}^2) = rank(\mathbf{A}) - \dim R(\mathbf{A}) \cap N(\mathbf{A})$, so $R(\mathbf{A}^2) = R(\mathbf{A}) \iff rank(\mathbf{A}^2) = rank(\mathbf{A}) \iff \dim R(\mathbf{A}) \cap N(\mathbf{A}) = 0$.

4.5.19. (a)   Since

$$\begin{pmatrix} \mathbf{A} \\ \mathbf{B} \end{pmatrix} (\mathbf{A} + \mathbf{B})(\mathbf{A} \mid \mathbf{B}) = \begin{pmatrix} \mathbf{A} \\ \mathbf{B} \end{pmatrix} (\mathbf{A} \mid \mathbf{B}) = \begin{pmatrix} \mathbf{A} & \mathbf{0} \\ \mathbf{0} & \mathbf{B} \end{pmatrix},$$

the result of Example 3.9.3 together with (4.5.2) insures

$$rank\,(\mathbf{A}) + rank\,(\mathbf{B}) \leq rank\,(\mathbf{A} + \mathbf{B}).$$

Couple this with the fact that $rank\,(\mathbf{A} + \mathbf{B}) \leq rank\,(\mathbf{A}) + rank\,(\mathbf{B})$ (see Example 4.4.8) to conclude $rank\,(\mathbf{A} + \mathbf{B}) = rank\,(\mathbf{A}) + rank\,(\mathbf{B})$.

(b)   Verify that if $\mathbf{B} = \mathbf{I} - \mathbf{A}$, then $\mathbf{B}^2 = \mathbf{B}$ and $\mathbf{AB} = \mathbf{BA} = \mathbf{0}$, and apply the result of part (a).

4.5.20. (a) $\mathbf{B}^T\mathbf{A}\mathbf{C}^T = \mathbf{B}^T\mathbf{B}\mathbf{C}\mathbf{C}^T$. The products $\mathbf{B}^T\mathbf{B}$ and $\mathbf{C}\mathbf{C}^T$ are each nonsingular because they are $r \times r$ with

$$rank\,\big(\mathbf{B}^T\mathbf{B}\big) = rank\,(\mathbf{B}) = r \quad \text{and} \quad rank\,\big(\mathbf{C}\mathbf{C}^T\big) = rank\,(\mathbf{C}) = r.$$

(b)   Notice that $\mathbf{A}^\dagger = \mathbf{C}^T\big(\mathbf{B}^T\mathbf{B}\mathbf{C}\mathbf{C}^T\big)^{-1}\mathbf{B}^T = \mathbf{C}^T\big(\mathbf{C}\mathbf{C}^T\big)^{-1}\big(\mathbf{B}^T\mathbf{B}\big)^{-1}\mathbf{B}^T$, so

$$\mathbf{A}^T\mathbf{A}\mathbf{A}^\dagger\mathbf{b} = \mathbf{C}^T\mathbf{B}^T\mathbf{B}\mathbf{C}\mathbf{C}^T\big(\mathbf{C}\mathbf{C}^T\big)^{-1}\big(\mathbf{B}^T\mathbf{B}\big)^{-1}\mathbf{B}^T\mathbf{b} = \mathbf{C}^T\mathbf{B}^T\mathbf{b} = \mathbf{A}^T\mathbf{b}.$$

If $\mathbf{Ax} = \mathbf{b}$ is consistent, then its solution set agrees with the solution set for the normal equations.

(c) $\mathbf{A}\mathbf{A}^\dagger\mathbf{A} = \mathbf{B}\mathbf{C}\mathbf{C}^T\big(\mathbf{C}\mathbf{C}^T\big)^{-1}\big(\mathbf{B}^T\mathbf{B}\big)^{-1}\mathbf{B}^T\mathbf{B}\mathbf{C} = \mathbf{B}\mathbf{C} = \mathbf{A}$. Now,

$$\begin{aligned} \mathbf{x} \in R\,\big(\mathbf{I} - \mathbf{A}^\dagger\mathbf{A}\big) &\implies \mathbf{x} = \big(\mathbf{I} - \mathbf{A}^\dagger\mathbf{A}\big)\mathbf{y} \quad \text{for some } \mathbf{y} \\ &\implies \mathbf{Ax} = \big(\mathbf{A} - \mathbf{A}\mathbf{A}^\dagger\mathbf{A}\big)\mathbf{y} = \mathbf{0} \implies \mathbf{x} \in N\,(\mathbf{A}). \end{aligned}$$

Conversely,

$$\mathbf{x} \in N\,(\mathbf{A}) \implies \mathbf{Ax} = \mathbf{0} \implies \mathbf{x} = \big(\mathbf{I} - \mathbf{A}^\dagger\mathbf{A}\big)\mathbf{x} \implies \mathbf{x} \in R\,\big(\mathbf{I} - \mathbf{A}^\dagger\mathbf{A}\big),$$

so $R\,\big(\mathbf{I} - \mathbf{A}^\dagger\mathbf{A}\big) = N\,(\mathbf{A})$. As $\mathbf{h}$ ranges over all of $\Re^{n \times 1}$, the expression $\big(\mathbf{I} - \mathbf{A}^\dagger\mathbf{A}\big)\mathbf{h}$ generates $R\,\big(\mathbf{I} - \mathbf{A}^\dagger\mathbf{A}\big) = N\,(\mathbf{A})$. Since $\mathbf{A}^\dagger\mathbf{b}$ is a particular solution of $\mathbf{A}^T\mathbf{Ax} = \mathbf{A}^T\mathbf{b}$, the general solution is

$$\mathbf{x} = \mathbf{A}^\dagger\mathbf{b} + N\,(\mathbf{A}) = \mathbf{A}^\dagger\mathbf{b} + \big(\mathbf{I} - \mathbf{A}^\dagger\mathbf{A}\big)\mathbf{h}.$$

(d)   If $r = n$, then $\mathbf{B} = \mathbf{A}$ and $\mathbf{C} = \mathbf{I}_n$.

(e)   If $\mathbf{A}$ is nonsingular, then so is $\mathbf{A}^T$, and

$$\mathbf{A}^\dagger = \big(\mathbf{A}^T\mathbf{A}\big)^{-1}\mathbf{A}^T = \mathbf{A}^{-1}\big(\mathbf{A}^T\big)^{-1}\mathbf{A}^T = \mathbf{A}^{-1}.$$

(f)   Follow along the same line as indicated in the solution to part (c) for the case $\mathbf{A}\mathbf{A}^\dagger\mathbf{A} = \mathbf{A}$.

Solutions for exercises in section 4. 6

4.6.1. $\mathbf{A} = \begin{pmatrix} 5 \\ 7 \\ 8 \\ 10 \\ 12 \end{pmatrix}$ and $\mathbf{b} = \begin{pmatrix} 11.1 \\ 15.4 \\ 17.5 \\ 22.0 \\ 26.3 \end{pmatrix}$, so $\mathbf{A}^T\mathbf{A} = 382$ and $\mathbf{A}^T\mathbf{b} = 838.9$. Thus the least squares estimate for k is $838.9/382 = 2.196$.

4.6.2. This is essentially the same problem as Exercise 4.6.1. Because it must pass through the origin, the equation of the least squares line is $y = mx$, and hence

$$\mathbf{A} = \begin{pmatrix} x_1 \\ x_2 \\ \vdots \\ x_n \end{pmatrix} \quad \text{and} \quad \mathbf{b} = \begin{pmatrix} y_1 \\ y_2 \\ \vdots \\ y_n \end{pmatrix}, \quad \text{so } \mathbf{A}^T\mathbf{A} = \sum_i x_i^2 \text{ and } \mathbf{A}^T\mathbf{b} = \sum_i x_i y_i.$$

4.6.3. Look for the line $p = \alpha + \beta t$ that comes closest to the data in the least squares sense. That is, find the least squares solution for the system $\mathbf{Ax} = \mathbf{b}$, where

$$\mathbf{A} = \begin{pmatrix} 1 & 1 \\ 1 & 2 \\ 1 & 3 \end{pmatrix}, \quad \mathbf{x} = \begin{pmatrix} \alpha \\ \beta \end{pmatrix}, \quad \text{and} \quad \mathbf{b} = \begin{pmatrix} 7 \\ 4 \\ 3 \end{pmatrix}.$$

Set up normal equations $\mathbf{A}^T\mathbf{Ax} = \mathbf{A}^T\mathbf{b}$ to get

$$\begin{pmatrix} 3 & 6 \\ 6 & 14 \end{pmatrix}\begin{pmatrix} \alpha \\ \beta \end{pmatrix} = \begin{pmatrix} 14 \\ 24 \end{pmatrix} \implies \begin{pmatrix} \alpha \\ \beta \end{pmatrix} = \begin{pmatrix} 26/3 \\ -2 \end{pmatrix} \implies p = (26/3) - 2t.$$

Setting $p = 0$ gives $t = 13/3$. In other words, we expect the company to begin losing money on May 1 of year five.

4.6.4. The associated linear system $\mathbf{Ax} = \mathbf{b}$ is

$$\begin{array}{ll} \text{Year 1:} & \alpha + \beta = 1 \\ \text{Year 2:} & 2\alpha = 1 \\ \text{Year 3:} & -\beta = 1 \end{array} \quad \text{or} \quad \begin{pmatrix} 1 & 1 \\ 2 & 0 \\ 0 & -1 \end{pmatrix}\begin{pmatrix} \alpha \\ \beta \end{pmatrix} = \begin{pmatrix} 1 \\ 1 \\ 1 \end{pmatrix}.$$

The least squares solution to this inconsistent system is obtained from the system of normal equations $\mathbf{A}^T\mathbf{Ax} = \mathbf{A}^T\mathbf{b}$ that is $\begin{pmatrix} 5 & 1 \\ 1 & 2 \end{pmatrix}\begin{pmatrix} \alpha \\ \beta \end{pmatrix} = \begin{pmatrix} 3 \\ 0 \end{pmatrix}$. The unique solution is $\begin{pmatrix} \alpha \\ \beta \end{pmatrix} = \begin{pmatrix} 2/3 \\ -1/3 \end{pmatrix}$, so the least squares estimate for the increase in bread prices is

$$B = \frac{2}{3}W - \frac{1}{3}M.$$

When $W = -1$ and $M = -1$, we estimate that $B = -1/3$.

4.6.5. (a) $\alpha_0 = .02$ and $\alpha_1 = .0983$. (b) 1.986 grams.

4.6.6. Use $\ln y = \ln \alpha_0 + \alpha_1 t$ to obtain the least squares estimates $\alpha_0 = 9.73$ and $\alpha_1 = .507$.

4.6.7. The least squares line is $y = 9.64 + .182x$ and for $\varepsilon_i = 9.64 + .182x_i - y_i$, the sum of the squares of these errors is $\sum_i \varepsilon_i^2 = 162.9$. The least squares quadratic is $y = 13.97 + .1818x - .4336x^2$, and the corresponding sum of squares of the errors is $\sum \varepsilon_i^2 = 1.622$. Therefore, we conclude that the quadratic provides a much better fit.

4.6.8. 230.7 min. $(\alpha_0 = 492.04, \alpha_1 = -23.435, \alpha_2 = -.076134, \alpha_3 = 1.8624)$

4.6.9. $\mathbf{x}_2$ is a least squares solution $\implies \mathbf{A}^T \mathbf{A} \mathbf{x}_2 = \mathbf{A}^T \mathbf{b} \implies \mathbf{0} = \mathbf{A}^T(\mathbf{b} - \mathbf{A}\mathbf{x}_2)$. If we set $\mathbf{x}_1 = \mathbf{b} - \mathbf{A}\mathbf{x}_2$, then

$$\begin{pmatrix} \mathbf{I}_{m\times m} & \mathbf{A} \\ \mathbf{A}^T & \mathbf{0}_{n\times n} \end{pmatrix} \begin{pmatrix} \mathbf{x}_1 \\ \mathbf{x}_2 \end{pmatrix} = \begin{pmatrix} \mathbf{I}_{m\times m} & \mathbf{A} \\ \mathbf{A}^T & \mathbf{0}_{n\times n} \end{pmatrix} \begin{pmatrix} \mathbf{b} - \mathbf{A}\mathbf{x}_2 \\ \mathbf{x}_2 \end{pmatrix} = \begin{pmatrix} \mathbf{b} \\ \mathbf{0} \end{pmatrix}.$$

The converse is true because

$$\begin{pmatrix} \mathbf{I}_{m\times m} & \mathbf{A} \\ \mathbf{A}^T & \mathbf{0}_{n\times n} \end{pmatrix} \begin{pmatrix} \mathbf{x}_1 \\ \mathbf{x}_2 \end{pmatrix} = \begin{pmatrix} \mathbf{b} \\ \mathbf{0} \end{pmatrix} \implies \mathbf{A}\mathbf{x}_2 = \mathbf{b} - \mathbf{x}_1 \quad \text{and} \quad \mathbf{A}^T \mathbf{x}_1 = \mathbf{0}$$

$$\implies \mathbf{A}^T \mathbf{A} \mathbf{x}_2 = \mathbf{A}^T \mathbf{b} - \mathbf{A}^T \mathbf{x}_1 = \mathbf{A}^T \mathbf{b}.$$

4.6.10. $\mathbf{t} \in R\left(\mathbf{A}^T\right) = R\left(\mathbf{A}^T \mathbf{A}\right) \implies \mathbf{t}^T = \mathbf{z}^T \mathbf{A}^T \mathbf{A}$ for some $\mathbf{z}$. For each $\mathbf{x}$ satisfying $\mathbf{A}^T \mathbf{A} \mathbf{x} = \mathbf{A}^T \mathbf{b}$, write

$$\hat{y} = \mathbf{t}^T \mathbf{x} = \mathbf{z}^T \mathbf{A}^T \mathbf{A} \mathbf{x} = \mathbf{z}^T \mathbf{A}^T \mathbf{b},$$

and notice that $\mathbf{z}^T \mathbf{A}^T \mathbf{b}$ is independent of $\mathbf{x}$.

Solutions for exercises in section 4. 7

4.7.1. (b) and (f)

4.7.2. (a), (c), and (d)

4.7.3. Use any $\mathbf{x}$ to write $\mathbf{T}(\mathbf{0}) = \mathbf{T}(\mathbf{x} - \mathbf{x}) = \mathbf{T}(\mathbf{x}) - \mathbf{T}(\mathbf{x}) = \mathbf{0}$.

4.7.4. (a)

4.7.5. (a) No   (b) Yes

4.7.6. $\mathbf{T}(\mathbf{u}_1) = (2, 2) = 2\mathbf{u}_1 + 0\mathbf{u}_2$ and $\mathbf{T}(\mathbf{u}_2) = (3, 6) = 0\mathbf{u}_1 + 3\mathbf{u}_2$ so that $[\mathbf{T}]_\mathcal{B} = \begin{pmatrix} 2 & 0 \\ 0 & 3 \end{pmatrix}$.

4.7.7. (a) $[\mathbf{T}]_{\mathcal{SS'}} = \begin{pmatrix} 1 & 3 \\ 0 & 0 \\ 2 & -4 \end{pmatrix}$ (b) $[\mathbf{T}]_{\mathcal{SS''}} = \begin{pmatrix} 2 & -4 \\ 0 & 0 \\ 1 & 3 \end{pmatrix}$

4.7.8. $[\mathbf{T}]_\mathcal{B} = \begin{pmatrix} 1 & -3/2 & 1/2 \\ -1 & 1/2 & 1/2 \\ 0 & 1/2 & -1/2 \end{pmatrix}$ and $[\mathbf{v}]_\mathcal{B} = \begin{pmatrix} 1 \\ 1 \\ 0 \end{pmatrix}$.

4.7.9. According to (4.7.4), the jth column of $[\mathbf{T}]_S$ is

$$[\mathbf{T}(\mathbf{e}_j)]_S = [\mathbf{Ae}_j]_S = [\mathbf{A}_{*j}]_S = \mathbf{A}_{*j}.$$

4.7.10. $[\mathbf{T}^k]_\mathcal{B} = [\mathbf{TT}\cdots\mathbf{T}]_\mathcal{B} = [\mathbf{T}]_\mathcal{B}[\mathbf{T}]_\mathcal{B}\cdots[\mathbf{T}]_\mathcal{B} = [\mathbf{T}]_\mathcal{B}^k$

4.7.11. (a) Sketch a picture to observe that $\mathbf{P}(\mathbf{e}_1) = \begin{pmatrix} x \\ x \end{pmatrix} = \mathbf{P}(\mathbf{e}_2)$ and that the vectors $\mathbf{e}_1$, $\mathbf{P}(\mathbf{e}_1)$, and $\mathbf{0}$ are vertices of a $45°$ right triangle (as are $\mathbf{e}_2$, $\mathbf{P}(\mathbf{v}_2)$, and $\mathbf{0}$). So, if $\|\star\|$ denotes length, the Pythagorean theorem may be applied to yield $1 = 2\|\mathbf{P}(\mathbf{e}_1)\|^2 = 4x^2$ and $1 = 2\|\mathbf{P}(\mathbf{e}_2)\|^2 = 4x^2$. Thus

$$\left\{ \begin{aligned} \mathbf{P}(\mathbf{e}_1) = \begin{pmatrix} 1/2 \\ 1/2 \end{pmatrix} = (1/2)\mathbf{e}_1 + (1/2)\mathbf{e}_2 \\ \mathbf{P}(\mathbf{e}_2) = \begin{pmatrix} 1/2 \\ 1/2 \end{pmatrix} = (1/2)\mathbf{e}_1 + (1/2)\mathbf{e}_2 \end{aligned} \right\} \implies [\mathbf{P}]_S = \begin{pmatrix} 1/2 & 1/2 \\ 1/2 & 1/2 \end{pmatrix}.$$

(b) $\mathbf{P}(\mathbf{v}) = \begin{pmatrix} \frac{\alpha+\beta}{2} \\ \frac{\alpha+\beta}{2} \end{pmatrix}$

4.7.12. (a) If $\mathbf{U}_1 = \begin{pmatrix} 1 & 0 \\ 0 & 0 \end{pmatrix}$, $\mathbf{U}_2 = \begin{pmatrix} 0 & 1 \\ 0 & 0 \end{pmatrix}$, $\mathbf{U}_3 = \begin{pmatrix} 0 & 0 \\ 1 & 0 \end{pmatrix}$, $\mathbf{U}_4 = \begin{pmatrix} 0 & 0 \\ 0 & 1 \end{pmatrix}$, then

$$\mathbf{T}(\mathbf{U}_1) = \mathbf{U}_1 + 0\mathbf{U}_2 + 0\mathbf{U}_3 + 0\mathbf{U}_4,$$

$$\mathbf{T}(\mathbf{U}_2) = \frac{1}{2}\begin{pmatrix} 0 & 1 \\ 1 & 0 \end{pmatrix} = 0\mathbf{U}_1 + 1/2\mathbf{U}_2 + 1/2\mathbf{U}_3 + 0\mathbf{U}_4,$$

$$\mathbf{T}(\mathbf{U}_3) = \frac{1}{2}\begin{pmatrix} 0 & 1 \\ 1 & 0 \end{pmatrix} = 0\mathbf{U}_1 + 1/2\mathbf{U}_2 + 1/2\mathbf{U}_3 + 0\mathbf{U}_4,$$

$$\mathbf{T}(\mathbf{U}_4) = 0\mathbf{U}_1 + 0\mathbf{U}_2 + 0\mathbf{U}_3 + \mathbf{U}_4,$$

so $[\mathbf{T}]_S = \begin{pmatrix} 1 & 0 & 0 & 0 \\ 0 & 1/2 & 1/2 & 0 \\ 0 & 1/2 & 1/2 & 0 \\ 0 & 0 & 0 & 1 \end{pmatrix}$. To verify $[\mathbf{T}(\mathbf{U})]_S = [\mathbf{T}]_S[\mathbf{U}]_S$, observe that

$$\mathbf{T}(\mathbf{U}) = \begin{pmatrix} a & (b+c)/2 \\ (b+c)/2 & d \end{pmatrix}, \quad [\mathbf{T}(\mathbf{U})]_S = \begin{pmatrix} a \\ (b+c)/2 \\ (b+c)/2 \\ d \end{pmatrix}, \quad [\mathbf{U}]_S = \begin{pmatrix} a \\ b \\ c \\ d \end{pmatrix}.$$

(b) For $\mathbf{U}_1$, $\mathbf{U}_2$, $\mathbf{U}_3$, and $\mathbf{U}_4$ as defined above,

$$\mathbf{T}(\mathbf{U}_1) = \begin{pmatrix} 0 & -1 \\ -1 & 0 \end{pmatrix} = 0\mathbf{U}_1 - \mathbf{U}_2 - \mathbf{U}_3 + 0\mathbf{U}_4,$$

$$\mathbf{T}(\mathbf{U}_2) = \begin{pmatrix} 1 & 2 \\ 0 & -1 \end{pmatrix} = \mathbf{U}_1 + 2\mathbf{U}_2 + 0\mathbf{U}_3 - \mathbf{U}_4,$$

$$\mathbf{T}(\mathbf{U}_3) = \begin{pmatrix} 1 & 0 \\ -2 & -1 \end{pmatrix} = \mathbf{U}_1 + 0\mathbf{U}_2 - 2\mathbf{U}_3 - 1\mathbf{U}_4,$$

$$\mathbf{T}(\mathbf{U}_4) = \begin{pmatrix} 0 & 1 \\ 1 & 0 \end{pmatrix} = 0\mathbf{U}_1 + \mathbf{U}_2 + \mathbf{U}_3 + 0\mathbf{U}_4,$$

so $[\mathbf{T}]_\mathcal{S} = \begin{pmatrix} 0 & 1 & 1 & 0 \\ -1 & 2 & 0 & 1 \\ -1 & 0 & -2 & 1 \\ 0 & -1 & -1 & 0 \end{pmatrix}$. To verify $[\mathbf{T}(\mathbf{U})]_\mathcal{S} = [\mathbf{T}]_\mathcal{S}[\mathbf{U}]_\mathcal{S}$, observe that

$$\mathbf{T}(\mathbf{U}) = \begin{pmatrix} c+b & -a+2b+d \\ -a-2c+d & -b-c \end{pmatrix} \text{ and } [\mathbf{T}(\mathbf{U})]_\mathcal{S} = \begin{pmatrix} c+b \\ -a+2b+d \\ -a-2c+d \\ -b-c \end{pmatrix}.$$

4.7.13. $[\mathbf{S}]_{\mathcal{B}\mathcal{B}'} = \begin{pmatrix} 0 & 0 & 0 \\ 1 & 0 & 0 \\ 0 & 1/2 & 0 \\ 0 & 0 & 1/3 \end{pmatrix}$

4.7.14. (a) $[\mathbf{R}\mathbf{Q}]_\mathcal{S} = [\mathbf{R}]_\mathcal{S}[\mathbf{Q}]_\mathcal{S} = \begin{pmatrix} 1 & 0 \\ 0 & -1 \end{pmatrix}\begin{pmatrix} \cos\theta & -\sin\theta \\ \sin\theta & \cos\theta \end{pmatrix} = \begin{pmatrix} \cos\theta & -\sin\theta \\ -\sin\theta & -\cos\theta \end{pmatrix}$

(b)

$$[\mathbf{Q}\mathbf{Q}]_\mathcal{S} = [\mathbf{Q}]_\mathcal{S}[\mathbf{Q}]_\mathcal{S} = \begin{pmatrix} \cos^2\theta - \sin^2\theta & -2\cos\theta\sin\theta \\ 2\cos\theta\sin\theta & \cos^2\theta - \sin^2\theta \end{pmatrix}$$

$$= \begin{pmatrix} \cos 2\theta & -\sin 2\theta \\ \sin 2\theta & \cos 2\theta \end{pmatrix}$$

4.7.15. (a) Let $\mathcal{B} = \{\mathbf{u}_i\}_{i=1}^n$, $\mathcal{B}' = \{\mathbf{v}_i\}_{i=1}^m$. If $[\mathbf{P}]_{\mathcal{B}\mathcal{B}'} = [\alpha_{ij}]$ and $[\mathbf{Q}]_{\mathcal{B}\mathcal{B}'} = [\beta_{ij}]$, then $\mathbf{P}(\mathbf{u}_j) = \sum_i \alpha_{ij}\mathbf{v}_i$ and $\mathbf{Q}(\mathbf{u}_j) = \sum_i \beta_{ij}\mathbf{v}_i$. Thus $(\mathbf{P}+\mathbf{Q})(\mathbf{u}_j) = \sum_i(\alpha_{ij} + \beta_{ij})\mathbf{v}_i$ and hence $[\mathbf{P}+\mathbf{Q}]_{\mathcal{B}\mathcal{B}'} = [\alpha_{ij} + \beta_{ij}] = [\alpha_{ij}] + [\beta_{ij}] = [\mathbf{P}]_{\mathcal{B}\mathcal{B}'} + [\mathbf{Q}]_{\mathcal{B}\mathcal{B}'}$. The proof of part (b) is similar.

4.7.16. (a) If $\mathcal{B} = \{\mathbf{x}_i\}_{i=1}^n$ is a basis, then $\mathbf{I}(\mathbf{x}_j) = 0\mathbf{x}_1 + 0\mathbf{x}_2 + \cdots + 1\mathbf{x}_j + \cdots + 0\mathbf{x}_n$ so that the jth column in $[\mathbf{I}]_\mathcal{B}$ is just the jth unit column.

(b) Suppose $\mathbf{x}_j = \sum_i \beta_{ij}\mathbf{y}_i$ so that $[\mathbf{x}_j]_{\mathcal{B}'} = \begin{pmatrix} \beta_{1j} \\ \vdots \\ \beta_{nj} \end{pmatrix}$. Then

$$\mathbf{I}(\mathbf{x}_j) = \mathbf{x}_j = \sum_i \beta_{ij}\mathbf{y}_i \implies [\mathbf{I}]_{\mathcal{B}\mathcal{B}'} = [\beta_{ij}] = \left([\mathbf{x}_1]_{\mathcal{B}'} \,\middle|\, [\mathbf{x}_2]_{\mathcal{B}'} \,\middle|\, \cdots \,\middle|\, [\mathbf{x}_n]_{\mathcal{B}'} \right).$$

Furthermore, $\mathbf{T}(\mathbf{y}_j) = \mathbf{x}_j = \sum_i \beta_{ij}\mathbf{y}_i \implies [\mathbf{T}]_{\mathcal{B}'} = [\beta_{ij}]$, and

$$\mathbf{T}(\mathbf{x}_j) = \mathbf{T}\left(\sum_i \beta_{ij}\mathbf{y}_i\right) = \sum_i \beta_{ij}\mathbf{T}(\mathbf{y}_i) = \sum_i \beta_{ij}\mathbf{x}_i \implies [\mathbf{T}]_\mathcal{B} = [\beta_{ij}].$$

$$(c) \quad \begin{pmatrix} 1 & -1 & 0 \\ 0 & 1 & -1 \\ 0 & 0 & 1 \end{pmatrix}$$

4.7.17. (a) $\quad \mathbf{T}^{-1}(x, y, z) = (x + y + z, \; x + 2y + 2z, \; x + 2y + 3z)$

(b) $\quad [\mathbf{T}^{-1}]_{\mathcal{S}} = \begin{pmatrix} 1 & 1 & 1 \\ 1 & 2 & 2 \\ 1 & 2 & 3 \end{pmatrix} = [\mathbf{T}]_{\mathcal{S}}^{-1}$

4.7.18. (1) $\implies$ (2): $\quad \mathbf{T}(\mathbf{x}) = \mathbf{T}(\mathbf{y}) \implies \mathbf{T}(\mathbf{x} - \mathbf{y}) = \mathbf{0} \implies (\mathbf{y} - \mathbf{x}) = \mathbf{T}^{-1}(\mathbf{0}) = \mathbf{0}$.

(2) $\implies$ (3): $\quad \mathbf{T}(\mathbf{x}) = \mathbf{0}$ and $\mathbf{T}(\mathbf{0}) = \mathbf{0} \implies \mathbf{x} = \mathbf{0}$.

(3) $\implies$ (4): If $\{\mathbf{u}_i\}_{i=1}^{n}$ is a basis for $\mathcal{V}$, show that $N(\mathbf{T}) = \{\mathbf{0}\}$ implies $\{\mathbf{T}(\mathbf{u}_i)\}_{i=1}^{n}$ is also a basis. Consequently, for each $\mathbf{v} \in \mathcal{V}$ there are coordinates ξ_i such that

$$\mathbf{v} = \sum_i \xi_i \mathbf{T}(\mathbf{u}_i) = \mathbf{T}\left(\sum_i \xi_i \mathbf{u}_i\right).$$

(4) $\implies$ (2): For each basis vector $\mathbf{u}_i$, there is a $\mathbf{v}_i$ such that $\mathbf{T}(\mathbf{v}_i) = \mathbf{u}_i$. Show that $\{\mathbf{v}_i\}_{i=1}^{n}$ is also a basis. If $\mathbf{T}(\mathbf{x}) = \mathbf{T}(\mathbf{y})$, then $\mathbf{T}(\mathbf{x} - \mathbf{y}) = \mathbf{0}$. Let $\mathbf{x} - \mathbf{y} = \sum_i \xi_i \mathbf{v}_i$ so that $\mathbf{0} = \mathbf{T}(\mathbf{x} - \mathbf{y}) = \mathbf{T}\left(\sum_i \xi_i \mathbf{v}_i\right) = \sum_i \xi_i \mathbf{T}(\mathbf{v}_i) = \sum_i \xi_i \mathbf{u}_i \implies$ each $\xi_i = 0 \implies \mathbf{x} - \mathbf{y} = \mathbf{0} \implies \mathbf{x} = \mathbf{y}$.

(4) and (2) $\implies$ (1): For each $\mathbf{y} \in \mathcal{V}$, show there is a *unique* $\mathbf{x}$ such that $\mathbf{T}(\mathbf{x}) = \mathbf{y}$. Let $\hat{\mathbf{T}}$ be the function defined by the rule $\hat{\mathbf{T}}(\mathbf{y}) = \mathbf{x}$. Clearly, $\mathbf{T}\hat{\mathbf{T}} = \hat{\mathbf{T}}\mathbf{T} = \mathbf{I}$. To show that $\hat{\mathbf{T}}$ is a linear function, consider $\alpha \mathbf{y}_1 + \mathbf{y}_2$, and let $\mathbf{x}_1$ and $\mathbf{x}_2$ be such that $\mathbf{T}(\mathbf{x}_1) = \mathbf{y}_1$, $\mathbf{T}(\mathbf{x}_2) = \mathbf{y}_2$. Now, $\mathbf{T}(\alpha \mathbf{x}_1 + \mathbf{x}_2) = \alpha \mathbf{y}_1 + \mathbf{y}_2$ so that $\hat{\mathbf{T}}(\alpha \mathbf{y}_1 + \mathbf{y}_2) = \alpha \mathbf{x}_1 + \mathbf{x}_2$. However, $\mathbf{x}_1 = \hat{\mathbf{T}}(\mathbf{y}_1)$, $\mathbf{x}_2 = \hat{\mathbf{T}}(\mathbf{y}_2)$ so that $\alpha \hat{\mathbf{T}}(\mathbf{y}_1) + \hat{\mathbf{T}}(\mathbf{y}_2) = \alpha \mathbf{x}_1 + \mathbf{x}_2 = \hat{\mathbf{T}}(\alpha \mathbf{y}_1 + \mathbf{y}_2)$. Therefore $\hat{\mathbf{T}} = \mathbf{T}^{-1}$.

4.7.19. (a) $\quad \mathbf{0} = \sum_i \alpha_i \mathbf{x}_i \iff \begin{pmatrix} 0 \\ \vdots \\ 0 \end{pmatrix} = [\mathbf{0}]_{\mathcal{B}} = \left[\sum_i \alpha_i \mathbf{x}_i\right]_{\mathcal{B}} = \sum_i [\alpha_i \mathbf{x}_i]_{\mathcal{B}} = \sum_i \alpha_i [\mathbf{x}_i]_{\mathcal{B}}$

(b) $\quad \mathcal{G} = \left\{\mathbf{T}(\mathbf{u}_1), \mathbf{T}(\mathbf{u}_2), \ldots, \mathbf{T}(\mathbf{u}_n)\right\}$ spans $R(\mathbf{T})$. From part (a), the set

$$\left\{\mathbf{T}(\mathbf{u}_{b_1}), \mathbf{T}(\mathbf{u}_{b_2}), \ldots, \mathbf{T}(\mathbf{u}_{b_r})\right\}$$

is a maximal independent subset of $\mathcal{G}$ if and only if the set

$$\left\{[\mathbf{T}(\mathbf{u}_{b_1})]_{\mathcal{B}}, [\mathbf{T}(\mathbf{u}_{b_2})]_{\mathcal{B}}, \ldots, [\mathbf{T}(\mathbf{u}_{b_r})]_{\mathcal{B}}\right\}$$

is a maximal linearly independent subset of

$$\left\{[\mathbf{T}(\mathbf{u}_1)]_{\mathcal{B}}, [\mathbf{T}(\mathbf{u}_2)]_{\mathcal{B}}, \ldots, [\mathbf{T}(\mathbf{u}_n)]_{\mathcal{B}}\right\},$$

which are the columns of $[\mathbf{T}]_{\mathcal{B}}$.

Solutions for exercises in section 4. 8

4.8.1. Multiplication by nonsingular matrices does not change rank.

4.8.2. $\mathbf{A} = \mathbf{Q}^{-1}\mathbf{B}\mathbf{Q}$ and $\mathbf{B} = \mathbf{P}^{-1}\mathbf{C}\mathbf{P} \implies \mathbf{A} = (\mathbf{PQ})^{-1}\mathbf{C}(\mathbf{PQ})$.

4.8.3. (a) $[\mathbf{A}]_{\mathcal{S}} = \begin{pmatrix} 1 & 2 & -1 \\ 0 & -1 & 0 \\ 1 & 0 & 7 \end{pmatrix}$

(b) $[\mathbf{A}]_{\mathcal{S}'} = \begin{pmatrix} 1 & 4 & 3 \\ -1 & -2 & -9 \\ 1 & 1 & 8 \end{pmatrix}$ and $\mathbf{Q} = \begin{pmatrix} 1 & 1 & 1 \\ 0 & 1 & 1 \\ 0 & 0 & 1 \end{pmatrix}$

4.8.4. Put the vectors from $\mathcal{B}$ into a matrix $\mathbf{Q}$ and compute

$$[\mathbf{A}]_{\mathcal{B}} = \mathbf{Q}^{-1}\mathbf{A}\mathbf{Q} = \begin{pmatrix} -2 & -3 & -7 \\ 7 & 9 & 12 \\ -2 & -1 & 0 \end{pmatrix}.$$

4.8.5. $[\mathbf{B}]_{\mathcal{S}} = \mathbf{B}$ and $[\mathbf{B}]_{\mathcal{S}'} = \mathbf{C}$. Therefore, $\mathbf{C} = \mathbf{Q}^{-1}\mathbf{B}\mathbf{Q}$, where $\mathbf{Q} = \begin{pmatrix} 2 & -3 \\ -1 & 2 \end{pmatrix}$ is the change of basis matrix from $\mathcal{S}'$ to $\mathcal{S}$.

4.8.6. If $\mathcal{B} = \{\mathbf{u}, \mathbf{v}\}$ is such a basis, then $\mathbf{T}(\mathbf{u}) = 2\mathbf{u}$ and $\mathbf{T}(\mathbf{v}) = 3\mathbf{v}$. For $\mathbf{u} = (u_1, u_2)$, $\mathbf{T}(\mathbf{u}) = 2\mathbf{u}$ implies

$$-7u_1 - 15u_2 = 2u_1$$
$$6u_1 + 12u_2 = 2u_2,$$

or
$$-9u_1 - 15u_2 = 0$$
$$6u_1 + 10u_2 = 0,$$

so $u_1 = (-5/3)u_2$ with u_2 being free. Letting $u_2 = -3$ produces $\mathbf{u} = (5, -3)$. Similarly, a solution to $\mathbf{T}(\mathbf{v}) = 3\mathbf{v}$ is $\mathbf{v} = (-3, 2)$. $[\mathbf{T}]_{\mathcal{S}} = \begin{pmatrix} -7 & -15 \\ 6 & 12 \end{pmatrix}$ and $[\mathbf{T}]_{\mathcal{B}} = \begin{pmatrix} 2 & 0 \\ 0 & 3 \end{pmatrix}$. For $\mathbf{Q} = \begin{pmatrix} 5 & -3 \\ -3 & 2 \end{pmatrix}$, $[\mathbf{T}]_{\mathcal{B}} = \mathbf{Q}^{-1}[\mathbf{T}]_{\mathcal{S}}\mathbf{Q}$.

4.8.7. If $\sin\theta = 0$, the result is trivial. Assume $\sin\theta \neq 0$. Notice that with respect to the standard basis $\mathcal{S}$, $[\mathbf{P}]_{\mathcal{S}} = \mathbf{R}$. This means that if $\mathbf{R}$ and $\mathbf{D}$ are to be similar, then there must exist a basis $\mathcal{B} = \{\mathbf{u}, \mathbf{v}\}$ such that $[\mathbf{P}]_{\mathcal{B}} = \mathbf{D}$, which implies that $\mathbf{P}(\mathbf{u}) = e^{i\theta}\mathbf{u}$ and $\mathbf{P}(\mathbf{v}) = e^{-i\theta}\mathbf{v}$. For $\mathbf{u} = (u_1, u_2)$, $\mathbf{P}(\mathbf{u}) = e^{i\theta}\mathbf{u}$ implies

$$u_1 \cos\theta - u_2 \sin\theta = e^{i\theta}u_1 = u_1 \cos\theta + iu_1 \sin\theta$$
$$u_1 \sin\theta + u_2 \cos\theta = e^{i\theta}u_2 = u_2 \cos\theta + iu_2 \sin\theta,$$

or
$$iu_1 + u_2 = 0$$
$$u_1 - iu_2 = 0,$$

so $u_1 = iu_2$ with u_2 being free. Letting $u_2 = 1$ produces $\mathbf{u} = (i, 1)$. Similarly, a solution to $\mathbf{P}(\mathbf{v}) = e^{-i\theta}\mathbf{v}$ is $\mathbf{v} = (1, i)$. Now, $[\mathbf{P}]_{\mathcal{S}} = \mathbf{R}$ and $[\mathbf{P}]_{\mathcal{B}} = \mathbf{D}$ so that $\mathbf{R}$ and $\mathbf{D}$ must be similar. The coordinate change matrix from $\mathcal{B}$ to $\mathcal{S}$ is $\mathbf{Q} = \begin{pmatrix} i & 1 \\ 1 & i \end{pmatrix}$, and therefore $\mathbf{D} = \mathbf{Q}^{-1}\mathbf{R}\mathbf{Q}$.

4.8.8. (a) $\mathbf{B} = \mathbf{Q}^{-1}\mathbf{C}\mathbf{Q} \implies (\mathbf{B} - \lambda\mathbf{I}) = \mathbf{Q}^{-1}\mathbf{C}\mathbf{Q} - \lambda\mathbf{Q}^{-1}\mathbf{Q} = \mathbf{Q}^{-1}(\mathbf{C} - \lambda\mathbf{I})\mathbf{Q}$. The result follows because multiplication by nonsingular matrices does not change rank.

(b) $\mathbf{B} = \mathbf{P}^{-1}\mathbf{D}\mathbf{P} \implies \mathbf{B} - \lambda_i\mathbf{I} = \mathbf{P}^{-1}(\mathbf{D} - \lambda_i\mathbf{I})\mathbf{P}$ and $(\mathbf{D} - \lambda_i\mathbf{I})$ is singular for each λ_i. Now use part (a).

4.8.9. $\mathbf{B} = \mathbf{P}^{-1}\mathbf{A}\mathbf{P} \implies \mathbf{B}^k = \mathbf{P}^{-1}\mathbf{A}\mathbf{P}\mathbf{P}^{-1}\mathbf{A}\mathbf{P}\cdots\mathbf{P}^{-1}\mathbf{A}\mathbf{P} = \mathbf{P}^{-1}\mathbf{A}\mathbf{A}\cdots\mathbf{A}\mathbf{P} = \mathbf{P}^{-1}\mathbf{A}^k\mathbf{P}$

4.8.10. (a) $\mathbf{Y}^T\mathbf{Y}$ is nonsingular because $rank\left(\mathbf{Y}^T\mathbf{Y}\right)_{n \times n} = rank\left(\mathbf{Y}\right) = n$. If

$$[\mathbf{v}]_{\mathcal{B}} = \begin{pmatrix} \alpha_1 \\ \vdots \\ \alpha_n \end{pmatrix} \quad \text{and} \quad [\mathbf{v}]_{\mathcal{B}'} = \begin{pmatrix} \beta_1 \\ \vdots \\ \beta_n \end{pmatrix},$$

then

$$\mathbf{v} = \sum_i \alpha_i\mathbf{x}_i = \mathbf{X}[\mathbf{v}]_{\mathcal{B}} \quad \text{and} \quad \mathbf{v} = \sum_i \beta_i\mathbf{y}_i = \mathbf{Y}[\mathbf{v}]_{\mathcal{B}'}$$

$$\implies \mathbf{X}[\mathbf{v}]_{\mathcal{B}} = \mathbf{Y}[\mathbf{v}]_{\mathcal{B}'} \implies \mathbf{Y}^T\mathbf{X}[\mathbf{v}]_{\mathcal{B}} = \mathbf{Y}^T\mathbf{Y}[\mathbf{v}]_{\mathcal{B}'}$$

$$\implies (\mathbf{Y}^T\mathbf{Y})^{-1}\mathbf{Y}^T\mathbf{X}[\mathbf{v}]_{\mathcal{B}} = [\mathbf{v}]_{\mathcal{B}'}.$$

(b) When $m = n$, $\mathbf{Y}$ is square and $(\mathbf{Y}^T\mathbf{Y})^{-1}\mathbf{Y}^T = \mathbf{Y}^{-1}$ so that $\mathbf{P} = \mathbf{Y}^{-1}\mathbf{X}$.

4.8.11. (a) Because $\mathcal{B}$ contains n vectors, you need only show that $\mathcal{B}$ is linearly independent. To do this, suppose $\sum_{i=0}^{n-1} \alpha_i\mathbf{N}^i(\mathbf{y}) = \mathbf{0}$ and apply $\mathbf{N}^{n-1}$ to both sides to get $\alpha_0\mathbf{N}^{n-1}(\mathbf{y}) = \mathbf{0} \implies \alpha_0 = 0$. Now $\sum_{i=1}^{n-1} \alpha_i\mathbf{N}^i(\mathbf{y}) = \mathbf{0}$. Apply $\mathbf{N}^{n-2}$ to both sides of this to conclude that $\alpha_1 = 0$. Continue this process until you have $\alpha_0 = \alpha_1 = \cdots = \alpha_{n-1} = 0$.

(b) Any $n \times n$ nilpotent matrix of index n can be viewed as a nilpotent operator of index n on $\Re^n$. Furthermore, $\mathbf{A} = [\mathbf{A}]_{\mathcal{S}}$ and $\mathbf{B} = [\mathbf{B}]_{\mathcal{S}}$, where $\mathcal{S}$ is the standard basis. According to part (a), there are bases $\mathcal{B}$ and $\mathcal{B}'$ such that $[\mathbf{A}]_{\mathcal{B}} = \mathbf{J}$ and $[\mathbf{B}]_{\mathcal{B}'} = \mathbf{J}$. Since $[\mathbf{A}]_{\mathcal{S}} \simeq [\mathbf{A}]_{\mathcal{B}}$, it follows that $\mathbf{A} \simeq \mathbf{J}$. Similarly $\mathbf{B} \simeq \mathbf{J}$, and hence $\mathbf{A} \simeq \mathbf{B}$ by Exercise 4.8.2.

(c) Trace and rank are similarity invariants, and part (a) implies that every $n \times n$ nilpotent matrix of index n is similar to $\mathbf{J}$, and $trace(\mathbf{J}) = 0$ and $rank(\mathbf{J}) = n - 1$.

4.8.12. (a) $\mathbf{x}_i \in R(\mathbf{E}) \implies \mathbf{x}_i = \mathbf{E}(\mathbf{v}_i)$ for some $\mathbf{v}_i \implies \mathbf{E}(\mathbf{x}_i) = \mathbf{E}^2(\mathbf{v}_i) = \mathbf{E}(\mathbf{v}_i) = \mathbf{x}_i$. Since $\mathcal{B}$ contains n vectors, you need only show that $\mathcal{B}$ is linearly independent. $\mathbf{0} = \sum_i \alpha_i\mathbf{x}_i + \beta_i\mathbf{y}_i \implies \mathbf{0} = \mathbf{E}(\mathbf{0}) = \sum_i \alpha_i\mathbf{E}(\mathbf{x}_i) + \beta_i\mathbf{E}(\mathbf{y}_i) = \sum_i \alpha_i\mathbf{x}_i \implies \alpha_i\text{'s} = 0 \implies \sum_i \beta_i\mathbf{y}_i = \mathbf{0} \implies \beta_i\text{'s} = 0$.

(b) Let $\mathcal{B} = \mathcal{X} \cup \mathcal{Y} = \{\mathbf{b}_1, \mathbf{b}_2, \ldots, \mathbf{b}_n\}$. For $j = 1, 2, \ldots, r$, the jth column of $[\mathbf{E}]_{\mathcal{B}}$ is $[\mathbf{E}(\mathbf{b}_j)]_{\mathcal{B}} = [\mathbf{E}(\mathbf{x}_j)]_{\mathcal{B}} = \mathbf{e}_j$. For $j = r+1, r+2, \ldots, n$, $[\mathbf{E}(\mathbf{b}_j)]_{\mathcal{B}} = [\mathbf{E}(\mathbf{y}_{j-r})]_{\mathcal{B}} = [\mathbf{0}]_{\mathcal{B}} = 0$.

(c) Suppose that $\mathbf{B}$ and $\mathbf{C}$ are two idempotent matrices of rank r. If you regard them as linear operators on $\Re^n$, then, with respect to the standard basis, $[\mathbf{B}]_{\mathcal{S}} = \mathbf{B}$ and $[\mathbf{C}]_{\mathcal{S}} = \mathbf{C}$. You know from part (b) that there are bases $\mathcal{U}$ and $\mathcal{V}$ such that $[\mathbf{B}]_{\mathcal{U}} = [\mathbf{C}]_{\mathcal{V}} = \begin{pmatrix} \mathbf{I}_r & \mathbf{0} \\ \mathbf{0} & \mathbf{0} \end{pmatrix} = \mathbf{P}$. This implies that $\mathbf{B} \simeq \mathbf{P}$, and $\mathbf{P} \simeq \mathbf{C}$. From Exercise 4.8.2, it follows that $\mathbf{B} \simeq \mathbf{C}$.

(d) It follows from part (c) that $\mathbf{F} \simeq \mathbf{P} = \begin{pmatrix} \mathbf{I}_r & \mathbf{0} \\ \mathbf{0} & \mathbf{0} \end{pmatrix}$. Since trace and rank are similarity invariants, $trace\,(\mathbf{F}) = trace\,(\mathbf{P}) = r = rank\,(\mathbf{P}) = rank\,(\mathbf{F})$.

Solutions for exercises in section 4. 9

4.9.1. (a)   Yes, because $\mathbf{T}(\mathbf{0}) = \mathbf{0}$.   (b)   Yes, because $\mathbf{x} \in \mathcal{V} \implies \mathbf{T}(\mathbf{x}) \in \mathcal{V}$.

4.9.2. Every subspace of $\mathcal{V}$ is invariant under $\mathbf{I}$.

4.9.3. (a) $\mathcal{X}$ is invariant because $\mathbf{x} \in \mathcal{X} \iff \mathbf{x} = (\alpha, \beta, 0, 0)$ for $\alpha, \beta \in \Re$, so

$$\mathbf{T}(\mathbf{x}) = \mathbf{T}(\alpha, \beta, 0, 0) = (\alpha + \beta,\ \beta, 0, 0) \in \mathcal{X}.$$

(b) $\left[\mathbf{T}_{/\mathcal{X}}\right]_{\{\mathbf{e}_1, \mathbf{e}_2\}} = \begin{pmatrix} 1 & 1 \\ 0 & 1 \end{pmatrix}$

(c) $[\mathbf{T}]_{\mathcal{B}} = \left(\begin{array}{cc|cc} 1 & 1 & * & * \\ 0 & 1 & * & * \\ \hline 0 & 0 & * & * \\ 0 & 0 & * & * \end{array} \right)$

4.9.4. (a) $\mathbf{Q}$ is nonsingular.   (b) $\mathcal{X}$ is invariant because

$$\mathbf{T}(\alpha_1 \mathbf{Q}_{*1} + \alpha_2 \mathbf{Q}_{*2}) = \alpha_1 \begin{pmatrix} 1 \\ 1 \\ -2 \\ 3 \end{pmatrix} + \alpha_2 \begin{pmatrix} 1 \\ 2 \\ -2 \\ 2 \end{pmatrix} = \alpha_1 \mathbf{Q}_{*1} + \alpha_2 (\mathbf{Q}_{*1} + \mathbf{Q}_{*2})$$

$$= (\alpha_1 + \alpha_2) \mathbf{Q}_{*1} + \alpha_2 \mathbf{Q}_{*2} \in span\,\{\mathbf{Q}_{*1},\ \mathbf{Q}_{*2}\}\,.$$

$\mathcal{Y}$ is invariant because

$$\mathbf{T}(\alpha_3 \mathbf{Q}_{*3} + \alpha_4 \mathbf{Q}_{*4}) = \alpha_3 \begin{pmatrix} 0 \\ 0 \\ 0 \\ 0 \end{pmatrix} + \alpha_4 \begin{pmatrix} 0 \\ 3 \\ 1 \\ -4 \end{pmatrix} = \alpha_4 \mathbf{Q}_{*3} \in span\,\{\mathbf{Q}_{*3},\ \mathbf{Q}_{*4}\}\,.$$

(c)   According to (4.9.10), $\mathbf{Q}^{-1}\mathbf{T}\mathbf{Q}$ should be block diagonal.

(d) $\quad \mathbf{Q}^{-1}\mathbf{T}\mathbf{Q} = \begin{pmatrix} 1 & 1 & 0 & 0 \\ 0 & 1 & 0 & 0 \\ \hline 0 & 0 & 0 & 1 \\ 0 & 0 & 0 & 0 \end{pmatrix} = \begin{pmatrix} \left[\mathbf{T}/_{\mathcal{X}}\right]_{\{\mathbf{Q}_{*1}, \mathbf{Q}_{*2}\}} & \mathbf{0} \\ \mathbf{0} & \left[\mathbf{T}/_{\mathcal{Y}}\right]_{\{\mathbf{Q}_{*3}, \mathbf{Q}_{*4}\}} \end{pmatrix}$

4.9.5. If $\mathbf{A} = [\alpha_{ij}]$ and $\mathbf{C} = [\gamma_{ij}]$, then

$$\mathbf{T}(\mathbf{u}_j) = \sum_{i=1}^{r} \alpha_{ij}\mathbf{u}_i \in \mathcal{U} \quad \text{and} \quad \mathbf{T}(\mathbf{w}_j) = \sum_{i=1}^{q} \gamma_{ij}\mathbf{w}_i \in \mathcal{W}.$$

4.9.6. If $\mathcal{S}$ is the standard basis for $\Re^{n \times 1}$, and if $\mathcal{B}$ is the basis consisting of the columns of $\mathbf{P}$, then

$$[\mathbf{T}]_{\mathcal{B}} = \mathbf{P}^{-1}[\mathbf{T}]_{\mathcal{S}}\mathbf{P} = \mathbf{P}^{-1}\mathbf{T}\mathbf{P} = \begin{pmatrix} \mathbf{A} & \mathbf{0} \\ \mathbf{0} & \mathbf{C} \end{pmatrix}.$$

(Recall Example 4.8.3.) The desired conclusion now follows from the result of Exercise 4.9.5.

4.9.7. $\mathbf{x} \in N(\mathbf{A} - \lambda\mathbf{I}) \implies (\mathbf{A} - \lambda\mathbf{I})\mathbf{x} = \mathbf{0} \implies \mathbf{A}\mathbf{x} = \lambda\mathbf{x} \in N(\mathbf{A} - \lambda\mathbf{I})$

4.9.8. (a) $(\mathbf{A} - \lambda\mathbf{I})$ is singular when $\lambda = -1$ and $\lambda = 3$.

(b) There are four invariant subspaces—the trivial space $\{\mathbf{0}\}$, the entire space $\Re^2$, and the two one-dimensional spaces

$$N(\mathbf{A} + \mathbf{I}) = span\left\{\begin{pmatrix} 1 \\ 2 \end{pmatrix}\right\} \quad \text{and} \quad N(\mathbf{A} - 3\mathbf{I}) = span\left\{\begin{pmatrix} 1 \\ 3 \end{pmatrix}\right\}.$$

(c) $\quad \mathbf{Q} = \begin{pmatrix} 1 & 1 \\ 2 & 3 \end{pmatrix}$

Clearly spoken, Mr. Fogg; you explain English by Greek.
— Benjamin Franklin (1706–1790)

Solutions for Chapter 5

Solutions for exercises in section 5. 1

5.1.1. (a) $\|\mathbf{x}\|_1 = 9, \quad \|\mathbf{x}\|_2 = 5, \quad \|\mathbf{x}\|_\infty = 4$

(b) $\|\mathbf{x}\|_1 = 5 + 2\sqrt{2}, \quad \|\mathbf{x}\|_2 = \sqrt{21}, \quad \|\mathbf{x}\|_\infty = 4$

5.1.2. (a) $\|\mathbf{u} - \mathbf{v}\| = \sqrt{31}$ (b) $\|\mathbf{u} + \mathbf{v}\| = \sqrt{27} \le 7 = \|\mathbf{u}\| + \|\mathbf{v}\|$

(c) $|\mathbf{u}^T\mathbf{v}| = 1 \le 10 = \|\mathbf{u}\|\,\|\mathbf{v}\|$

5.1.3. Use the CBS inequality with $\mathbf{x} = \begin{pmatrix} \alpha_1 \\ \alpha_2 \\ \vdots \\ \alpha_n \end{pmatrix}$ and $\mathbf{y} = \begin{pmatrix} 1 \\ 1 \\ \vdots \\ 1 \end{pmatrix}$.

5.1.4. (a) $\{\mathbf{x} \in \Re^n \mid \|\mathbf{x}\|_2 \le 1\}$ (b) $\{\mathbf{x} \in \Re^n \mid \|\mathbf{x} - \mathbf{c}\|_2 \le \rho\}$

5.1.5. $\|\mathbf{x} - \mathbf{y}\|^2 = \|\mathbf{x} + \mathbf{y}\|^2 \implies -2\mathbf{x}^T\mathbf{y} = 2\mathbf{x}^T\mathbf{y} \implies \mathbf{x}^T\mathbf{y} = 0.$

5.1.6. $\|\mathbf{x} - \mathbf{y}\| = \|(-1)(\mathbf{y} - \mathbf{x})\| = |(-1)|\,\|\mathbf{y} - \mathbf{x}\| = \|\mathbf{y} - \mathbf{x}\|$

5.1.7. $\mathbf{x} - \mathbf{y} = \sum_{i=1}^n (x_i - y_i)\mathbf{e}_i \implies \|\mathbf{x} - \mathbf{y}\| \le \sum_{i=1}^n |x_i - y_i|\,\|\mathbf{e}_i\| \le \nu \sum_{i=1}^n |x_i - y_i|$, where $\nu = \max_i \|\mathbf{e}_i\|$. For each $\epsilon > 0$, set $\delta = \epsilon/n\nu$. If $|x_i - y_i| < \delta$ for each i, then, using (5.1.6), $\big|\,\|\mathbf{x}\| - \|\mathbf{y}\|\,\big| \le \|\mathbf{x} - \mathbf{y}\| < \nu n\delta = \epsilon.$

5.1.8. To show that $\|\mathbf{x}\|_1 \le \sqrt{n}\,\|\mathbf{x}\|_2$, apply the CBS inequality to the standard inner product of a vector of all 1's with a vector whose components are the $|x_i|$'s.

5.1.9. If $\mathbf{y} = \alpha\mathbf{x}$, then $|\mathbf{x}^*\mathbf{y}| = |\alpha|\,\|\mathbf{x}\|^2 = \|\mathbf{x}\|\,\|\mathbf{y}\|$. Conversely, if $|\mathbf{x}^*\mathbf{y}| = \|\mathbf{x}\|\,\|\mathbf{y}\|$, then (5.1.4) implies that $\|\alpha\mathbf{x} - \mathbf{y}\| = 0$, and hence $\alpha\mathbf{x} - \mathbf{y} = \mathbf{0}$ —recall (5.1.1).

5.1.10. If $\mathbf{y} = \alpha\mathbf{x}$ for $\alpha > 0$, then $\|\mathbf{x} + \mathbf{y}\| = \|(1 + \alpha)\mathbf{x}\| = (1 + \alpha)\|\mathbf{x}\| = \|\mathbf{x}\| + \|\mathbf{y}\|$. Conversely, $\|\mathbf{x} + \mathbf{y}\| = \|\mathbf{x}\| + \|\mathbf{y}\| \implies (\|\mathbf{x}\| + \|\mathbf{y}\|)^2 = \|\mathbf{x} + \mathbf{y}\|^2 \implies$

$$\|\mathbf{x}\|^2 + 2\|\mathbf{x}\|\,\|\mathbf{y}\| + \|\mathbf{y}\|^2 = (\mathbf{x}^* + \mathbf{y}^*)(\mathbf{x} + \mathbf{y})$$
$$= \mathbf{x}^*\mathbf{x} + \mathbf{x}^*\mathbf{y} + \mathbf{y}^*\mathbf{x} + \mathbf{y}^*\mathbf{y}$$
$$= \|\mathbf{x}\|^2 + 2\,\mathrm{Re}(\mathbf{x}^*\mathbf{y}) + \|\mathbf{y}\|^2,$$

and hence $\|\mathbf{x}\|\,\|\mathbf{y}\| = \mathrm{Re}\,(\mathbf{x}^*\mathbf{y})$. But it's always true that $\mathrm{Re}\,(\mathbf{x}^*\mathbf{y}) \le |\mathbf{x}^*\mathbf{y}|$, so the CBS inequality yields

$$\|\mathbf{x}\|\,\|\mathbf{y}\| = \mathrm{Re}\,(\mathbf{x}^*\mathbf{y}) \le |\mathbf{x}^*\mathbf{y}| \le \|\mathbf{x}\|\,\|\mathbf{y}\|.$$

In other words, $|\mathbf{x}^*\mathbf{y}| = \|\mathbf{x}\|\,\|\mathbf{y}\|$. We know from Exercise 5.1.9 that equality in the CBS inequality implies $\mathbf{y} = \alpha\mathbf{x}$, where $\alpha = \mathbf{x}^*\mathbf{y}/\mathbf{x}^*\mathbf{x}$. We now need to show that this α is real and positive. Using $\mathbf{y} = \alpha\mathbf{x}$ in the equality $\|\mathbf{x} + \mathbf{y}\| =$

$\|\mathbf{x}\| + \|\mathbf{y}\|$ produces $|1 + \alpha| = 1 + |\alpha|$, or $|1 + \alpha|^2 = (1 + |\alpha|)^2$. Expanding this yields

$$(1 + \bar{\alpha})(1 + \alpha) = 1 + 2|\alpha| + |\alpha|^2$$
$$\implies \quad 1 + 2\operatorname{Re}(\alpha) + \bar{\alpha}\alpha = 1 + 2|\alpha| + \bar{\alpha}\alpha$$
$$\implies \quad \operatorname{Re}(\alpha) = |\alpha|,$$

which implies that α must be real. Furthermore, $\alpha = \operatorname{Re}(\alpha) = |\alpha| \geq 0$. Since $\mathbf{y} = \alpha\mathbf{x}$ and $\mathbf{y} \neq \mathbf{0}$, it follows that $\alpha \neq 0$, and therefore $\alpha > 0$.

5.1.11. This is a consequence of Hölder's inequality because

$$|\mathbf{x}^T\mathbf{y}| = |\mathbf{x}^T(\mathbf{y} - \alpha\mathbf{e})| \leq \|\mathbf{x}\|_1 \|\mathbf{y} - \alpha\mathbf{e}\|_\infty$$

for all α, and $\min_\alpha \|\mathbf{y} - \alpha\mathbf{e}\|_\infty = (y_{\max} - y_{\min})/2$ (with the minimum being attained at $\alpha = (y_{\max} + y_{\min})/2$).

5.1.12. (a) It's not difficult to see that $f'(t) < 0$ for $t < 1$, and $f'(t) > 0$ for $t > 1$, so we can conclude that $f(t) > f(1) = 0$ for $t \neq 1$. The desired inequality follows by setting $t = \alpha/\beta$.

(b) This inequality follows from the inequality of part (a) by setting

$$\alpha = |\hat{x}_i|^p, \quad \beta = |\hat{y}_i|^q, \quad \lambda = 1/p, \quad \text{and} \quad (1 - \lambda) = 1/q.$$

(c) Hölder's inequality results from part (b) by setting $\hat{x}_i = x_i/\|\mathbf{x}\|_p$ and $\hat{y}_i = y_i/\|\mathbf{y}\|_q$. To obtain the "vector form" of the inequality, use the triangle inequality for complex numbers to write

$$|\mathbf{x}^*\mathbf{y}| = \left|\sum_{i=1}^n \overline{x}_i y_i\right| \leq \sum_{i=1}^n |\overline{x}_i| |y_i| = \sum_{i=1}^n |x_i y_i| \leq \left(\sum_{i=1}^n |x_i|^p\right)^{1/p} \left(\sum_{i=1}^n |y_i|^q\right)^{1/q}$$
$$= \|\mathbf{x}\|_p \|\mathbf{y}\|_q.$$

5.1.13. For $p = 1$, Minkowski's inequality is a consequence of the triangle inequality for scalars. The inequality in the hint follows from the fact that $p = 1 + p/q$ together with the scalar triangle inequality, and it implies that

$$\sum_{i=1}^n |x_i + y_i|^p = \sum_{i=1}^n |x_i + y_i| |x_i + y_i|^{p/q} \leq \sum_{i=1}^n |x_i| |x_i + y_i|^{p/q} + \sum_{i=1}^n |y_i| |x_i + y_i|^{p/q}.$$

Application of Hölder's inequality produces

$$\sum_{i=1}^n |x_i| |x_i + y_i|^{p/q} \leq \left(\sum_{i=1}^n |x_i|^p\right)^{1/p} \left(\sum_{i=1}^n |x_i + y_i|^p\right)^{1/q}$$
$$= \left(\sum_{i=1}^n |x_i|^p\right)^{1/p} \left(\sum_{i=1}^n |x_i + y_i|^p\right)^{(p-1)/p}$$
$$= \|\mathbf{x}\|_p \|\mathbf{x} + \mathbf{y}\|_p^{p-1}.$$

Similarly, $\sum_{i=1}^{n} |y_i|\, |x_i + y_i|^{p/q} \leq \|\mathbf{y}\|_p \|\mathbf{x} + \mathbf{y}\|_p^{p-1}$, and therefore

$$\|\mathbf{x} + \mathbf{y}\|_p^p \leq \left(\|\mathbf{x}\|_p + \|\mathbf{y}\|_p\right) \|\mathbf{x} + \mathbf{y}\|_p^{p-1} \implies \|\mathbf{x} + \mathbf{y}\|_p \leq \|\mathbf{x}\|_p + \|\mathbf{y}\|_p .$$

Solutions for exercises in section 5. 2

5.2.1. $\|\mathbf{A}\|_F = \left[\sum_{i,j} |a_{ij}|^2\right]^{1/2} = [trace\,(\mathbf{A}^*\mathbf{A})]^{1/2} = \sqrt{10},$

$\|\mathbf{B}\|_F = \sqrt{3}$, and $\|\mathbf{C}\|_F = 9$.

5.2.2. (a) $\|\mathbf{A}\|_1 = $ max absolute column sum $= 4$, and $\|\mathbf{A}\|_\infty = $ max absolute row sum $= 3$. $\|\mathbf{A}\|_2 = \sqrt{\lambda_{\max}}$, where $\lambda_{\max}$ is the largest value of λ for which $\mathbf{A}^T\mathbf{A} - \lambda\mathbf{I}$ is singular. Determine these λ's by row reduction.

$$\mathbf{A}^T\mathbf{A} - \lambda\mathbf{I} = \begin{pmatrix} 2 - -\lambda & -4 \\ -4 & 8 - \lambda \end{pmatrix} \longrightarrow \begin{pmatrix} -4 & 8 - \lambda \\ 2 - \lambda & -4 \end{pmatrix}$$

$$\longrightarrow \begin{pmatrix} -4 & 8 - \lambda \\ 0 & -4 + \frac{2-\lambda}{4}(8 - \lambda) \end{pmatrix}$$

This matrix is singular if and only if the second pivot is zero, so we must have $(2 - \lambda)(8 - \lambda) - 16 = 0 \implies \lambda^2 - 10\lambda = 0 \implies \lambda = 0, \lambda = 10$, and therefore $\|\mathbf{A}\|_2 = \sqrt{10}$.

(b) Use the same technique to get $\|\mathbf{B}\|_1 = \|\mathbf{B}\|_2 = \|\mathbf{B}\|_\infty = 1$, and

(c) $\|\mathbf{C}\|_1 = \|\mathbf{C}\|_\infty = 10$ and $\|\mathbf{C}\|_2 = 9$.

5.2.3. (a) $\|\mathbf{I}\| = \max_{\|\mathbf{x}\|=1} \|\mathbf{Ix}\| = \max_{\|\mathbf{x}\|=1} \|\mathbf{x}\| = 1$.

(b) $\|\mathbf{I}_{n \times n}\|_F = \left[trace\,(\mathbf{I}^T\mathbf{I})\right]^{1/2} = \sqrt{n}$.

5.2.4. Use the fact that $trace\,(\mathbf{AB}) = trace\,(\mathbf{BA})$ (recall Example 3.6.5) to write

$$\|\mathbf{A}\|_F^2 = trace\,(\mathbf{A}^*\mathbf{A}) = trace\,(\mathbf{AA}^*) = \|\mathbf{A}^*\|_F^2 .$$

5.2.5. (a) For $\mathbf{x} = \mathbf{0}$, the statement is trivial. For $\mathbf{x} \neq \mathbf{0}$, we have $\|(\mathbf{x}/\|\mathbf{x}\|)\| = 1$, so for any particular $\mathbf{x}_0 \neq \mathbf{0}$,

$$\|\mathbf{A}\| = \max_{\|\mathbf{x}\|=1} \|\mathbf{Ax}\| = \max_{\mathbf{x}\neq 0} \left\|\mathbf{A}\frac{\mathbf{x}}{\|\mathbf{x}\|}\right\| \geq \frac{\|\mathbf{Ax}_0\|}{\|\mathbf{x}_0\|} \implies \|\mathbf{Ax}_0\| \leq \|\mathbf{A}\| \|\mathbf{x}_0\| .$$

(b) Let $\mathbf{x}_0$ be a vector such that $\|\mathbf{x}_0\| = 1$ and

$$\|\mathbf{ABx}_0\| = \max_{\|\mathbf{x}\|=1} \|\mathbf{ABx}\| = \|\mathbf{AB}\| .$$

Make use of the result of part (a) to write

$$\|\mathbf{AB}\| = \|\mathbf{ABx}_0\| \leq \|\mathbf{A}\| \|\mathbf{Bx}_0\| \leq \|\mathbf{A}\| \|\mathbf{B}\| \|\mathbf{x}_0\| = \|\mathbf{A}\| \|\mathbf{B}\| .$$

(c) $\|\mathbf{A}\| = \max\limits_{\|\mathbf{x}\|=1} \|\mathbf{Ax}\| \leq \max\limits_{\|\mathbf{x}\|\leq 1} \|\mathbf{Ax}\|$ because $\{\mathbf{x}\,|\,\|\mathbf{x}\| = 1\} \subset \{\mathbf{x}\,|\,\|\mathbf{x}\| \leq 1\}\,.$
If there would exist a vector $\mathbf{x}_0$ such that $\|\mathbf{x}_0\| < 1$ and $\|\mathbf{A}\| < \|\mathbf{Ax}_0\|$,
then part (a) would insure that $\|\mathbf{A}\| < \|\mathbf{Ax}_0\| \leq \|\mathbf{A}\|\,\|\mathbf{x}_0\| < \|\mathbf{A}\|$, which is
impossible.

5.2.6. (a)   Applying the CBS inequality yields

$$|\mathbf{y}^*\mathbf{Ax}| \leq \|\mathbf{y}\|_2\,\|\mathbf{Ax}\|_2 \quad\Longrightarrow\quad \max_{\substack{\|\mathbf{x}\|_2=1 \\ \|\mathbf{y}\|_2=1}} |\mathbf{y}^*\mathbf{Ax}| \leq \max_{\|\mathbf{x}\|_2=1} \|\mathbf{Ax}\|_2 = \|\mathbf{A}\|_2\,.$$

Now show that equality is actually attained for some pair $\mathbf{x}$ and $\mathbf{y}$ on the unit
2-sphere. To do so, notice that if $\mathbf{x}_0$ is a vector of unit length such that

$$\|\mathbf{Ax}_0\|_2 = \max_{\|\mathbf{x}\|_2=1} \|\mathbf{Ax}\|_2 = \|\mathbf{A}\|_2\,, \quad \text{and if} \quad \mathbf{y}_0 = \frac{\mathbf{Ax}_0}{\|\mathbf{Ax}_0\|_2} = \frac{\mathbf{Ax}_0}{\|\mathbf{A}\|_2},$$

then

$$\mathbf{y}_0^*\mathbf{Ax}_0 = \frac{\mathbf{x}_0^*\mathbf{A}^*\mathbf{Ax}_0}{\|\mathbf{A}\|_2} = \frac{\|\mathbf{Ax}_0\|_2^2}{\|\mathbf{A}\|_2} = \frac{\|\mathbf{A}\|_2^2}{\|\mathbf{A}\|_2} = \|\mathbf{A}\|_2\,.$$

(b)   This follows directly from the result of part (a) because

$$\|\mathbf{A}\|_2 = \max_{\substack{\|\mathbf{x}\|_2=1 \\ \|\mathbf{y}\|_2=1}} |\mathbf{y}^*\mathbf{Ax}| = \max_{\substack{\|\mathbf{x}\|_2=1 \\ \|\mathbf{y}\|_2=1}} |(\mathbf{y}^*\mathbf{Ax})^*| = \max_{\substack{\|\mathbf{x}\|_2=1 \\ \|\mathbf{y}\|_2=1}} |\mathbf{x}^*\mathbf{A}^*\mathbf{y}| = \|\mathbf{A}^*\|_2\,.$$

(c)   Use part (a) with the CBS inequality to write

$$\|\mathbf{A}^*\mathbf{A}\|_2 = \max_{\substack{\|\mathbf{x}\|_2=1 \\ \|\mathbf{y}\|_2=1}} |\mathbf{y}^*\mathbf{A}^*\mathbf{Ax}| \leq \max_{\substack{\|\mathbf{x}\|_2=1 \\ \|\mathbf{y}\|_2=1}} \|\mathbf{Ay}\|_2\,\|\mathbf{Ax}\|_2 = \|\mathbf{A}\|_2^2\,.$$

To see that equality is attained, let $\mathbf{x} = \mathbf{y} = \mathbf{x}_0$, where $\mathbf{x}_0$ is a vector of unit
length such that $\|\mathbf{Ax}_0\|_2 = \max_{\|\mathbf{x}\|_2=1} \|\mathbf{Ax}\|_2 = \|\mathbf{A}\|_2$, and observe

$$|\mathbf{x}_0^*\mathbf{A}^*\mathbf{Ax}_0| = \mathbf{x}_0^*\mathbf{A}^*\mathbf{Ax}_0 = \|\mathbf{Ax}_0\|_2^2 = \|\mathbf{A}\|_2^2\,.$$

(d)   Let $\mathbf{D} = \begin{pmatrix} \mathbf{A} & \mathbf{0} \\ \mathbf{0} & \mathbf{B} \end{pmatrix}$. We know from (5.2.7) that $\|\mathbf{D}\|_2^2$ is the largest value
λ such that $\mathbf{D}^T\mathbf{D} - \lambda\mathbf{I}$ is singular. But $\mathbf{D}^T\mathbf{D} - \lambda\mathbf{I}$ is singular if and only if
$\mathbf{A}^T\mathbf{A} - \lambda\mathbf{I}$ or $\mathbf{B}^T\mathbf{B} - \lambda\mathbf{I}$ is singular, so $\lambda_{\max}(\mathbf{D}) = \max\left\{\lambda_{\max}(\mathbf{A}),\,\lambda_{\max}(\mathbf{B})\right\}.$

(e)   If $\mathbf{UU}^* = \mathbf{I}$, then $\|\mathbf{U}^*\mathbf{Ax}\|_2^2 = \mathbf{x}^*\mathbf{A}^*\mathbf{UU}^*\mathbf{Ax} = \mathbf{x}^*\mathbf{A}^*\mathbf{Ax} = \|\mathbf{Ax}\|_2^2$, so
$\|\mathbf{U}^*\mathbf{A}\|_2 = \max_{\|\mathbf{x}\|_2=1} \|\mathbf{U}^*\mathbf{Ax}\|_2 = \max_{\|\mathbf{x}\|_2=1} \|\mathbf{Ax}\|_2 = \|\mathbf{A}\|_2$. Now, if $\mathbf{V}^*\mathbf{V} =$
$\mathbf{I}$, use what was just established with part (b) to write

$$\|\mathbf{AV}\|_2 = \|(\mathbf{AV})^*\|_2 = \|\mathbf{V}^*\mathbf{A}^*\|_2 = \|\mathbf{A}^*\|_2 = \|\mathbf{A}\|_2 \quad\Longrightarrow\quad \|\mathbf{U}^*\mathbf{AV}\|_2 = \|\mathbf{A}\|_2.$$

5.2.7. Proceed as follows.

$$\frac{1}{\min_{\|\mathbf{x}\|=1} \left\|\mathbf{A}^{-1}\mathbf{x}\right\|} = \max_{\|\mathbf{x}\|=1}\left\{\frac{1}{\left\|\mathbf{A}^{-1}\mathbf{x}\right\|}\right\} = \max_{\mathbf{y}\neq\mathbf{0}}\left\{\frac{1}{\left\|\mathbf{A}^{-1}\frac{(\mathbf{Ay})}{\|\mathbf{Ay}\|}\right\|}\right\}$$

$$= \max_{\mathbf{y}\neq\mathbf{0}}\frac{\|\mathbf{Ay}\|}{\left\|\mathbf{A}^{-1}(\mathbf{Ay})\right\|} = \max_{\mathbf{y}\neq\mathbf{0}}\frac{\|\mathbf{Ay}\|}{\|\mathbf{y}\|} = \max_{\mathbf{y}\neq\mathbf{0}}\left\|\mathbf{A}\left(\frac{\mathbf{y}}{\|\mathbf{y}\|}\right)\right\|$$

$$= \max_{\|\mathbf{x}\|=1}\|\mathbf{Ax}\| = \|\mathbf{A}\|.$$

5.2.8. Use (5.2.6) on p. 280 to write $\|(z\mathbf{I}-\mathbf{A})^{-1}\| = (1/\min_{\|\mathbf{x}\|=1}\|(z\mathbf{I}-\mathbf{A})\mathbf{x}\|)$, and let $\mathbf{w}$ be a vector for which $\|\mathbf{w}\| = 1$ and $\|(z\mathbf{I}-\mathbf{A})\mathbf{w}\| = \min_{\|\mathbf{x}\|=1}\|(z\mathbf{I}-\mathbf{A})\mathbf{x}\|$. Use $\|\mathbf{Aw}\| \le \|\mathbf{A}\| < |z|$ together with the "backward triangle inequality" from Example 5.1.1 (p. 273) to write

$$\|(z\mathbf{I}-\mathbf{A})\mathbf{w}\| = \|z\mathbf{w} - \mathbf{Aw}\| \ge \big|\|z\mathbf{w}\| - \|\mathbf{Aw}\|\big| = \big||z| - \|\mathbf{Aw}\|\big|$$
$$= |z| - \|\mathbf{Aw}\| \ge |z| - \|\mathbf{A}\|.$$

Consequently, $\min_{\|\mathbf{x}\|=1}\|(z\mathbf{I}-\mathbf{A})\mathbf{x}\| = \|(z\mathbf{I}-\mathbf{A})\mathbf{w}\| \ge |z| - \|\mathbf{A}\|$ implies that

$$\|(z\mathbf{I}-\mathbf{A})^{-1}\| = \frac{1}{\min_{\|\mathbf{x}\|=1}\|(z\mathbf{I}-\mathbf{A})\mathbf{x}\|} \le \frac{1}{|z| - \|\mathbf{A}\|}.$$

Solutions for exercises in section 5. 3

5.3.1. Only (c) is an inner product. The expressions in (a) and (b) each fail the first condition of the definition (5.3.1), and (d) fails the second.

5.3.2. (a) $\langle\mathbf{x}|\mathbf{y}\rangle = 0 \ \forall \ \mathbf{x} \in \mathcal{V} \implies \langle\mathbf{y}|\mathbf{y}\rangle = 0 \implies \mathbf{y} = \mathbf{0}.$

(b) $\langle\alpha\mathbf{x}|\mathbf{y}\rangle = \overline{\langle\mathbf{y}|\alpha\mathbf{x}\rangle} = \overline{\alpha\langle\mathbf{y}|\mathbf{x}\rangle} = \overline{\alpha}\overline{\langle\mathbf{y}|\mathbf{x}\rangle} = \overline{\alpha}\langle\mathbf{x}|\mathbf{y}\rangle$

(c) $\langle\mathbf{x}+\mathbf{y}|\mathbf{z}\rangle = \overline{\langle\mathbf{z}|\mathbf{x}+\mathbf{y}\rangle} = \overline{\langle\mathbf{z}|\mathbf{x}\rangle + \langle\mathbf{z}|\mathbf{y}\rangle} = \overline{\langle\mathbf{z}|\mathbf{x}\rangle} + \overline{\langle\mathbf{z}|\mathbf{y}\rangle} = \langle\mathbf{x}|\mathbf{z}\rangle + \langle\mathbf{y}|\mathbf{z}\rangle$

5.3.3. The first property in (5.2.3) holds because $\langle\mathbf{x}|\mathbf{x}\rangle \ge 0$ for all $\mathbf{x} \in \mathcal{V}$ implies $\|\mathbf{x}\| = \sqrt{\langle\mathbf{x}|\mathbf{x}\rangle} \ge 0$, and $\|\mathbf{x}\| = 0 \iff \langle\mathbf{x}|\mathbf{x}\rangle = 0 \iff \mathbf{x} = \mathbf{0}$. The second property in (5.2.3) holds because

$$\|\alpha\mathbf{x}\|^2 = \langle\alpha\mathbf{x}|\alpha\mathbf{x}\rangle = \alpha\langle\alpha\mathbf{x}|\mathbf{x}\rangle = \alpha\overline{\langle\mathbf{x}|\alpha\mathbf{x}\rangle} = \alpha\overline{\alpha}\overline{\langle\mathbf{x}|\mathbf{x}\rangle} = |\alpha|^2\langle\mathbf{x}|\mathbf{x}\rangle = |\alpha|^2\|\mathbf{x}\|^2.$$

5.3.4. $0 \le \|\mathbf{x}-\mathbf{y}\|^2 = \langle\mathbf{x}-\mathbf{y}|\mathbf{x}-\mathbf{y}\rangle = \langle\mathbf{x}|\mathbf{x}\rangle - 2\langle\mathbf{x}|\mathbf{y}\rangle + \langle\mathbf{y}|\mathbf{y}\rangle = \|\mathbf{x}\|^2 - 2\langle\mathbf{x}|\mathbf{y}\rangle + \|\mathbf{y}\|^2$

5.3.5. (a) Use the CBS inequality with the Frobenius matrix norm and the standard inner product as illustrated in Example 5.3.3, and set $\mathbf{A} = \mathbf{I}$.

(b) Proceed as in part (a), but this time set $\mathbf{A} = \mathbf{B}^T$ (recall from Example 3.6.5 that $trace\left(\mathbf{B}^T\mathbf{B}\right) = trace\left(\mathbf{B}\mathbf{B}^T\right)$).

(c) Use the result of Exercise 5.3.4 with the Frobenius matrix norm and the inner product for matrices.

5.3.6. Suppose that parallelogram identity holds, and verify that (5.3.10) satisfies the four conditions in (5.3.1). The first condition follows because $\langle \mathbf{x} | \mathbf{x} \rangle_r = \| \mathbf{x} \|^2$ and $\langle \mathbf{i}\mathbf{x} | \mathbf{x} \rangle_r = 0$ combine to yield $\langle \mathbf{x} | \mathbf{x} \rangle = \| \mathbf{x} \|^2$. The second condition (for real α) and third condition hold by virtue of the argument for (5.3.7). We will prove the fourth condition and then return to show that the second holds for complex α. By observing that $\langle \mathbf{x} | \mathbf{y} \rangle_r = \langle \mathbf{y} | \mathbf{x} \rangle_r$ and $\langle \mathbf{i}\mathbf{x} | \mathbf{i}\mathbf{y} \rangle_r = \langle \mathbf{x} | \mathbf{y} \rangle_r$, we have

$$\langle \mathbf{i}\mathbf{y} | \mathbf{x} \rangle_r = \left\langle \mathbf{i}\mathbf{y} \middle| -\mathbf{i}^2 \mathbf{x} \right\rangle_r = \langle \mathbf{y} | -\mathbf{i}\mathbf{x} \rangle_r = -\langle \mathbf{y} | \mathbf{i}\mathbf{x} \rangle_r = -\langle \mathbf{i}\mathbf{x} | \mathbf{y} \rangle_r ,$$

and hence

$$\langle \mathbf{y} | \mathbf{x} \rangle = \langle \mathbf{y} | \mathbf{x} \rangle_r + \mathbf{i} \langle \mathbf{i}\mathbf{y} | \mathbf{x} \rangle_r = \langle \mathbf{y} | \mathbf{x} \rangle_r - \mathbf{i} \langle \mathbf{i}\mathbf{x} | \mathbf{y} \rangle_r = \langle \mathbf{x} | \mathbf{y} \rangle_r - \mathbf{i} \langle \mathbf{i}\mathbf{x} | \mathbf{y} \rangle_r = \overline{\langle \mathbf{x} | \mathbf{y} \rangle}.$$

Now prove that $\langle \mathbf{x} | \alpha \mathbf{y} \rangle = \alpha \langle \mathbf{x} | \mathbf{y} \rangle$ for all complex α. Begin by showing it is true for $\alpha = \mathbf{i}$.

$$\begin{aligned}
\langle \mathbf{x} | \mathbf{i}\mathbf{y} \rangle &= \langle \mathbf{x} | \mathbf{i}\mathbf{y} \rangle_r + \mathbf{i} \langle \mathbf{i}\mathbf{x} | \mathbf{i}\mathbf{y} \rangle_r = \langle \mathbf{x} | \mathbf{i}\mathbf{y} \rangle_r + \mathbf{i} \langle \mathbf{x} | \mathbf{y} \rangle_r = \langle \mathbf{i}\mathbf{y} | \mathbf{x} \rangle_r + \mathbf{i} \langle \mathbf{x} | \mathbf{y} \rangle_r \\
&= -\langle \mathbf{i}\mathbf{x} | \mathbf{y} \rangle_r + \mathbf{i} \langle \mathbf{x} | \mathbf{y} \rangle_r = \mathbf{i} (\langle \mathbf{x} | \mathbf{y} \rangle_r + \mathbf{i} \langle \mathbf{i}\mathbf{x} | \mathbf{y} \rangle_r) \\
&= \mathbf{i} \langle \mathbf{x} | \mathbf{y} \rangle
\end{aligned}$$

For $\alpha = \xi + i\eta$,

$$\langle \mathbf{x} | \alpha \mathbf{y} \rangle = \langle \mathbf{x} | \xi \mathbf{y} + i\eta \mathbf{y} \rangle = \langle \mathbf{x} | \xi \mathbf{y} \rangle + \langle \mathbf{x} | i\eta \mathbf{y} \rangle = \xi \langle \mathbf{x} | \mathbf{y} \rangle + i\eta \langle \mathbf{x} | \mathbf{y} \rangle = \alpha \langle \mathbf{x} | \mathbf{y} \rangle .$$

Conversely, if $\langle \star | \star \rangle$ is any inner product on $\mathcal{V}$, then with $\| \star \|^2 = \langle \star | \star \rangle$ we have

$$\begin{aligned}
\| \mathbf{x} + \mathbf{y} \|^2 + \| \mathbf{x} - \mathbf{y} \|^2 &= \langle \mathbf{x} + \mathbf{y} | \mathbf{x} + \mathbf{y} \rangle + \langle \mathbf{x} - \mathbf{y} | \mathbf{x} - \mathbf{y} \rangle \\
&= \| \mathbf{x} \|^2 + 2\mathrm{Re} \langle \mathbf{x} | \mathbf{y} \rangle + \| \mathbf{y} \|^2 + \| \mathbf{x} \|^2 - 2\mathrm{Re} \langle \mathbf{x} | \mathbf{y} \rangle + \| \mathbf{y} \|^2 \\
&= 2 \left(\| \mathbf{x} \|^2 + \| \mathbf{y} \|^2 \right).
\end{aligned}$$

5.3.7. The parallelogram identity (5.3.7) fails to hold for all $\mathbf{x}, \mathbf{y} \in \mathcal{C}^n$. For example, if $\mathbf{x} = \mathbf{e}_1$ and $\mathbf{y} = \mathbf{e}_2$, then

$$\| \mathbf{e}_1 + \mathbf{e}_2 \|_\infty^2 + \| \mathbf{e}_1 - \mathbf{e}_2 \|_\infty^2 = 2, \quad \text{but} \quad 2 (\| \mathbf{e}_1 \|_\infty^2 + \| \mathbf{e}_2 \|_\infty^2) = 4.$$

5.3.8. (a) As shown in Example 5.3.2, the Frobenius matrix norm $\mathcal{C}^{n \times n}$ is generated by the standard matrix inner product (5.3.2), so the result on p. 290 guarantees that $\| \star \|_F$ satisfies the parallelogram identity.

5.3.9. No, because the parallelogram inequality (5.3.7) doesn't hold. To see that $\| \mathbf{X} + \mathbf{Y} \|^2 + \| \mathbf{X} - \mathbf{Y} \|^2 = 2 (\| \mathbf{X} \|^2 + \| \mathbf{Y} \|^2)$ is not valid for all $\mathbf{X}, \mathbf{Y} \in \mathcal{C}^{n \times n}$, let $\mathbf{X} = \mathrm{diag}(1, 0, \ldots, 0)$ and $\mathbf{Y} = \mathrm{diag}(0, 1, \ldots, 0)$. For $\star = 1, 2$, or ∞,

$$\| \mathbf{X} + \mathbf{Y} \|_\star^2 + \| \mathbf{X} - \mathbf{Y} \|_\star^2 = 1 + 1 = 2, \quad \text{but} \quad 2 (\| \mathbf{X} \|_\star^2 + \| \mathbf{Y} \|_\star^2) = 4.$$

Solutions for exercises in section 5. 4

5.4.1. (a), (b), and (e) are orthogonal pairs.

5.4.2. First find $\mathbf{v} = \begin{pmatrix} \alpha_1 \\ \alpha_2 \end{pmatrix}$ such that $3\alpha_1 - 2\alpha_2 = 0$, and then normalize $\mathbf{v}$. The second must be the negative of $\mathbf{v}$.

5.4.3. (a) Simply verify that $\mathbf{x}_i^T \mathbf{x}_j = 0$ for $i \neq j$.

(b) Let $\mathbf{x}_4^T = (\alpha_1 \ \ \alpha_2 \ \ \alpha_3 \ \ \alpha_4)$, and notice that $\mathbf{x}_i^T \mathbf{x}_4 = 0$ for $i = 1, 2, 3$ is three homogeneous equations in four unknowns

$$\begin{pmatrix} 1 & -1 & 0 & 2 \\ 1 & 1 & 1 & 0 \\ -1 & -1 & 2 & 0 \end{pmatrix} \begin{pmatrix} \alpha_1 \\ \alpha_2 \\ \alpha_3 \\ \alpha_4 \end{pmatrix} = \begin{pmatrix} 0 \\ 0 \\ 0 \end{pmatrix} \implies \begin{pmatrix} \alpha_1 \\ \alpha_2 \\ \alpha_3 \\ \alpha_4 \end{pmatrix} = \beta \begin{pmatrix} -1 \\ 1 \\ 0 \\ 1 \end{pmatrix}.$$

(c) Simply normalize the set by dividing each vector by its norm.

5.4.4. The Fourier coefficients are

$$\xi_1 = \langle \mathbf{u}_1 | \mathbf{x} \rangle = \frac{1}{\sqrt{2}}, \quad \xi_2 = \langle \mathbf{u}_2 | \mathbf{x} \rangle = \frac{-1}{\sqrt{3}}, \quad \xi_3 = \langle \mathbf{u}_3 | \mathbf{x} \rangle = \frac{-5}{\sqrt{6}},$$

so

$$\mathbf{x} = \xi_1 \mathbf{u}_1 + \xi_2 \mathbf{u}_2 + \xi_3 \mathbf{u}_3 = \frac{1}{2} \begin{pmatrix} 1 \\ -1 \\ 0 \end{pmatrix} - \frac{1}{3} \begin{pmatrix} 1 \\ 1 \\ 1 \end{pmatrix} - \frac{5}{6} \begin{pmatrix} -1 \\ -1 \\ 2 \end{pmatrix}.$$

5.4.5. If $\mathbf{U}_1$, $\mathbf{U}_2$, $\mathbf{U}_3$, and $\mathbf{U}_4$ denote the elements of $\mathcal{B}$, verify they constitute an orthonormal set by showing that

$$\langle \mathbf{U}_i | \mathbf{U}_j \rangle = trace(\mathbf{U}_i^T \mathbf{U}_j) = 0 \text{ for } i \neq j \quad \text{and} \quad \|\mathbf{U}_i\| = \sqrt{trace(\mathbf{U}_i^T \mathbf{U}_i)} = 1.$$

Consequently, $\mathcal{B}$ is linearly independent—recall (5.4.2)—and therefore $\mathcal{B}$ is a basis because it is a *maximal* independent set—part (b) of Exercise 4.4.4 insures $\dim \Re^{2 \times 2} = 4$. The Fourier coefficients $\langle \mathbf{U}_i | \mathbf{A} \rangle = trace(\mathbf{U}_i^T \mathbf{A})$ are

$$\langle \mathbf{U}_1 | \mathbf{A} \rangle = \frac{2}{\sqrt{2}}, \quad \langle \mathbf{U}_2 | \mathbf{A} \rangle = 0, \quad \langle \mathbf{U}_3 | \mathbf{A} \rangle = 1, \quad \langle \mathbf{U}_4 | \mathbf{A} \rangle = 1,$$

so the Fourier expansion of $\mathbf{A}$ is $\mathbf{A} = (2/\sqrt{2})\mathbf{U}_1 + \mathbf{U}_3 + \mathbf{U}_4$.

5.4.6. $\cos \theta = \mathbf{x}^T \mathbf{y} / \|\mathbf{x}\| \|\mathbf{y}\| = 1/2$, so $\theta = \pi/3$.

5.4.7. This follows because each vector has a unique representation in terms of a basis—see Exercise 4.4.8 or the discussion of coordinates in §4.7.

5.4.8. If the columns of $\mathbf{U} = [\mathbf{u}_1 | \mathbf{u}_2 | \cdots | \mathbf{u}_n]$ are an orthonormal basis for $\mathcal{C}^n$, then

$$[\mathbf{U}^* \mathbf{U}]_{ij} = \mathbf{u}_i^* \mathbf{u}_j = \begin{cases} 1 & \text{when } i = j, \\ 0 & \text{when } i \neq j, \end{cases} \tag{$\ddagger$}$$

and, therefore, $\mathbf{U}^*\mathbf{U} = \mathbf{I}$. Conversely, if $\mathbf{U}^*\mathbf{U} = \mathbf{I}$, then ($\ddagger$) holds, so the columns of $\mathbf{U}$ are orthonormal—they are a basis for $\mathcal{C}^n$ because orthonormal sets are always linearly independent.

5.4.9. Equations (4.5.5) and (4.5.6) guarantee that

$$R(\mathbf{A}) = R(\mathbf{A}\mathbf{A}^*) \quad \text{and} \quad N(\mathbf{A}) = N(\mathbf{A}^*\mathbf{A}),$$

and consequently $\mathbf{r} \in R(\mathbf{A}) = R(\mathbf{A}\mathbf{A}^*) \implies \mathbf{r} = \mathbf{A}\mathbf{A}^*\mathbf{x}$ for some $\mathbf{x}$, and $\mathbf{n} \in N(\mathbf{A}) = N(\mathbf{A}^*\mathbf{A}) \implies \mathbf{A}^*\mathbf{A}\mathbf{n} = \mathbf{0}$. Therefore,

$$\langle \mathbf{r} | \mathbf{n} \rangle = \mathbf{r}^*\mathbf{n} = \mathbf{x}^*\mathbf{A}\mathbf{A}^*\mathbf{n} = \mathbf{x}^*\mathbf{A}^*\mathbf{A}\mathbf{n} = \mathbf{0}.$$

5.4.10. (a) $\pi/4$ (b) $\pi/2$

5.4.11. The number $\mathbf{x}^T\mathbf{y}$ or $\mathbf{x}^*\mathbf{y}$ will in general be complex. In order to guarantee that we end up with a real number, we should take

$$\cos\theta = \frac{|\operatorname{Re}(\mathbf{x}^*\mathbf{y})|}{\|\mathbf{x}\|\,\|\mathbf{y}\|}.$$

5.4.12. Use the Fourier expansion $\mathbf{y} = \sum_i \langle \mathbf{u}_i | \mathbf{y} \rangle \mathbf{u}_i$ together with the various properties of an inner product to write

$$\langle \mathbf{x} | \mathbf{y} \rangle = \left\langle \mathbf{x} \left| \sum_i \langle \mathbf{u}_i | \mathbf{y} \rangle \mathbf{u}_i \right\rangle \right. = \sum_i \langle \mathbf{x} | \langle \mathbf{u}_i | \mathbf{y} \rangle \mathbf{u}_i \rangle = \sum_i \langle \mathbf{u}_i | \mathbf{y} \rangle \langle \mathbf{x} | \mathbf{u}_i \rangle.$$

5.4.13. In a real space, $\langle \mathbf{x} | \mathbf{y} \rangle = \langle \mathbf{y} | \mathbf{x} \rangle$, so the third condition in the definition (5.3.1) of an inner product and Exercise 5.3.2(c) produce

$$
\begin{aligned}
\langle \mathbf{x} + \mathbf{y} | \mathbf{x} - \mathbf{y} \rangle &= \langle \mathbf{x} + \mathbf{y} | \mathbf{x} \rangle - \langle \mathbf{x} + \mathbf{y} | \mathbf{y} \rangle \\
&= \langle \mathbf{x} | \mathbf{x} \rangle + \langle \mathbf{y} | \mathbf{x} \rangle - \langle \mathbf{x} | \mathbf{y} \rangle - \langle \mathbf{y} | \mathbf{y} \rangle \\
&= \|\mathbf{x}\|^2 - \|\mathbf{y}\|^2 = 0.
\end{aligned}
$$

5.4.14. (a) In a real space, $\langle \mathbf{x} | \mathbf{y} \rangle = \langle \mathbf{y} | \mathbf{x} \rangle$, so the third condition in the definition (5.3.1) of an inner product and Exercise 5.3.2(c) produce

$$
\begin{aligned}
\|\mathbf{x} + \mathbf{y}\|^2 = \langle \mathbf{x} + \mathbf{y} | \mathbf{x} + \mathbf{y} \rangle &= \langle \mathbf{x} + \mathbf{y} | \mathbf{x} \rangle + \langle \mathbf{x} + \mathbf{y} | \mathbf{y} \rangle \\
&= \langle \mathbf{x} | \mathbf{x} \rangle + \langle \mathbf{y} | \mathbf{x} \rangle + \langle \mathbf{x} | \mathbf{y} \rangle + \langle \mathbf{y} | \mathbf{y} \rangle \\
&= \|\mathbf{x}\|^2 + 2\langle \mathbf{x} | \mathbf{y} \rangle + \|\mathbf{y}\|^2,
\end{aligned}
$$

and hence $\langle \mathbf{x} | \mathbf{y} \rangle = 0$ if and only if $\|\mathbf{x} + \mathbf{y}\|^2 = \|\mathbf{x}\|^2 + \|\mathbf{y}\|^2$.

(b) In a complex space, $\mathbf{x} \perp \mathbf{y} \implies \|\mathbf{x} + \mathbf{y}\|^2 = \|\mathbf{x}\|^2 + \|\mathbf{y}\|^2$, but the converse is not valid—e.g., consider $\mathcal{C}^2$ with the standard inner product, and let $\mathbf{x} = \begin{pmatrix} -i \\ 1 \end{pmatrix}$ and $\mathbf{y} = \begin{pmatrix} 1 \\ i \end{pmatrix}$.

(c)   Again, using the properties of a general inner product, derive the expansion

$$\|\alpha\mathbf{x} + \beta\mathbf{y}\|^2 = \langle \alpha\mathbf{x} + \beta\mathbf{y} \,|\, \alpha\mathbf{x} + \beta\mathbf{y} \rangle$$
$$= \langle \alpha\mathbf{x} | \alpha\mathbf{x} \rangle + \langle \alpha\mathbf{x} | \beta\mathbf{y} \rangle + \langle \beta\mathbf{y} | \alpha\mathbf{x} \rangle + \langle \beta\mathbf{y} | \beta\mathbf{y} \rangle$$
$$= \|\alpha\mathbf{x}\|^2 + \overline{\alpha}\beta \langle \mathbf{x} | \mathbf{y} \rangle + \overline{\beta}\alpha \langle \mathbf{y} | \mathbf{x} \rangle + \|\beta\mathbf{y}\|^2 .$$

Clearly, $\mathbf{x} \perp \mathbf{y} \implies \|\alpha\mathbf{x} + \beta\mathbf{y}\|^2 = \|\alpha\mathbf{x}\|^2 + \|\beta\mathbf{y}\|^2 \;\forall\; \alpha, \beta$. Conversely, if $\|\alpha\mathbf{x} + \beta\mathbf{y}\|^2 = \|\alpha\mathbf{x}\|^2 + \|\beta\mathbf{y}\|^2 \;\forall\; \alpha, \beta$, then $\overline{\alpha}\beta \langle \mathbf{x} | \mathbf{y} \rangle + \overline{\beta}\alpha \langle \mathbf{y} | \mathbf{x} \rangle = 0 \;\forall\; \alpha, \beta$. Letting $\alpha = \langle \mathbf{x} | \mathbf{y} \rangle$ and $\beta = 1$ produces the conclusion that $2| \langle \mathbf{x} | \mathbf{y} \rangle |^2 = 0$, and thus $\langle \mathbf{x} | \mathbf{y} \rangle = 0$.

5.4.15. (a) $\cos \theta_i = \langle \mathbf{u}_i | \mathbf{x} \rangle / \|\mathbf{u}_i\| \, \|\mathbf{x}\| = \langle \mathbf{u}_i | \mathbf{x} \rangle / \|\mathbf{x}\| = \xi_i / \|\mathbf{x}\|$

(b)   Use the Pythagorean theorem (Exercise 5.4.14) to write

$$\|\mathbf{x}\|^2 = \|\xi_1 \mathbf{u}_1 + \xi_2 \mathbf{u}_2 + \cdots + \xi_n \mathbf{u}_n\|^2$$
$$= \|\xi_1 \mathbf{u}_1\|^2 + \|\xi_2 \mathbf{u}_2\|^2 + \cdots + \|\xi_n \mathbf{u}_n\|^2$$
$$= |\xi_1|^2 + |\xi_2|^2 + \cdots + |\xi_n|^2.$$

5.4.16. Use the properties of an inner product to write

$$\left\| \mathbf{x} - \sum_{i=1}^{k} \xi_i \mathbf{u}_i \right\|^2 = \left\langle \mathbf{x} - \sum_{i=1}^{k} \xi_i \mathbf{u}_i \,\middle|\, \mathbf{x} - \sum_{i=1}^{k} \xi_i \mathbf{u}_i \right\rangle$$
$$= \langle \mathbf{x} | \mathbf{x} \rangle - 2 \sum_i |\xi_i|^2 + \left\langle \sum_{i=1}^{k} \xi_i \mathbf{u}_i \,\middle|\, \sum_{i=1}^{k} \xi_i \mathbf{u}_i \right\rangle$$
$$= \|\mathbf{x}\|^2 - 2 \sum_i |\xi_i|^2 + \left\| \sum_{i=1}^{k} \xi_i \mathbf{u}_i \right\|^2 ,$$

and then invoke the Pythagorean theorem (Exercise 5.4.14) to conclude

$$\left\| \sum_{i=1}^{k} \xi_i \mathbf{u}_i \right\|^2 = \sum_i \|\xi_i \mathbf{u}_i\|^2 = \sum_i |\xi_i|^2.$$

Consequently,

$$0 \leq \left\| \mathbf{x} - \sum_{i=1}^{k} \xi_i \mathbf{u}_i \right\|^2 = \|\mathbf{x}\|^2 - \sum_i |\xi_i|^2 \implies \sum_{i=1}^{k} |\xi_i|^2 \leq \|\mathbf{x}\|^2 . \qquad (\ddagger)$$

If $\mathbf{x} \in span \{\mathbf{u}_1, \mathbf{u}_2, \ldots, \mathbf{u}_k\}$, then the Fourier expansion of $\mathbf{x}$ with respect to the $\mathbf{u}_i$'s is $\mathbf{x} = \sum_{i=1}^{k} \xi_i \mathbf{u}_i$, and hence equality holds in $(\ddagger)$. Conversely, if equality holds in $(\ddagger)$, then $\mathbf{x} - \sum_{i=1}^{k} \xi_i \mathbf{u}_i = \mathbf{0}$.

5.4.17. Choose any unit vector $\mathbf{e}_i$ for $\mathbf{y}$. The angle between $\mathbf{e}$ and $\mathbf{e}_i$ approaches $\pi/2$ as $n \to \infty$, but $\mathbf{e}^T \mathbf{e}_i = 1$ for all n.

5.4.18. If $\mathbf{y}$ is negatively correlated to $\mathbf{x}$, then $\mathbf{z_x} = -\mathbf{z_y}$, but $\|\mathbf{z_x} - \mathbf{z_y}\|_2 = 2\sqrt{n}$ gives no indication of the fact that $\mathbf{z_x}$ and $\mathbf{z_y}$ are on the same line. Continuity therefore dictates that when $\mathbf{y} \approx \beta_0 \mathbf{e} + \beta_1 \mathbf{x}$ with $\beta_1 < 0$, then $\mathbf{z_x} \approx -\mathbf{z_y}$, but $\|\mathbf{z_x} - \mathbf{z_y}\|_2 \approx 2\sqrt{n}$ gives no hint that $\mathbf{z_x}$ and $\mathbf{z_y}$ are almost on the same line. If we want to use norms to gauge linear correlation, we should use

$$\min\left\{\|\mathbf{z_x} - \mathbf{z_y}\|_2, \|\mathbf{z_x} + \mathbf{z_y}\|_2\right\}.$$

5.4.19. (a) $\cos\theta = 1 \implies \langle \mathbf{x}|\mathbf{y}\rangle = \|\mathbf{x}\|\,\|\mathbf{y}\| > 0$, and the straightforward extension of Exercise 5.1.9 guarantees that

$$\mathbf{y} = \frac{\langle \mathbf{x}|\mathbf{y}\rangle}{\|\mathbf{x}\|^2}\mathbf{x}, \quad \text{and clearly} \quad \frac{\langle \mathbf{x}|\mathbf{y}\rangle}{\|\mathbf{x}\|^2} > 0.$$

Conversely, if $\mathbf{y} = \alpha\mathbf{x}$ for $\alpha > 0$, then $\langle \mathbf{x}|\mathbf{y}\rangle = \alpha\|\mathbf{x}\|^2 \implies \cos\theta = 1$.

(b) $\cos\theta = -1 \implies \langle \mathbf{x}|\mathbf{y}\rangle = -\|\mathbf{x}\|\,\|\mathbf{y}\| < 0$, so the generalized version of Exercise 5.1.9 guarantees that

$$\mathbf{y} = \frac{\langle \mathbf{x}|\mathbf{y}\rangle}{\|\mathbf{x}\|^2}\mathbf{x}, \quad \text{and in this case} \quad \frac{\langle \mathbf{x}|\mathbf{y}\rangle}{\|\mathbf{x}\|^2} < 0.$$

Conversely, if $\mathbf{y} = \alpha\mathbf{x}$ for $\alpha < 0$, then $\langle \mathbf{x}|\mathbf{y}\rangle = \alpha\|\mathbf{x}\|^2$, so

$$\cos\theta = \frac{\alpha\|\mathbf{x}\|^2}{|\alpha|\,\|\mathbf{x}\|^2} = -1.$$

5.4.20. $F(t) = \sum_n^\infty (-1)^n \frac{2}{n}\sin nt$.

Solutions for exercises in section 5. 5

5.5.1. (a)

$$\mathbf{u}_1 = \frac{1}{2}\begin{pmatrix} 1 \\ 1 \\ 1 \\ -1 \end{pmatrix}, \quad \mathbf{u}_2 = \frac{1}{2\sqrt{3}}\begin{pmatrix} 3 \\ -1 \\ -1 \\ 1 \end{pmatrix}, \quad \mathbf{u}_3 = \frac{1}{\sqrt{6}}\begin{pmatrix} 0 \\ 1 \\ 1 \\ 2 \end{pmatrix}$$

(b) First verify this is an orthonormal set by showing $\mathbf{u}_i^T\mathbf{u}_j = \begin{cases} 1 & \text{when } i = j, \\ 0 & \text{when } i \neq j. \end{cases}$
To show that the $\mathbf{x}_i$'s and the $\mathbf{u}_i$'s span the same space, place the $\mathbf{x}_i$'s as rows in a matrix $\mathbf{A}$, and place the $\mathbf{u}_i$'s as rows in a matrix $\mathbf{B}$, and then verify that $\mathbf{E_A} = \mathbf{E_B}$—recall Example 4.2.2.

(c) The result should be the same as in part (a).

5.5.2. First reduce $\mathbf{A}$ to $\mathbf{E_A}$ to determine a "regular" basis for each space.

$$R\left(\mathbf{A}\right) = span\left\{\begin{pmatrix} 1 \\ 2 \\ 3 \end{pmatrix}\right\} \qquad N\left(\mathbf{A}^T\right) = span\left\{\begin{pmatrix} -2 \\ 1 \\ 0 \end{pmatrix}, \begin{pmatrix} -3 \\ 0 \\ 1 \end{pmatrix}\right\}$$

$$R\left(\mathbf{A}^T\right) = span\left\{\begin{pmatrix} 1 \\ -2 \\ 3 \\ -1 \end{pmatrix}\right\} \qquad N\left(\mathbf{A}\right) = span\left\{\begin{pmatrix} 2 \\ 1 \\ 0 \\ 0 \end{pmatrix}, \begin{pmatrix} -3 \\ 0 \\ 1 \\ 0 \end{pmatrix}, \begin{pmatrix} 1 \\ 0 \\ 0 \\ 1 \end{pmatrix}\right\}$$

Now apply Gram–Schmidt to each of these.

$$R\left(\mathbf{A}\right) = span\left\{\frac{1}{\sqrt{14}}\begin{pmatrix} 1 \\ 2 \\ 3 \end{pmatrix}\right\} \quad N\left(\mathbf{A}^T\right) = span\left\{\frac{1}{\sqrt{5}}\begin{pmatrix} -2 \\ 1 \\ 0 \end{pmatrix}, \frac{1}{\sqrt{70}}\begin{pmatrix} -3 \\ -6 \\ 5 \end{pmatrix}\right\}$$

$$R\left(\mathbf{A}^T\right) = span\left\{\frac{1}{\sqrt{15}}\begin{pmatrix} 1 \\ -2 \\ 3 \\ -1 \end{pmatrix}\right\}$$

$$N\left(\mathbf{A}\right) = span\left\{\frac{1}{\sqrt{5}}\begin{pmatrix} 2 \\ 1 \\ 0 \\ 0 \end{pmatrix}, \frac{1}{\sqrt{70}}\begin{pmatrix} -3 \\ 6 \\ 5 \\ 0 \end{pmatrix}, \frac{1}{\sqrt{210}}\begin{pmatrix} 1 \\ -2 \\ 3 \\ 14 \end{pmatrix}\right\}$$

5.5.3.

$$\mathbf{u}_1 = \frac{1}{\sqrt{3}}\begin{pmatrix} i \\ i \\ i \end{pmatrix}, \quad \mathbf{u}_2 = \frac{1}{\sqrt{6}}\begin{pmatrix} -2i \\ i \\ i \end{pmatrix}, \quad \mathbf{u}_3 = \frac{1}{\sqrt{2}}\begin{pmatrix} 0 \\ -i \\ i \end{pmatrix}$$

5.5.4. Nothing! The resulting orthonormal set is the same as the original.

5.5.5. It breaks down at the first vector such that $\mathbf{x}_k \in span\left\{\mathbf{x}_1, \mathbf{x}_2, \ldots, \mathbf{x}_{k-1}\right\}$ because if

$$\mathbf{x}_k \in span\left\{\mathbf{x}_1, \mathbf{x}_2, \ldots, \mathbf{x}_{k-1}\right\} = span\left\{\mathbf{u}_1, \mathbf{u}_2, \ldots, \mathbf{u}_{k-1}\right\},$$

then the Fourier expansion of $\mathbf{x}_k$ with respect to $span\left\{\mathbf{u}_1, \mathbf{u}_2, \ldots, \mathbf{u}_{k-1}\right\}$ is

$$\mathbf{x}_k = \sum_{i=1}^{k-1} \langle \mathbf{u}_i | \mathbf{x}_k \rangle \mathbf{u}_i,$$

and therefore

$$\mathbf{u}_k = \frac{\left(\mathbf{x}_k - \sum_{i=1}^{k-1} \langle \mathbf{u}_i | \mathbf{x}_k \rangle \mathbf{u}_i\right)}{\left\|\left(\mathbf{x}_k - \sum_{i=1}^{k-1} \langle \mathbf{u}_i | \mathbf{x}_k \rangle \mathbf{u}_i\right)\right\|} = \frac{\mathbf{0}}{\|\mathbf{0}\|}$$

is not defined.

5.5.6. (a)   The rectangular QR factors are

$$\mathbf{Q} = \begin{pmatrix} 1/\sqrt{3} & -1/\sqrt{3} & 1/\sqrt{6} \\ 1/\sqrt{3} & 1/\sqrt{3} & 1/\sqrt{6} \\ 1/\sqrt{3} & 0 & -2/\sqrt{6} \\ 0 & 1/\sqrt{3} & 0 \end{pmatrix} \quad \text{and} \quad \mathbf{R} = \begin{pmatrix} \sqrt{3} & \sqrt{3} & -\sqrt{3} \\ 0 & \sqrt{3} & \sqrt{3} \\ 0 & 0 & \sqrt{6} \end{pmatrix}.$$

(b)   Following Example 5.5.3, solve $\mathbf{Rx} = \mathbf{Q}^T\mathbf{b}$ to get $\mathbf{x} = \begin{pmatrix} 2/3 \\ 1/3 \\ 0 \end{pmatrix}$.

5.5.7. For $k = 1$, there is nothing to prove. For $k > 1$, assume that $\mathcal{O}_k$ is an orthonormal basis for $\mathcal{S}_k$. First establish that $\mathcal{O}_{k+1}$ must be an orthonormal set. Orthogonality follows because for each $j < k + 1$,

$$\begin{aligned} \langle \mathbf{u}_j | \mathbf{u}_{k+1} \rangle &= \left\langle \mathbf{u}_j \left| \frac{1}{\nu_{k+1}} \left(\mathbf{x}_{k+1} - \sum_{i=1}^{k} \langle \mathbf{u}_i | \mathbf{x}_{k+1} \rangle \mathbf{u}_i \right) \right. \right\rangle \\ &= \frac{1}{\nu_{k+1}} \left(\langle \mathbf{u}_j | \mathbf{x}_{k+1} \rangle - \left\langle \mathbf{u}_j \left| \sum_{i=1}^{k} \langle \mathbf{u}_i | \mathbf{x}_{k+1} \rangle \mathbf{u}_i \right. \right\rangle \right) \\ &= \frac{1}{\nu_{k+1}} \left(\langle \mathbf{u}_j | \mathbf{x}_{k+1} \rangle - \sum_{i=1}^{k} \langle \mathbf{u}_i | \mathbf{x}_{k+1} \rangle \langle \mathbf{u}_j | \mathbf{u}_i \rangle \right) \\ &= \frac{1}{\nu_{k+1}} \left(\langle \mathbf{u}_j | \mathbf{x}_{k+1} \rangle - \langle \mathbf{u}_j | \mathbf{x}_{k+1} \rangle \right) = 0. \end{aligned}$$

This together with the fact that each $\mathbf{u}_i$ has unit norm means that $\mathcal{O}_{k+1}$ is an orthonormal set. Now assume $\mathcal{O}_k$ is a basis for $\mathcal{S}_k$, and prove that $\mathcal{O}_{k+1}$ is a basis for $\mathcal{S}_{k+1}$. If $\mathbf{x} \in \mathcal{S}_{k+1}$, then $\mathbf{x}$ can be written as a combination

$$\mathbf{x} = \sum_{i=1}^{k+1} \alpha_i \mathbf{x}_i = \left(\sum_{i=1}^{k} \alpha_i \mathbf{x}_i \right) + \alpha_{k+1} \mathbf{x}_{k+1},$$

where $\sum_{i=1}^{k} \alpha_i \mathbf{x}_i \in \mathcal{S}_k = span\,(\mathcal{O}_k) \subset span\,(\mathcal{O}_{k+1})$. Couple this together with the fact that

$$\mathbf{x}_{k+1} = \nu_{k+1}\mathbf{u}_{k+1} + \sum_{i=1}^{k} \langle \mathbf{u}_i | \mathbf{x}_{k+1} \rangle \mathbf{u}_i \in span\,(\mathcal{O}_{k+1})$$

to conclude that $\mathbf{x} \in span\,(\mathcal{O}_{k+1})$. Consequently, $\mathcal{O}_{k+1}$ spans $\mathcal{S}_{k+1}$, and therefore $\mathcal{O}_{k+1}$ is a basis for $\mathcal{S}_{k+1}$ because orthonormal sets are always linearly independent.

5.5.8. If $\mathbf{A} = \mathbf{Q}_1\mathbf{R}_1 = \mathbf{Q}_2\mathbf{R}_2$ are two rectangular QR factorizations, then (5.5.6) implies $\mathbf{A}^T\mathbf{A} = \mathbf{R}_1^T\mathbf{R}_1 = \mathbf{R}_2^T\mathbf{R}_2$. It follows from Example 3.10.7 that $\mathbf{A}^T\mathbf{A}$ is positive definite, and $\mathbf{R}_1 = \mathbf{R}_2$ because the Cholesky factorization of a positive definite matrix is unique. Therefore, $\mathbf{Q}_1 = \mathbf{A}\mathbf{R}_1^{-1} = \mathbf{A}\mathbf{R}_2^{-1} = \mathbf{Q}_2$.

5.5.9. (a) **Step 1:** $fl\,\|\mathbf{x}_1\| = 1$, so $\mathbf{u}_1 \leftarrow \mathbf{x}_1$.

Step 2: $\mathbf{u}_1^T\mathbf{x}_2 = 1$, so

$$\mathbf{u}_2 \leftarrow \mathbf{x}_2 - \left(\mathbf{u}_1^T\mathbf{x}_2\right)\mathbf{u}_1 = \begin{pmatrix} 0 \\ 0 \\ -10^{-3} \end{pmatrix} \quad \text{and} \quad \mathbf{u}_2 \leftarrow \frac{\mathbf{u}_2}{\|\mathbf{u}_2\|} = \begin{pmatrix} 0 \\ 0 \\ -1 \end{pmatrix}.$$

Step 3: $\mathbf{u}_1^T\mathbf{x}_3 = 1$ and $\mathbf{u}_2^T\mathbf{x}_3 = 0$, so

$$\mathbf{u}_3 \leftarrow \mathbf{x}_3 - \left(\mathbf{u}_1^T\mathbf{x}_3\right)\mathbf{u}_1 - \left(\mathbf{u}_2^T\mathbf{x}_3\right)\mathbf{u}_2 = \begin{pmatrix} 0 \\ 10^{-3} \\ -10^{-3} \end{pmatrix} \quad \text{and} \quad \mathbf{u}_3 \leftarrow \frac{\mathbf{u}_3}{\|\mathbf{u}_3\|} = \begin{pmatrix} 0 \\ .709 \\ -.709 \end{pmatrix}.$$

Therefore, the result of the classical Gram–Schmidt algorithm using 3-digit arithmetic is

$$\mathbf{u}_1 = \begin{pmatrix} 1 \\ 0 \\ 10^{-3} \end{pmatrix}, \quad \mathbf{u}_2 = \begin{pmatrix} 0 \\ 0 \\ -1 \end{pmatrix}, \quad \mathbf{u}_3 = \begin{pmatrix} 0 \\ .709 \\ -.709 \end{pmatrix},$$

which is not very good because $\mathbf{u}_2$ and $\mathbf{u}_3$ are not even close to being orthogonal.

(b) **Step 1:** $fl\,\|\mathbf{x}_1\| = 1$, so

$$\{\mathbf{u}_1, \mathbf{u}_2, \mathbf{u}_3\} \leftarrow \{\mathbf{x}_1, \mathbf{x}_2, \mathbf{x}_3\}.$$

Step 2: $\mathbf{u}_1^T\mathbf{u}_2 = 1$ and $\mathbf{u}_1^T\mathbf{u}_3 = 1$, so

$$\mathbf{u}_2 \leftarrow \mathbf{u}_2 - \left(\mathbf{u}_1^T\mathbf{u}_2\right)\mathbf{u}_1 = \begin{pmatrix} 0 \\ 0 \\ -10^{-3} \end{pmatrix}, \quad \mathbf{u}_3 \leftarrow \mathbf{u}_3 - \left(\mathbf{u}_1^T\mathbf{u}_3\right)\mathbf{u}_1 = \begin{pmatrix} 0 \\ 10^{-3} \\ -10^{-3} \end{pmatrix},$$

and then

$$\mathbf{u}_2 \leftarrow \frac{\mathbf{u}_2}{\|\mathbf{u}_2\|} = \begin{pmatrix} 0 \\ 0 \\ -1 \end{pmatrix}.$$

Step 3: $\mathbf{u}_2^T\mathbf{u}_3 = 10^{-3}$, so

$$\mathbf{u}_3 \leftarrow \mathbf{u}_3 - \left(\mathbf{u}_2^T\mathbf{u}_3\right)\mathbf{u}_2 = \begin{pmatrix} 0 \\ 10^{-3} \\ 0 \end{pmatrix} \quad \text{and} \quad \mathbf{u}_3 \leftarrow \frac{\mathbf{u}_3}{\|\mathbf{u}_3\|} = \begin{pmatrix} 0 \\ 1 \\ 0 \end{pmatrix}.$$

Thus the modified Gram–Schmidt algorithm produces

$$\mathbf{u}_1 = \begin{pmatrix} 1 \\ 0 \\ 10^{-3} \end{pmatrix}, \quad \mathbf{u}_2 = \begin{pmatrix} 0 \\ 0 \\ -1 \end{pmatrix}, \quad \mathbf{u}_3 = \begin{pmatrix} 0 \\ 1 \\ 0 \end{pmatrix},$$

which is as close to being an orthonormal set as one could reasonably hope to obtain by using 3-digit arithmetic.

5.5.10. Yes. In both cases r_{ij} is the (i,j)-entry in the upper-triangular matrix R in the QR factorization.

5.5.11. $p_0(x) = 1/\sqrt{2}, \quad p_1(x) = \sqrt{3/2}\, x, \quad p_2(x) = \sqrt{5/8}\, (3x^2 - 1)$

Solutions for exercises in section 5. 6

5.6.1. (a), (c), and (d).

5.6.2. Yes, because $\mathbf{U}^*\mathbf{U} = \begin{pmatrix} 1 & 0 \\ 0 & 1 \end{pmatrix}.$

5.6.3. (a) Eight: $\mathbf{D} = \begin{pmatrix} \pm 1 & 0 & 0 \\ 0 & \pm 1 & 0 \\ 0 & 0 & \pm 1 \end{pmatrix}$ (b) 2^n : $\mathbf{D} = \begin{pmatrix} \pm 1 & 0 & \cdots & 0 \\ 0 & \pm 1 & \cdots & 0 \\ \vdots & \vdots & \ddots & \vdots \\ 0 & 0 & \cdots & \pm 1 \end{pmatrix}$

(c) There are infinitely many because each diagonal entry can be any point on the unit circle in the complex plane—these matrices have the form given in part (d) of Exercise 5.6.1.

5.6.4. (a) When $\alpha^2 + \beta^2 = 1/2$. (b) When $\alpha^2 + \beta^2 = 1$.

5.6.5. (a) $(\mathbf{UV})^*(\mathbf{UV}) = \mathbf{V}^*\mathbf{U}^*\mathbf{UV} = \mathbf{V}^*\mathbf{V} = \mathbf{I}.$

(b) Consider $\mathbf{I} + (-\mathbf{I}) = \mathbf{0}.$

(c)

$$\begin{pmatrix} \mathbf{U} & \mathbf{0} \\ \mathbf{0} & \mathbf{V} \end{pmatrix}^* \begin{pmatrix} \mathbf{U} & \mathbf{0} \\ \mathbf{0} & \mathbf{V} \end{pmatrix} = \begin{pmatrix} \mathbf{U}^* & \mathbf{0} \\ \mathbf{0} & \mathbf{V}^* \end{pmatrix} \begin{pmatrix} \mathbf{U} & \mathbf{0} \\ \mathbf{0} & \mathbf{V} \end{pmatrix}$$
$$= \begin{pmatrix} \mathbf{U}^*\mathbf{U} & \mathbf{0} \\ \mathbf{0} & \mathbf{V}^*\mathbf{V} \end{pmatrix}$$
$$= \begin{pmatrix} \mathbf{I} & \mathbf{0} \\ \mathbf{0} & \mathbf{I} \end{pmatrix}.$$

5.6.6. Recall from (3.7.8) or (4.2.10) that $(\mathbf{I}+\mathbf{A})^{-1}$ exists if and only if $N(\mathbf{I}+\mathbf{A}) = \mathbf{0}$, and write $\mathbf{x} \in N(\mathbf{I}+\mathbf{A}) \implies \mathbf{x} = -\mathbf{A}\mathbf{x} \implies \mathbf{x}^*\mathbf{x} = -\mathbf{x}^*\mathbf{A}\mathbf{x}$. But taking the conjugate transpose of both sides yields $\mathbf{x}^*\mathbf{x} = -\mathbf{x}^*\mathbf{A}^*\mathbf{x} = \mathbf{x}^*\mathbf{A}\mathbf{x}$, so $\mathbf{x}^*\mathbf{x} = 0$, and thus $\mathbf{x} = \mathbf{0}$. Replacing $\mathbf{A}$ by $-\mathbf{A}$ in Exercise 3.7.6 gives $\mathbf{A}(\mathbf{I}+\mathbf{A})^{-1} = (\mathbf{I}+\mathbf{A})^{-1}\mathbf{A}$, so

$$(\mathbf{I} - \mathbf{A})(\mathbf{I}+\mathbf{A})^{-1} = (\mathbf{I}+\mathbf{A})^{-1} - \mathbf{A}(\mathbf{I}+\mathbf{A})^{-1}$$
$$= (\mathbf{I}+\mathbf{A})^{-1} - (\mathbf{I}+\mathbf{A})^{-1}\mathbf{A} = (\mathbf{I}+\mathbf{A})^{-1}(\mathbf{I} - \mathbf{A}).$$

These results together with the fact that $\mathbf{A}$ is skew hermitian produce

$$\begin{aligned} \mathbf{U}^*\mathbf{U} &= (\mathbf{I}+\mathbf{A})^{-1^*}(\mathbf{I}-\mathbf{A})^*(\mathbf{I}-\mathbf{A})(\mathbf{I}+\mathbf{A})^{-1} \\ &= (\mathbf{I}+\mathbf{A})^{*-1}(\mathbf{I}-\mathbf{A})^*(\mathbf{I}-\mathbf{A})(\mathbf{I}+\mathbf{A})^{-1} \\ &= (\mathbf{I}-\mathbf{A})^{-1}(\mathbf{I}+\mathbf{A})(\mathbf{I}-\mathbf{A})(\mathbf{I}+\mathbf{A})^{-1} = \mathbf{I}. \end{aligned}$$

5.6.7. (a) Yes—because if $\mathbf{R} = \mathbf{I} - 2\mathbf{u}\mathbf{u}^*$, where $\|\mathbf{u}\| = 1$, then

$$\begin{pmatrix} \mathbf{I} & \mathbf{0} \\ \mathbf{0} & \mathbf{R} \end{pmatrix} = \mathbf{I} - 2 \begin{pmatrix} \mathbf{0} \\ \mathbf{u} \end{pmatrix} \begin{pmatrix} \mathbf{0} & \mathbf{u}^* \end{pmatrix} \quad \text{and} \quad \left\| \begin{pmatrix} \mathbf{0} \\ \mathbf{u} \end{pmatrix} \right\| = 1.$$

(b) No—Suppose $\mathbf{R} = \mathbf{I} - 2\mathbf{u}\mathbf{u}^*$ and $\mathbf{S} = \mathbf{I} - 2\mathbf{v}\mathbf{v}^*$, where $\|\mathbf{u}\| = 1$ and $\|\mathbf{v}\| = 1$ so that

$$\begin{pmatrix} \mathbf{R} & \mathbf{0} \\ \mathbf{0} & \mathbf{S} \end{pmatrix} = \mathbf{I} - 2 \begin{pmatrix} \mathbf{u}\mathbf{u}^* & \mathbf{0} \\ \mathbf{0} & \mathbf{v}\mathbf{v}^* \end{pmatrix}.$$

If we could find a vector $\mathbf{w}$ such that $\|\mathbf{w}\| = 1$ and

$$\begin{pmatrix} \mathbf{R} & \mathbf{0} \\ \mathbf{0} & \mathbf{S} \end{pmatrix} = \mathbf{I} - 2\mathbf{w}\mathbf{w}^*, \quad \text{then} \quad \mathbf{w}\mathbf{w}^* = \begin{pmatrix} \mathbf{u}\mathbf{u}^* & \mathbf{0} \\ \mathbf{0} & \mathbf{v}\mathbf{v}^* \end{pmatrix}.$$

But this is impossible because (recall Example 3.9.3)

$$rank\,(\mathbf{w}\mathbf{w}^*) = 1 \quad \text{and} \quad rank \begin{pmatrix} \mathbf{u}\mathbf{u}^* & \mathbf{0} \\ \mathbf{0} & \mathbf{v}\mathbf{v}^* \end{pmatrix} = 2.$$

5.6.8. (a) $\mathbf{u}^*\mathbf{v} = (\mathbf{U}\mathbf{x})^*\mathbf{U}\mathbf{y} = \mathbf{x}^*\mathbf{U}^*\mathbf{U}\mathbf{y} = \mathbf{x}^*\mathbf{y}$

(b) The fact that $\mathbf{P}$ is an isometry means $\|\mathbf{u}\| = \|\mathbf{x}\|$ and $\|\mathbf{v}\| = \|\mathbf{y}\|$. Use this together with part (a) and the definition of cosine given in (5.4.1) to obtain

$$\cos\theta_{\mathbf{u},\mathbf{v}} = \frac{\mathbf{u}^T\mathbf{v}}{\|\mathbf{u}\|\,\|\mathbf{v}\|} = \frac{\mathbf{x}^T\mathbf{y}}{\|\mathbf{x}\|\,\|\mathbf{y}\|} = \cos\theta_{\mathbf{x},\mathbf{y}}.$$

5.6.9. (a) Since $\mathbf{U}_{m\times r}$ has orthonormal columns, we have $\mathbf{U}^*\mathbf{U} = \mathbf{I}_r$ so that

$$\|\mathbf{U}\|_2^2 = \max_{\|\mathbf{x}\|_2=1} \mathbf{x}^*\mathbf{U}^*\mathbf{U}\mathbf{x} = \max_{\|\mathbf{x}\|_2=1} \mathbf{x}^*\mathbf{x} = 1.$$

This together with $\|\mathbf{A}\|_2 = \|\mathbf{A}^*\|_2$—recall (5.2.10)—implies $\|\mathbf{V}\|_2 = 1$. For the Frobenius norm we have

$$\|\mathbf{U}\|_F = [trace\,(\mathbf{U}^*\mathbf{U})]^{1/2} = [trace\,(\mathbf{I})]^{1/2} = \sqrt{r}.$$

$trace\,(\mathbf{AB}) = trace\,(\mathbf{BA})$ (Example 3.6.5) and $\mathbf{VV}^* = \mathbf{I}_k \implies \|\mathbf{V}\|_F = \sqrt{k}.$

(b)   First show that $\|\mathbf{U}\mathbf{A}\|_2 = \|\mathbf{A}\|_2$ by writing

$$\|\mathbf{U}\mathbf{A}\|_2^2 = \max_{\|\mathbf{x}\|_2=1} \|\mathbf{U}\mathbf{A}\mathbf{x}\|_2^2 = \max_{\|\mathbf{x}\|_2=1} \mathbf{x}^*\mathbf{A}^*\mathbf{U}^*\mathbf{U}\mathbf{A}\mathbf{x} = \max_{\|\mathbf{x}\|_2=1} \mathbf{x}^*\mathbf{A}^*\mathbf{A}\mathbf{x}$$

$$= \max_{\|\mathbf{x}\|_2=1} \|\mathbf{A}\mathbf{x}\|_2^2 = \|\mathbf{A}\|_2^2 \, .$$

Now use this together with $\|\mathbf{A}\|_2 = \|\mathbf{A}^*\|_2$ to observe that

$$\|\mathbf{A}\mathbf{V}\|_2 = \|\mathbf{V}^*\mathbf{A}^*\|_2 = \|\mathbf{A}^*\|_2 = \|\mathbf{A}\|_2 \, .$$

Therefore, $\|\mathbf{U}\mathbf{A}\mathbf{V}\|_2 = \|\mathbf{U}(\mathbf{A}\mathbf{V})\|_2 = \|\mathbf{A}\mathbf{V}\|_2 = \|\mathbf{A}\|_2 \, .$

(c)   Use $trace\,(\mathbf{A}\mathbf{B}) = trace\,(\mathbf{B}\mathbf{A})$ with $\mathbf{U}^*\mathbf{U} = \mathbf{I}_r$ and $\mathbf{V}\mathbf{V}^* = \mathbf{I}_k$ to write

$$\|\mathbf{U}\mathbf{A}\mathbf{V}\|_F^2 = trace\,((\mathbf{U}\mathbf{A}\mathbf{V})^*\mathbf{U}\mathbf{A}\mathbf{V}) = trace\,(\mathbf{V}^*\mathbf{A}^*\mathbf{U}^*\mathbf{U}\mathbf{A}\mathbf{V})$$

$$= trace\,(\mathbf{V}^*\mathbf{A}^*\mathbf{A}\mathbf{V}) = trace\,(\mathbf{A}^*\mathbf{A}\mathbf{V}\mathbf{V}^*)$$

$$= trace\,(\mathbf{A}^*\mathbf{A}) = \|\mathbf{A}\|_F^2 \, .$$

5.6.10. Use (5.6.6) to compute the following quantities.

(a) $\dfrac{\mathbf{v}\mathbf{v}^T}{\mathbf{v}^T\mathbf{v}}\mathbf{u} = \left(\dfrac{\mathbf{v}^T\mathbf{u}}{\mathbf{v}^T\mathbf{v}}\right)\mathbf{v} = \dfrac{1}{6}\mathbf{v} = \dfrac{1}{6}\begin{pmatrix} 1 \\ 4 \\ 0 \\ -1 \end{pmatrix}$

(b) $\dfrac{\mathbf{u}\mathbf{u}^T}{\mathbf{u}^T\mathbf{u}}\mathbf{v} = \left(\dfrac{\mathbf{u}^T\mathbf{v}}{\mathbf{u}^T\mathbf{u}}\right)\mathbf{u} = \dfrac{1}{5}\mathbf{u} = \dfrac{1}{5}\begin{pmatrix} -2 \\ 1 \\ 3 \\ -1 \end{pmatrix}$

(c) $\left(\mathbf{I} - \dfrac{\mathbf{v}\mathbf{v}^T}{\mathbf{v}^T\mathbf{v}}\right)\mathbf{u} = \mathbf{u} - \left(\dfrac{\mathbf{v}^T\mathbf{u}}{\mathbf{v}^T\mathbf{v}}\right)\mathbf{v} = \mathbf{u} - \dfrac{1}{6}\mathbf{v} = \dfrac{1}{6}\begin{pmatrix} -13 \\ 2 \\ 18 \\ -5 \end{pmatrix}$

(d) $\left(\mathbf{I} - \dfrac{\mathbf{u}\mathbf{u}^T}{\mathbf{u}^T\mathbf{u}}\right)\mathbf{v} = \mathbf{v} - \left(\dfrac{\mathbf{u}^T\mathbf{v}}{\mathbf{u}^T\mathbf{u}}\right)\mathbf{u} = \mathbf{v} - \dfrac{1}{5}\mathbf{u} = \dfrac{1}{5}\begin{pmatrix} 7 \\ 19 \\ -3 \\ -4 \end{pmatrix}$

5.6.11. (a) $N\,(\mathbf{Q}) \neq \{\mathbf{0}\}$ because $\mathbf{Q}\mathbf{u} = \mathbf{0}$ and $\|\mathbf{u}\| = 1 \implies \mathbf{u} \neq \mathbf{0}$, so $\mathbf{Q}$ must be singular by (4.2.10).

(b)   The result of Exercise 4.4.10 insures that $n - 1 \le rank\,(\mathbf{Q})$, and the result of part (a) says $rank\,(\mathbf{Q}) \le n - 1$, and therefore $rank\,(\mathbf{Q}) = n - 1$.

5.6.12. Use (5.6.5) in conjunction with the CBS inequality given in (5.1.3) to write

$$\|\mathbf{p}\| = |\mathbf{u}^*\mathbf{x}| \le \|\mathbf{u}\| \, \|\mathbf{x}\| = \|\mathbf{x}\| \, .$$

The fact that equality holds if and only if $\mathbf{x}$ is a scalar multiple of $\mathbf{u}$ follows from the result of Exercise 5.1.9.

5.6.13. (a) Set $\mathbf{u} = \mathbf{x} - \|\mathbf{x}\| \, \mathbf{e}_1 = -2/3 \begin{pmatrix} 1 \\ 1 \\ 1 \end{pmatrix}$, and compute

$$\mathbf{R} = \mathbf{I} - \frac{2\mathbf{u}\mathbf{u}^T}{\mathbf{u}^T\mathbf{u}} = \frac{1}{3} \begin{pmatrix} 1 & -2 & -2 \\ -2 & 1 & -2 \\ -2 & -2 & 1 \end{pmatrix}.$$

(You could also use $\mathbf{u} = \mathbf{x} + \|\mathbf{x}\| \, \mathbf{e}_1$.)

(b) Verify that $\mathbf{R} = \mathbf{R}^T$, $\mathbf{R}^T\mathbf{R} = \mathbf{I}$, and $\mathbf{R}^2 = \mathbf{I}$.

(c) The columns of the reflector $\mathbf{R}$ computed in part (a) do the job.

5.6.14. $\mathbf{R}\mathbf{x} = \mathbf{x} \implies 2\mathbf{u}\mathbf{u}^*\mathbf{x} = \mathbf{0} \implies \mathbf{u}^*\mathbf{x} = 0$ because $\mathbf{u} \neq \mathbf{0}$.

5.6.15. If $\mathbf{R}\mathbf{x} = \mathbf{y}$ in Figure 5.6.2, then the line segment between $\mathbf{x} - \mathbf{y}$ is parallel to the line determined by $\mathbf{u}$, so $\mathbf{x} - \mathbf{y}$ itself must be a scalar multiple of $\mathbf{u}$. If $\mathbf{x} - \mathbf{y} = \alpha\mathbf{u}$, then

$$\mathbf{u} = \frac{\mathbf{x} - \mathbf{y}}{\alpha} = \frac{\mathbf{x} - \mathbf{y}}{\|\mathbf{x} - \mathbf{y}\|}.$$

It is straightforward to verify that this choice of $\mathbf{u}$ produces the desired reflector.

5.6.16. You can verify by direct multiplication that $\mathbf{P}^T\mathbf{P} = \mathbf{I}$ and $\mathbf{U}^*\mathbf{U} = \mathbf{I}$, but you can also recognize that $\mathbf{P}$ and $\mathbf{U}$ are elementary reflectors that come from Example 5.6.3 in the sense that

$$\mathbf{P} = \mathbf{I} - 2\frac{\mathbf{u}\mathbf{u}^T}{\mathbf{u}^T\mathbf{u}}, \quad \text{where} \quad \mathbf{u} = \mathbf{x} - \mathbf{e}_1 = \begin{pmatrix} x_1 - 1 \\ \tilde{\mathbf{x}} \end{pmatrix}$$

and

$$\mathbf{U} = \mu \left(\mathbf{I} - 2\frac{\mathbf{u}\mathbf{u}^*}{\mathbf{u}^*\mathbf{u}} \right), \quad \text{where} \quad \mathbf{u} = \mathbf{x} - \mu\mathbf{e}_1 = \begin{pmatrix} x_1 - \mu \\ \tilde{\mathbf{x}} \end{pmatrix}.$$

5.6.17. The final result is

$$\mathbf{v}_3 = \begin{pmatrix} -\sqrt{2}/2 \\ \sqrt{6}/2 \\ 1 \end{pmatrix}$$

and

$$\mathbf{Q} = \mathbf{P}_z(\pi/6)\mathbf{P}_y(-\pi/2)\mathbf{P}_x(\pi/4) = \frac{1}{4} \begin{pmatrix} 0 & -\sqrt{6} - \sqrt{2} & -\sqrt{6} + \sqrt{2} \\ 0 & \sqrt{6} - \sqrt{2} & -\sqrt{6} - \sqrt{2} \\ 4 & 0 & 0 \end{pmatrix}.$$

5.6.18. It matters because the rotation matrices given on p. 328 generally do not commute with each other (this is easily verified by direct multiplication). For example, this means that it is generally the case that

$$\mathbf{P}_y(\phi)\mathbf{P}_x(\theta)\mathbf{v} \neq \mathbf{P}_x(\theta)\mathbf{P}_y(\phi)\mathbf{v}.$$

5.6.19. As pointed out in Example 5.6.2, $\mathbf{u}^{\perp} = (\mathbf{u}/\|\mathbf{u}\|)^{\perp}$, so we can assume without any loss of generality that $\mathbf{u}$ has unit norm. We also know that any vector of unit norm can be extended to an orthonormal basis for $\mathcal{C}^n$—Examples 5.6.3 and 5.6.6 provide two possible ways to accomplish this. Let $\{\mathbf{u}, \mathbf{v}_1, \mathbf{v}_2, \ldots, \mathbf{v}_{n-1}\}$ be such an orthonormal basis for $\mathcal{C}^n$.

Claim: $span\{\mathbf{v}_1, \mathbf{v}_2, \ldots, \mathbf{v}_{n-1}\} = \mathbf{u}^{\perp}$.

> *Proof.* $\mathbf{x} \in span\{\mathbf{v}_1, \mathbf{v}_2, \ldots, \mathbf{v}_{n-1}\} \implies \mathbf{x} = \sum_i \alpha_i \mathbf{v}_i \implies \mathbf{u}^*\mathbf{x} = \sum_i \alpha_i \mathbf{u}^*\mathbf{v}_i = 0 \implies \mathbf{x} \in \mathbf{u}^{\perp}$, and thus $span\{\mathbf{v}_1, \mathbf{v}_2, \ldots, \mathbf{v}_{n-1}\} \subseteq \mathbf{u}^{\perp}$. To establish the reverse inclusion, write $\mathbf{x} = \alpha_0 \mathbf{u} + \sum_i \alpha_i \mathbf{v}_i$, and then note that $\mathbf{x} \perp \mathbf{u} \implies 0 = \mathbf{u}^*\mathbf{x} = \alpha_0 \implies \mathbf{x} \in span\{\mathbf{v}_1, \mathbf{v}_2, \ldots, \mathbf{v}_{n-1}\}$, and hence $\implies \mathbf{u}^{\perp} \subseteq span\{\mathbf{v}_1, \mathbf{v}_2, \ldots, \mathbf{v}_{n-1}\}$.

Consequently, $\{\mathbf{v}_1, \mathbf{v}_2, \ldots, \mathbf{v}_{n-1}\}$ is a basis for $\mathbf{u}^{\perp}$ because it is a spanning set that is linearly independent—recall (4.3.14)—and thus $\dim \mathbf{u}^{\perp} = n - 1$.

5.6.20. The relationship between the matrices in (5.6.6) and (5.6.7) on p. 324 suggests that if $\mathbf{P}$ is a projector, then $\mathbf{A} = \mathbf{I} - 2\mathbf{P}$ is an involution—and indeed this is true because $\mathbf{A}^2 = (\mathbf{I} - 2\mathbf{P})^2 = \mathbf{I} - 4\mathbf{P} + 4\mathbf{P}^2 = \mathbf{I}$. Similarly, if $\mathbf{A}$ is an involution, then $\mathbf{P} = (\mathbf{I} - \mathbf{A})/2$ is easily verified to be a projector. Thus each projector uniquely defines an involution, and vice versa.

5.6.21. The outside of the face is visible from the perspective indicated in Figure 5.6.6 if and only if the angle θ between $\mathbf{n}$ and the positive x-axis is between $-90°$ and $+90°$. This is equivalent to saying that the cosine between $\mathbf{n}$ and $\mathbf{e}_1$ is positive, so the desired conclusion follows from the fact that

$$\cos\theta > 0 \iff \frac{\mathbf{n}^T\mathbf{e}_1}{\|\mathbf{n}\|\,\|\mathbf{e}_1\|} > 0 \iff \mathbf{n}^T\mathbf{e}_1 > 0 \iff n_1 > 0.$$

Solutions for exercises in section 5.7

5.7.1. (a) Householder reduction produces

$$\mathbf{R}_2\mathbf{R}_1\mathbf{A} = \begin{pmatrix} 1 & 0 & 0 \\ 0 & -3/5 & 4/5 \\ 0 & 4/5 & 3/5 \end{pmatrix} \begin{pmatrix} 1/3 & -2/3 & 2/3 \\ -2/3 & 1/3 & 2/3 \\ 2/3 & 2/3 & 1/3 \end{pmatrix} \begin{pmatrix} 1 & 19 & -34 \\ -2 & -5 & 20 \\ 2 & 8 & 37 \end{pmatrix}$$

$$= \begin{pmatrix} 3 & 15 & 0 \\ 0 & 15 & -30 \\ 0 & 0 & 45 \end{pmatrix} = \mathbf{R},$$

so

$$\mathbf{Q} = (\mathbf{R}_2\mathbf{R}_1)^T = \begin{pmatrix} 1/3 & 14/15 & -2/15 \\ -2/3 & 1/3 & 2/3 \\ 2/3 & -2/15 & 11/15 \end{pmatrix}.$$

(b)   Givens reduction produces $\mathbf{P}_{23}\mathbf{P}_{13}\mathbf{P}_{12}\mathbf{A} = \mathbf{R}$, where

$$\mathbf{P}_{12} = \begin{pmatrix} 1/\sqrt{5} & -2/\sqrt{5} & 0 \\ 2/\sqrt{5} & 1/\sqrt{5} & 0 \\ 0 & 0 & 1 \end{pmatrix} \quad \mathbf{P}_{13} = \begin{pmatrix} \sqrt{5}/3 & 0 & 2/3 \\ 0 & 1 & 0 \\ -2/3 & 0 & \sqrt{5}/3 \end{pmatrix}$$

$$\mathbf{P}_{23} = \begin{pmatrix} 1 & 0 & 0 \\ 0 & 11/5\sqrt{5} & -2/5\sqrt{5} \\ 0 & 2/5\sqrt{5} & 11/5\sqrt{5} \end{pmatrix}$$

5.7.2. Since $\mathbf{P}$ is an orthogonal matrix, so is $\mathbf{P}^T$, and hence the columns of $\mathbf{X}$ are an orthonormal set. By writing

$$\mathbf{A} = \mathbf{P}^T\mathbf{T} = [\mathbf{X}\,|\,\mathbf{Y}]\begin{pmatrix} \mathbf{R} \\ \mathbf{0} \end{pmatrix} = \mathbf{XR},$$

and by using the fact that $rank\,(\mathbf{A}) = n \implies rank\,(\mathbf{R}) = n$, it follows that $R\,(\mathbf{A}) = R\,(\mathbf{XR}) = R\,(\mathbf{X})$—recall Exercise 4.5.12. Since every orthonormal set is linearly independent, the columns of $\mathbf{X}$ are a linearly independent spanning set for $R\,(\mathbf{A})$, and thus the columns of $\mathbf{X}$ are an orthonormal basis for $R\,(\mathbf{A})$. Notice that when the diagonal entries of $\mathbf{R}$ are positive, $\mathbf{A} = \mathbf{XR}$ is the "rectangular" QR factorization for $\mathbf{A}$ introduced on p. 311, and the columns of $\mathbf{X}$ are the same columns as those produced by the Gram–Schmidt procedure.

5.7.3. According to (5.7.1), set $\mathbf{u} = \mathbf{A}_{*1} - \|\mathbf{A}_{*1}\|\,\mathbf{e}_1 = \begin{pmatrix} -1 \\ 2 \\ -2 \\ 1 \end{pmatrix}$, so

$$\mathbf{R}_1 = \mathbf{I} - 2\frac{\mathbf{uu}^*}{\mathbf{u}^*\mathbf{u}} = \frac{1}{5}\begin{pmatrix} 4 & 2 & -2 & 1 \\ 2 & 1 & 4 & -2 \\ -2 & 4 & 1 & 2 \\ 1 & -2 & 2 & 4 \end{pmatrix} \quad \text{and} \quad \mathbf{R}_1\mathbf{A} = \begin{pmatrix} 5 & -15 & 5 \\ 0 & 10 & -5 \\ 0 & -10 & 2 \\ 0 & 5 & 14 \end{pmatrix}.$$

Next use $\mathbf{u} = \begin{pmatrix} 10 \\ -10 \\ 5 \end{pmatrix} - \begin{pmatrix} 15 \\ 0 \\ 0 \end{pmatrix} = \begin{pmatrix} -5 \\ -10 \\ 5 \end{pmatrix}$ to build

$$\hat{\mathbf{R}}_2 = \mathbf{I} - 2\frac{\mathbf{uu}^*}{\mathbf{u}^*\mathbf{u}} = \frac{1}{3}\begin{pmatrix} 2 & -2 & 1 \\ -2 & -1 & 2 \\ 1 & 2 & 2 \end{pmatrix} \quad \text{and} \quad \mathbf{R}_2 = \frac{1}{3}\begin{pmatrix} 3 & 0 & 0 & 0 \\ 0 & 2 & -2 & 1 \\ 0 & -2 & -1 & 2 \\ 0 & 1 & 2 & 2 \end{pmatrix},$$

so

$$\mathbf{R}_2\mathbf{R}_1\mathbf{A} = \begin{pmatrix} 5 & -15 & 5 \\ 0 & 15 & 0 \\ 0 & 0 & 12 \\ 0 & 0 & 9 \end{pmatrix}.$$

Finally, with $\mathbf{u} = \begin{pmatrix} 12 \\ 9 \end{pmatrix} - \begin{pmatrix} 15 \\ 0 \end{pmatrix} = \begin{pmatrix} -3 \\ 9 \end{pmatrix}$, build

$$\hat{\mathbf{R}}_3 = \frac{1}{5}\begin{pmatrix} 4 & 3 \\ 3 & -4 \end{pmatrix} \quad \text{and} \quad \mathbf{R}_3 = \frac{1}{5}\begin{pmatrix} 5 & 0 & 0 & 0 \\ 0 & 5 & 0 & 0 \\ 0 & 0 & 4 & 3 \\ 0 & 0 & 3 & -4 \end{pmatrix},$$

so that

$$\mathbf{R}_3\mathbf{R}_2\mathbf{R}_1\mathbf{A} = \begin{pmatrix} 5 & -15 & 5 \\ 0 & 15 & 0 \\ 0 & 0 & 15 \\ 0 & 0 & 0 \end{pmatrix}.$$

Therefore, $\mathbf{PA} = \mathbf{T} = \begin{pmatrix} \mathbf{R} \\ \mathbf{0} \end{pmatrix}$, where

$$\mathbf{P} = \mathbf{R}_3\mathbf{R}_2\mathbf{R}_1 = \frac{1}{15}\begin{pmatrix} 12 & 6 & -6 & 3 \\ 9 & -8 & 8 & -4 \\ 0 & -5 & 2 & 14 \\ 0 & -10 & -11 & -2 \end{pmatrix} \quad \text{and} \quad \mathbf{R} = \begin{pmatrix} 5 & -15 & 5 \\ 0 & 15 & 0 \\ 0 & 0 & 15 \end{pmatrix}.$$

The result of Exercise 5.7.2 insures that the first three columns in

$$\mathbf{P}^T = \mathbf{R}_1\mathbf{R}_2\mathbf{R}_3 = \frac{1}{15}\begin{pmatrix} 12 & 9 & 0 & 0 \\ 6 & -8 & -5 & -10 \\ -6 & 8 & 2 & -11 \\ 3 & -4 & 14 & -2 \end{pmatrix}$$

are an orthonormal basis for $R(\mathbf{A})$. Since the diagonal entries of $\mathbf{R}$ are positive,

$$\frac{1}{15}\begin{pmatrix} 12 & 9 & 0 \\ 6 & -8 & -5 \\ -6 & 8 & 2 \\ 3 & -4 & 14 \end{pmatrix}\begin{pmatrix} 5 & -15 & 5 \\ 0 & 15 & 0 \\ 0 & 0 & 15 \end{pmatrix} = \mathbf{A}$$

is the "rectangular" QR factorization for $\mathbf{A}$ discussed on p. 311.

5.7.4. If $\mathbf{A}$ has full column rank, and if $\mathbf{P}$ is an orthogonal matrix such that

$$\mathbf{PA} = \mathbf{T} = \begin{pmatrix} \mathbf{R} \\ \mathbf{0} \end{pmatrix} \quad \text{and} \quad \mathbf{Pb} = \begin{pmatrix} \mathbf{c} \\ \mathbf{d} \end{pmatrix},$$

where $\mathbf{R}$ is an upper-triangular matrix, then the results of Example 5.7.3 insure that the least squares solution of $\mathbf{Ax} = \mathbf{b}$ can be obtained by solving the triangular system $\mathbf{Rx} = \mathbf{c}$. The matrices $\mathbf{P}$ and $\mathbf{R}$ were computed in Exercise 5.7.3, so the least squares solution of $\mathbf{Ax} = \mathbf{b}$ is the solution to

$$\begin{pmatrix} 5 & -15 & 5 \\ 0 & 15 & 0 \\ 0 & 0 & 15 \end{pmatrix}\begin{pmatrix} x_1 \\ x_2 \\ x_3 \end{pmatrix} = \begin{pmatrix} 4 \\ 3 \\ 33 \end{pmatrix} \implies \mathbf{x} = \frac{1}{5}\begin{pmatrix} -4 \\ 1 \\ 11 \end{pmatrix}.$$

5.7.5. $\|\mathbf{A}\|_F = \|\mathbf{QR}\|_F = \|\mathbf{R}\|_F$ because orthogonal matrices are norm preserving transformations—recall Exercise 5.6.9.

5.7.6. Follow the procedure outlined in Example 5.7.4 to compute the reflector

$$\hat{\mathbf{R}} = \begin{pmatrix} -3/5 & 4/5 \\ 4/5 & 3/5 \end{pmatrix}, \quad \text{and then set} \quad \mathbf{R} = \begin{pmatrix} 1 & 0 & 0 \\ 0 & -3/5 & 4/5 \\ 0 & 4/5 & 3/5 \end{pmatrix}.$$

Since $\mathbf{A}$ is 3×3, there is only one step, so $\mathbf{P} = \mathbf{R}$ and

$$\mathbf{P}^T \mathbf{A} \mathbf{P} = \mathbf{H} = \begin{pmatrix} -2 & -5 & 0 \\ -5 & -41 & 38 \\ 0 & 38 & 41 \end{pmatrix}.$$

5.7.7. First argue that the product of an upper-Hessenberg matrix with an upper-triangular matrix must be upper Hessenberg—regardless of which side the triangular factor appears. This implies that $\mathbf{Q}$ is upper Hessenberg because $\mathbf{Q} = \mathbf{HR}^{-1}$ and $\mathbf{R}^{-1}$ is upper triangular—recall Exercise 3.7.4. This in turn means that $\mathbf{RQ}$ must be upper Hessenberg.

5.7.8. From the structure of the matrices in Example 5.7.5, it can be seen that $\mathbf{P}_{12}$ requires $4n$ multiplications, $\mathbf{P}_{23}$ requires $4(n-1)$ multiplications, etc. Use the formula $1 + 2 + \cdots + n = n(n+1)/2$ to obtain the total as

$$4[n + (n-1) + (n-2) + \cdots + 2] = 4\left(\frac{n^2 + n}{2} - 1\right) \approx 2n^2.$$

Solutions for exercises in section 5. 8

5.8.1. (a) $\begin{pmatrix} 4 \\ 13 \\ 28 \\ 27 \\ 18 \\ 0 \end{pmatrix}$ (b) $\begin{pmatrix} -1 \\ 0 \\ 2 \\ 0 \\ -1 \\ 0 \end{pmatrix}$ (c) $\begin{pmatrix} \alpha_0 \\ \alpha_0 + \alpha_1 \\ \alpha_0 + \alpha_1 + \alpha_2 \\ \alpha_1 + \alpha_2 \\ \alpha_2 \\ 0 \end{pmatrix}$

5.8.2. (a) $\begin{pmatrix} 0 \\ 0 \\ 0 \\ 4 \end{pmatrix}$ (b) $\begin{pmatrix} 0 \\ 0 \\ 0 \\ 1 \end{pmatrix}$

5.8.3. $\mathbf{F}_2 = \begin{pmatrix} 1 & 1 \\ 1 & -1 \end{pmatrix}$, $\mathbf{D}_2 = \begin{pmatrix} 1 & 0 \\ 0 & -i \end{pmatrix}$, and

$$\mathbf{F}_4 \mathbf{P}_4^T = \begin{pmatrix} 1 & 1 & 1 & 1 \\ 1 & -i & -1 & i \\ 1 & -1 & 1 & -1 \\ 1 & i & -1 & -i \end{pmatrix} \mathbf{P}_4^T = \begin{pmatrix} 1 & 1 & 1 & 1 \\ 1 & -1 & -i & i \\ 1 & 1 & -1 & -1 \\ 1 & -1 & i & -i \end{pmatrix}$$

$$= \left(\begin{array}{cc|cc} \begin{matrix} 1 & 1 \\ 1 & -1 \end{matrix} & & \begin{pmatrix} 1 & 0 \\ 0 & -i \end{pmatrix}\begin{pmatrix} 1 & 1 \\ 1 & -1 \end{pmatrix} \\ \hline \begin{matrix} 1 & 1 \\ 1 & -1 \end{matrix} & & -\begin{pmatrix} 1 & 0 \\ 0 & -i \end{pmatrix}\begin{pmatrix} 1 & 1 \\ 1 & -1 \end{pmatrix} \end{array} \right) = \begin{pmatrix} \mathbf{F}_2 & \mathbf{D}_2\mathbf{F}_2 \\ \mathbf{F}_2 & -\mathbf{D}_2\mathbf{F}_2 \end{pmatrix}.$$

5.8.4. (a) $\mathbf{a} \odot \mathbf{b} = \begin{pmatrix} \alpha_0\beta_0 \\ \alpha_0\beta_1 + \alpha_1\beta_0 \\ \alpha_1\beta_1 \\ 0 \end{pmatrix}$

$$\mathbf{F}_4(\mathbf{a} \odot \mathbf{b}) = \begin{pmatrix} \alpha_0\beta_0 + \alpha_0\beta_1 + \alpha_1\beta_0 + \alpha_1\beta_1 \\ \alpha_0\beta_0 - i\alpha_0\beta_1 - i\alpha_1\beta_0 - \alpha_1\beta_1 \\ \alpha_0\beta_0 - \alpha_0\beta_1 - \alpha_1\beta_0 + \alpha_1\beta_1 \\ \alpha_0\beta_0 + i\alpha_0\beta_1 + i\alpha_1\beta_0 - \alpha_1\beta_1 \end{pmatrix} = (\mathbf{F}_4\hat{\mathbf{a}}) \times (\mathbf{F}_4\hat{\mathbf{b}})$$

(b) $\mathbf{F}_4^{-1}\left[(\mathbf{F}_4\hat{\mathbf{a}}) \times (\mathbf{F}_4\hat{\mathbf{b}})\right] = \mathbf{a} \odot \mathbf{b}$

5.8.5. $p(x)q(x) = \gamma_0 + \gamma_1 x + \gamma_2 x^2 + \gamma_3 x^3,$ where

$$\begin{pmatrix} \gamma_0 \\ \gamma_1 \\ \gamma_2 \\ \gamma_3 \end{pmatrix} = \mathbf{F}_4^{-1}\left[(\mathbf{F}_4\hat{\mathbf{a}}) \times (\mathbf{F}_4\hat{\mathbf{b}})\right]$$

$$= \mathbf{F}_4^{-1}\left[\begin{pmatrix} 1 & 1 & 1 & 1 \\ 1 & -i & -1 & i \\ 1 & -1 & 1 & -1 \\ 1 & i & -1 & -i \end{pmatrix}\begin{pmatrix} -3 \\ 2 \\ 0 \\ 0 \end{pmatrix} \times \begin{pmatrix} 1 & 1 & 1 & 1 \\ 1 & -i & -1 & i \\ 1 & -1 & 1 & -1 \\ 1 & i & -1 & -i \end{pmatrix}\begin{pmatrix} -4 \\ 3 \\ 0 \\ 0 \end{pmatrix} \right]$$

$$= \mathbf{F}_4^{-1}\left[\begin{pmatrix} -1 \\ -3 - 2i \\ -5 \\ -3 + 2i \end{pmatrix} \times \begin{pmatrix} -1 \\ -4 - 3i \\ -7 \\ -4 + 3i \end{pmatrix} \right]$$

$$= \frac{1}{4}\begin{pmatrix} 1 & 1 & 1 & 1 \\ 1 & i & -1 & -i \\ 1 & -1 & 1 & -1 \\ 1 & -i & -1 & i \end{pmatrix}\begin{pmatrix} 1 \\ 6 + 17i \\ 35 \\ 6 - 17i \end{pmatrix} = \begin{pmatrix} 12 \\ -17 \\ 6 \\ 0 \end{pmatrix}.$$

5.8.6. (a) $\begin{pmatrix} 3 \\ 4 \end{pmatrix} \odot \begin{pmatrix} 1 \\ 2 \end{pmatrix} = \begin{pmatrix} 3 \\ 10 \\ 8 \\ 0 \end{pmatrix},$ so

$$43_{10} \times 21_{10} = (8 \times 10^2) + (10 \times 10^1) + (3 \times 10^0)$$
$$= (9 \times 10^2) + (0 \times 10^1) + (3 \times 10^0) = 903.$$

(b)
$$\begin{pmatrix} 3 \\ 2 \\ 1 \end{pmatrix} \odot \begin{pmatrix} 1 \\ 0 \\ 6 \end{pmatrix} = \begin{pmatrix} 3 \\ 2 \\ 19 \\ 12 \\ 6 \\ 0 \end{pmatrix}, \text{ so}$$

$$123_8 \times 601_8 = (6 \times 8^4) + (12 \times 8^3) + (19 \times 8^2) + (2 \times 8^1) + (3 \times 8^0).$$

Since
$$12 = 8 + 4 \implies 12 \times 8^3 = (8 + 4) \times 8^3 = 8^4 + (4 \times 8^3)$$
$$19 = (2 \times 8) + 3 \implies 19 \times 8^2 = (2 \times 8^3) + (3 \times 8^2),$$

we have that

$$123_8 \times 601_8 = (7 \times 8^4) + (6 \times 8^3) + (3 \times 8^2) + (2 \times 8^1) + (3 \times 8^0) = 76323_8.$$

(c)
$$\begin{pmatrix} 0 \\ 1 \\ 0 \\ 1 \end{pmatrix} \odot \begin{pmatrix} 1 \\ 0 \\ 1 \\ 1 \end{pmatrix} = \begin{pmatrix} 0 \\ 1 \\ 0 \\ 2 \\ 1 \\ 1 \\ 1 \\ 0 \end{pmatrix}, \text{ so}$$

$$1010_2 \times 1101_2 = (1 \times 2^6) + (1 \times 2^5) + (1 \times 2^4) + (2 \times 2^3) + (0 \times 2^2) + (1 \times 2^1) + (0 \times 2^0).$$

Substituting $2 \times 2^3 = 1 \times 2^4$ in this expression and simplifying yields

$$1010_2 \times 1101_2 = (1 \times 2^7) + (0 \times 2^6) + (0 \times 2^5) + (0 \times 2^4)$$
$$+ (0 \times 2^3) + (0 \times 2^2) + (1 \times 2^1) + (0 \times 2^0)$$
$$= 10000010_2.$$

5.8.7. (a) The number of multiplications required by the definition is

$$1 + 2 + \cdots + (n - 1) + n + (n - 1) + \cdots + 2 + 1$$
$$= 2\Big(1 + 2 + \cdots + (n - 1)\Big) + n$$
$$= (n - 1)n + n = n^2.$$

(b) In the formula $\mathbf{a}_{n \times 1} \odot \mathbf{b}_{n \times 1} = \mathbf{F}_{2n}^{-1}\big[(\mathbf{F}_{2n}\hat{\mathbf{a}}) \times (\mathbf{F}_{2n}\hat{\mathbf{b}})\big]$, using the FFT to compute $\mathbf{F}_{2n}\hat{\mathbf{a}}$ and $\mathbf{F}_{2n}\hat{\mathbf{b}}$ requires $(2n/2)\log_2 2n = n(1 + \log_2 n)$ multiplications for each term, and an additional $2n$ multiplications are needed to form the product $(\mathbf{F}_{2n}\hat{\mathbf{a}}) \times (\mathbf{F}_{2n}\hat{\mathbf{b}})$. Using the FFT in conjunction with the procedure described in Example 5.8.2 to apply $\mathbf{F}^{-1}$ to $(\mathbf{F}_{2n}\hat{\mathbf{a}}) \times (\mathbf{F}_{2n}\hat{\mathbf{b}})$ requires another $(2n/2)\log_2 2n = n(1 + \log_2 n)$ multiplications to compute $\overline{\mathbf{F}_{2n}\overline{\mathbf{x}}}$ followed by $2n$ more multiplications to produce $(1/2n)\overline{\mathbf{F}_{2n}\overline{\mathbf{x}}} = \mathbf{F}_{2n}^{-1}\mathbf{x}$. Therefore, the total count is $3n(1 + \log_2 n) + 4n = 3n\log_2 n + 7n$.

5.8.8. Recognize that $\mathbf{y}$ is of the form

$$\mathbf{y} = 1(\mathbf{e}_2 + \mathbf{e}_6) + 4(\mathbf{e}_3 + \mathbf{e}_5) + 5i(-\mathbf{e}_1 + \mathbf{e}_7) + 3i(-\mathbf{e}_2 + \mathbf{e}_6).$$

The real part says that there are two cosines—one with amplitude 1 and frequency 2, and the other with amplitude 4 and frequency 3. The imaginary part says there are two sines—one with amplitude 5 and frequency 1, and the other with amplitude 3 and frequency 2. Therefore,

$$x(\tau) = \cos 4\pi\tau + 4\cos 6\pi\tau + 5\sin 2\pi\tau + 3\sin 4\pi\tau.$$

5.8.9. Use (5.8.12) to write $\mathbf{a} \odot \mathbf{b} = \mathbf{F}^{-1}\big[(\mathbf{F}\hat{\mathbf{a}}) \times (\mathbf{F}\hat{\mathbf{b}})\big] = \mathbf{F}^{-1}\big[(\mathbf{F}\hat{\mathbf{b}}) \times (\mathbf{F}\hat{\mathbf{a}})\big] = \mathbf{a} \odot \mathbf{b}$.

5.8.10. This is a special case of the result given in Example 4.3.5. The Fourier matrix $\mathbf{F}_n$ is a special case of the Vandermonde matrix—simply let x_k's that define the Vandermonde matrix be the nth roots of unity.

5.8.11. The result of Exercise 5.8.10 implies that if

$$\hat{\mathbf{a}} = \begin{pmatrix} \alpha_0 \\ \vdots \\ \alpha_{n-1} \\ 0 \\ \vdots \\ 0 \end{pmatrix}_{2n \times 1} \quad \text{and} \quad \hat{\mathbf{b}} = \begin{pmatrix} \beta_0 \\ \vdots \\ \beta_{n-1} \\ 0 \\ \vdots \\ 0 \end{pmatrix}_{2n \times 1},$$

then $\mathbf{F}_{2n}\hat{\mathbf{a}} = \mathbf{p}$ and $\mathbf{F}_{2n}\hat{\mathbf{b}} = \mathbf{q}$, and we know from (5.8.11) that the γ_k's are given by $\gamma_k = [\mathbf{a} \odot \mathbf{b}]_k$. Therefore, the convolution theorem guarantees

$$\begin{pmatrix} \gamma_0 \\ \gamma_1 \\ \gamma_2 \\ \vdots \end{pmatrix} = \mathbf{a} \odot \mathbf{b} = \mathbf{F}_{2n}^{-1}\big[(\mathbf{F}_{2n}\hat{\mathbf{a}}) \times (\mathbf{F}_{2n}\hat{\mathbf{b}})\big] = \mathbf{F}_{2n}^{-1}[\mathbf{p} \times \mathbf{q}] = \mathbf{F}_{2n}^{-1}\begin{pmatrix} p(1)q(1) \\ p(\xi)q(\xi) \\ p(\xi^2)q(\xi^2) \\ \vdots \end{pmatrix}.$$

5.8.12. (a) This follows from the observation that $\mathbf{Q}^k$ has 1's on the kth subdiagonal and 1's on the $(n-k)^{th}$ superdiagonal. For example, if $n = 8$, then

$$\mathbf{Q}^3 = \begin{pmatrix} 0 & 0 & 0 & 0 & 0 & 1 & 0 & 0 \\ 0 & 0 & 0 & 0 & 0 & 0 & 1 & 0 \\ 0 & 0 & 0 & 0 & 0 & 0 & 0 & 1 \\ 1 & 0 & 0 & 0 & 0 & 0 & 0 & 0 \\ 0 & 1 & 0 & 0 & 0 & 0 & 0 & 0 \\ 0 & 0 & 1 & 0 & 0 & 0 & 0 & 0 \\ 0 & 0 & 0 & 1 & 0 & 0 & 0 & 0 \\ 0 & 0 & 0 & 0 & 1 & 0 & 0 & 0 \end{pmatrix}.$$

(b) If the rows of $\mathbf{F}$ are indexed from 0 to $n-1$, then they satisfy the relationships $\mathbf{F}_{k*}\mathbf{Q} = \xi^k \mathbf{F}_{k*}$ for each k (verifying this for $n=4$ will indicate why it is true in general). This means that $\mathbf{FQ} = \mathbf{DF}$, which in turn implies $\mathbf{FQF}^{-1} = \mathbf{D}$.

(c) Couple parts (a) and (b) with $\mathbf{FQ}^k\mathbf{F}^{-1} = (\mathbf{FQF}^{-1})^k = \mathbf{D}^k$ to write

$$
\begin{aligned}
\mathbf{FCF}^{-1} &= \mathbf{F}p(\mathbf{Q})\mathbf{F}^{-1} \\
&= \mathbf{F}(c_0\mathbf{I} + c_1\mathbf{Q} + \cdots + c_{n-1}\mathbf{Q}^{n-1})\mathbf{F}^{-1} \\
&= c_0\mathbf{I} + c_1\mathbf{FQF}^{-1} + \cdots + c_{n-1}\mathbf{FQ}^{n-1}\mathbf{F}^{-1} \\
&= c_0\mathbf{I} + c_1\mathbf{D} + \cdots + c_{n-1}\mathbf{D}^{n-1} \\
&= \begin{pmatrix} p(1) & 0 & \cdots & 0 \\ 0 & p(\xi) & \cdots & 0 \\ \vdots & \vdots & \ddots & \vdots \\ 0 & 0 & \cdots & p(\xi^{n-1}) \end{pmatrix}.
\end{aligned}
$$

(d) $\mathbf{FC}_1\mathbf{F}^{-1} = \mathbf{D}_1$ and $\mathbf{FC}_2\mathbf{F}^{-1} = \mathbf{D}_2$, where $\mathbf{D}_1$ and $\mathbf{D}_2$ are diagonal matrices, and therefore

$$
\begin{aligned}
\mathbf{C}_1\mathbf{C}_2 &= \mathbf{F}^{-1}\mathbf{D}_1\mathbf{FF}^{-1}\mathbf{D}_2\mathbf{F} = \mathbf{F}^{-1}\mathbf{D}_1\mathbf{D}_2\mathbf{F} = \mathbf{F}^{-1}\mathbf{D}_2\mathbf{D}_1\mathbf{F} = \mathbf{F}^{-1}\mathbf{D}_2\mathbf{FF}^{-1}\mathbf{D}_1\mathbf{F} \\
&= \mathbf{C}_2\mathbf{C}_1.
\end{aligned}
$$

5.8.13. (a) According to Exercise 5.8.12,

$$
\mathbf{C} = \begin{pmatrix} \sigma_0 & \sigma_{n-1} & \cdots & \sigma_1 \\ \sigma_1 & \sigma_0 & \cdots & \sigma_2 \\ \vdots & \vdots & \ddots & \vdots \\ \sigma_{n-1} & \sigma_{n-2} & \cdots & \sigma_0 \end{pmatrix} = \mathbf{F}^{-1} \begin{pmatrix} p(1) & 0 & \cdots & 0 \\ 0 & p(\xi) & \cdots & 0 \\ \vdots & \vdots & \ddots & \vdots \\ 0 & 0 & \cdots & p(\xi^{n-1}) \end{pmatrix} \mathbf{F} = \mathbf{F}^{-1}\mathbf{DF}
$$

in which $p(x) = \sigma_0 + \sigma_1 x + \cdots + \sigma_{n-1}x^{n-1}$. Therefore, $\mathbf{x} = \mathbf{C}^{-1}\mathbf{b} = \mathbf{F}^{-1}\mathbf{D}^{-1}\mathbf{Fb}$, so we can execute the following computations.

(i) $\quad \begin{pmatrix} p(0) \\ p(\xi) \\ \vdots \\ p(\xi^{n-1}) \end{pmatrix} \longleftarrow \mathbf{F} \begin{pmatrix} \sigma_0 \\ \sigma_1 \\ \vdots \\ \sigma_{n-1} \end{pmatrix} \quad$ using the FFT

(ii) $\quad \mathbf{x} \longleftarrow \mathbf{Fb} \quad$ using the FFT

(iii) $\quad x_k \longleftarrow x_k/p(\xi^k) \quad$ for $\quad k = 0, 1, \ldots, n-1$

(iv) $\quad \mathbf{x} \longleftarrow \mathbf{F}^{-1}\mathbf{x} \quad$ using the FFT as described in Example 5.8.2

(b)   Use the same techniques described in part (a) to compute the kth column of $\mathbf{C}^{-1}$ from the formula

$$[\mathbf{C}^{-1}]_{*k} = \mathbf{C}^{-1}\mathbf{e}_k = \mathbf{F}^{-1}\mathbf{D}^{-1}\mathbf{F}\mathbf{e}_k$$
$$= \mathbf{F}^{-1}\left(\mathbf{D}^{-1}[\mathbf{F}]_{*k}\right)$$
$$= \mathbf{F}^{-1}\begin{pmatrix} 1/p(1) \\ \xi^k/p(\xi) \\ \xi^{2k}/p(\xi^2) \\ \vdots \\ \xi^{n-k}/p(\xi^{n-1}) \end{pmatrix}.$$

(c)   The kth column of $\mathbf{P} = \mathbf{C}_1\mathbf{C}_2$ is given by

$$\mathbf{P}_{*k} = \mathbf{P}\mathbf{e}_k = \mathbf{F}^{-1}\mathbf{D}_1\mathbf{F}\mathbf{F}^{-1}\mathbf{D}_2\mathbf{F}\mathbf{e}_k = \mathbf{F}^{-1}\left(\mathbf{D}_1\mathbf{D}_2[\mathbf{F}]_{*k}\right).$$

If $(\sigma_0 \quad \sigma_1 \quad \cdots \quad \sigma_{n-1})$ and $(\eta_0 \quad \eta_1 \quad \cdots \quad \eta_{n-1})$ are the first rows in $\mathbf{C}_1$ and $\mathbf{C}_2$, respectively, and if $p(x) = \sum_{k=0}^{n-1}\sigma_k x^k$ and $q(x) = \sum_{k=0}^{n-1}\eta_k x^k$, then first compute

$$\mathbf{p} = \begin{pmatrix} p(0) \\ p(\xi) \\ \vdots \\ p(\xi^{n-1}) \end{pmatrix} \longleftarrow \mathbf{F}\begin{pmatrix} \sigma_0 \\ \sigma_1 \\ \vdots \\ \sigma_{n-1} \end{pmatrix} \quad \text{and} \quad \mathbf{q} = \begin{pmatrix} q(0) \\ q(\xi) \\ \vdots \\ q(\xi^{n-1}) \end{pmatrix} \longleftarrow \mathbf{F}\begin{pmatrix} \eta_0 \\ \eta_1 \\ \vdots \\ \eta_{n-1} \end{pmatrix}.$$

The kth column of the product can now be obtained from

$$\mathbf{P}_{*k} \longleftarrow \mathbf{F}^{-1}\left(\mathbf{p} \times \mathbf{q} \times \mathbf{F}_{*k}\right) \quad \text{for} \quad k = 0, 1, \dots, n-1.$$

5.8.14. (a)   For $n = 3$ we have

$$\mathbf{C}\hat{\mathbf{b}} = \begin{pmatrix} \alpha_0 & 0 & 0 & 0 & \alpha_2 & \alpha_1 \\ \alpha_1 & \alpha_0 & 0 & 0 & 0 & \alpha_2 \\ \alpha_2 & \alpha_1 & \alpha_0 & 0 & 0 & 0 \\ 0 & \alpha_2 & \alpha_1 & \alpha_0 & 0 & 0 \\ 0 & 0 & \alpha_2 & \alpha_1 & \alpha_0 & 0 \\ 0 & 0 & 0 & \alpha_2 & \alpha_1 & \alpha_0 \end{pmatrix}\begin{pmatrix} \beta_0 \\ \beta_1 \\ \beta_2 \\ 0 \\ 0 \\ 0 \end{pmatrix} = \begin{pmatrix} \alpha_0\beta_0 \\ \alpha_1\beta_0 + \alpha_0\beta_1 \\ \alpha_2\beta_0 + \alpha_1\beta_1 + \alpha_0\beta_2 \\ \alpha_2\beta_1 + \alpha_1\beta_2 \\ \alpha_2\beta_2 \\ 0 \end{pmatrix}.$$

Use this as a model to write the expression for $\mathbf{C}\hat{\mathbf{b}}$, where n is arbitrary.

(b)   We know from part (c) of Exercise 5.8.12 that if $\mathbf{F}$ is the Fourier matrix of order $2n$, then $\mathbf{F}\mathbf{C}\mathbf{F}^{-1} = \mathbf{D}$, where

$$\mathbf{D} = \begin{pmatrix} p(1) & 0 & \cdots & 0 \\ 0 & p(\xi) & \cdots & 0 \\ \vdots & \vdots & \ddots & \vdots \\ 0 & 0 & \cdots & p(\xi^{2n-1}) \end{pmatrix} \quad \text{(the } \xi^k \text{'s are the } 2n^{th} \text{ roots of unity)}$$

in which $p(x) = \alpha_0 + \alpha_1 x + \cdots + \alpha_{n-1} x^{n-1}$. Therefore, from part (a),

$$\mathbf{F}(\mathbf{a} \odot \mathbf{b}) = \mathbf{F}\mathbf{C}\hat{\mathbf{b}} = \mathbf{F}\mathbf{C}\mathbf{F}^{-1}\mathbf{F}\hat{\mathbf{b}} = \mathbf{D}\mathbf{F}\hat{\mathbf{b}}.$$

According to Exercise 5.8.10, we also know that

$$\mathbf{F}\hat{\mathbf{a}} = \begin{pmatrix} p(1) \\ p(\xi) \\ \vdots \\ p(\xi^{2n-1}) \end{pmatrix},$$

and hence

$$\mathbf{F}(\mathbf{a} \odot \mathbf{b}) = \mathbf{D}\mathbf{F}\hat{\mathbf{b}} = (\mathbf{F}\hat{\mathbf{a}}) \times (\mathbf{F}\hat{\mathbf{b}}).$$

5.8.15. (a) $\mathbf{P}_n \mathbf{x}$ performs an even–odd permutation to all components of $\mathbf{x}$. The matrix

$$(\mathbf{I}_2 \otimes \mathbf{P}_{n/2}) = \begin{pmatrix} \mathbf{P}_{n/2} & \mathbf{0} \\ \mathbf{0} & \mathbf{P}_{n/2} \end{pmatrix} \mathbf{x}$$

performs an even–odd permutation to the top half of $\mathbf{x}$ and then does the same to the bottom half of $\mathbf{x}$. The matrix

$$(\mathbf{I}_4 \otimes \mathbf{P}_{n/4}) = \begin{pmatrix} \mathbf{P}_{n/4} & \mathbf{0} & \mathbf{0} & \mathbf{0} \\ \mathbf{0} & \mathbf{P}_{n/4} & \mathbf{0} & \mathbf{0} \\ \mathbf{0} & \mathbf{0} & \mathbf{P}_{n/4} & \mathbf{0} \\ \mathbf{0} & \mathbf{0} & \mathbf{0} & \mathbf{P}_{n/4} \end{pmatrix} \mathbf{x}$$

performs an even–odd permutation to each individual quarter of $\mathbf{x}$. As this pattern is continued, the product

$$\mathbf{R}_n = (\mathbf{I}_{2^{r-1}} \otimes \mathbf{P}_{2^1})(\mathbf{I}_{2^{r-2}} \otimes \mathbf{P}_{2^2}) \cdots (\mathbf{I}_{2^1} \otimes \mathbf{P}_{2^{r-1}})(\mathbf{I}_{2^0} \otimes \mathbf{P}_{2^r})\mathbf{x}$$

produces the bit-reversing permutation. For example, when $n = 8$,

$$\mathbf{R}_8 \mathbf{x} = (\mathbf{I}_4 \otimes \mathbf{P}_2)(\mathbf{I}_2 \otimes \mathbf{P}_4)(\mathbf{I}_1 \otimes \mathbf{P}_8)\mathbf{x}$$

$$= \begin{pmatrix} \mathbf{P}_2 & \mathbf{0} & \mathbf{0} & \mathbf{0} \\ \mathbf{0} & \mathbf{P}_2 & \mathbf{0} & \mathbf{0} \\ \mathbf{0} & \mathbf{0} & \mathbf{P}_2 & \mathbf{0} \\ \mathbf{0} & \mathbf{0} & \mathbf{0} & \mathbf{P}_2 \end{pmatrix} \begin{pmatrix} \mathbf{P}_4 & \mathbf{0} \\ \mathbf{0} & \mathbf{P}_4 \end{pmatrix} \mathbf{P}_8 \begin{pmatrix} x_0 \\ x_1 \\ x_2 \\ x_3 \\ x_4 \\ x_5 \\ x_6 \\ x_7 \end{pmatrix}$$

$$= \begin{pmatrix} \mathbf{P}_2 & \mathbf{0} & \mathbf{0} & \mathbf{0} \\ \mathbf{0} & \mathbf{P}_2 & \mathbf{0} & \mathbf{0} \\ \mathbf{0} & \mathbf{0} & \mathbf{P}_2 & \mathbf{0} \\ \mathbf{0} & \mathbf{0} & \mathbf{0} & \mathbf{P}_2 \end{pmatrix} \begin{pmatrix} \mathbf{P}_4 & \mathbf{0} \\ \mathbf{0} & \mathbf{P}_4 \end{pmatrix} \begin{pmatrix} x_0 \\ x_2 \\ x_4 \\ x_6 \\ \overline{x_1} \\ x_3 \\ x_5 \\ x_7 \end{pmatrix}$$

$$= \begin{pmatrix} \mathbf{P}_2 & \mathbf{0} & \mathbf{0} & \mathbf{0} \\ \mathbf{0} & \mathbf{P}_2 & \mathbf{0} & \mathbf{0} \\ \mathbf{0} & \mathbf{0} & \mathbf{P}_2 & \mathbf{0} \\ \mathbf{0} & \mathbf{0} & \mathbf{0} & \mathbf{P}_2 \end{pmatrix} \begin{pmatrix} x_0 \\ x_4 \\ x_2 \\ x_6 \\ \overline{x_1} \\ x_5 \\ x_3 \\ x_7 \end{pmatrix} = \begin{pmatrix} x_0 \\ x_4 \\ x_2 \\ x_6 \\ x_1 \\ x_5 \\ x_3 \\ x_7 \end{pmatrix} \quad \text{because} \quad \mathbf{P}_2 = \begin{pmatrix} 1 & 0 \\ 0 & 1 \end{pmatrix}.$$

(b)   To prove that $\mathbf{I}_{2^{r-k}} \otimes \mathbf{F}_{2^k} = \mathbf{L}_{2^k}\mathbf{R}_{2^k}$ using induction, note first that for $k = 1$ we have

$$\mathbf{L}_2 = (\mathbf{I}_{2^{r-1}} \otimes \mathbf{B}_2)1 = \mathbf{I}_{2^{r-1}} \otimes \mathbf{F}_2 \quad \text{and} \quad \mathbf{R}_2 = \mathbf{I}_n(\mathbf{I}_{2^{r-1}} \otimes \mathbf{P}_2) = \mathbf{I}_n\mathbf{I}_n = \mathbf{I}_n,$$

so $\mathbf{L}_2\mathbf{R}_2 = \mathbf{I}_{2^{r-1}} \otimes \mathbf{F}_2$. Now assume that the result holds for $k = j$—i.e., assume

$$\mathbf{I}_{2^{r-j}} \otimes \mathbf{F}_{2^j} = \mathbf{L}_{2^j}\mathbf{R}_{2^j}.$$

Prove that the result is true for $k = j + 1$—i.e., prove

$$\mathbf{I}_{2^{r-(j+1)}} \otimes \mathbf{F}_{2^{j+1}} = \mathbf{L}_{2^{j+1}}\mathbf{R}_{2^{j+1}}.$$

Use the fact that $\mathbf{F}_{2^{j+1}} = \mathbf{B}_{2^{j+1}}(\mathbf{I}_2 \otimes \mathbf{F}_j)\mathbf{P}_{2^{j+1}}$ along with the two basic properties of the tensor product given in the introduction of this exercise to write

$$
\begin{aligned}
\mathbf{I}_{2^{r-(j+1)}} \otimes \mathbf{F}_{2^{j+1}} &= \mathbf{I}_{2^{r-(j+1)}} \otimes \mathbf{B}_{2^{j+1}}(\mathbf{I}_2 \otimes \mathbf{F}_{2^j})\mathbf{P}_{2^{j+1}} \\
&= \Big(\mathbf{I}_{2^{r-(j+1)}} \otimes \mathbf{B}_{2^{j+1}}(\mathbf{I}_2 \otimes \mathbf{F}_{2^j})\Big)\Big(\mathbf{I}_{2^{r-(j+1)}} \otimes \mathbf{P}_{2^{j+1}}\Big) \\
&= (\mathbf{I}_{2^{r-(j+1)}} \otimes \mathbf{B}_{2^{j+1}})(\mathbf{I}_{2^{r-(j+1)}} \otimes \mathbf{I}_2 \otimes \mathbf{F}_{2^j})(\mathbf{I}_{2^{r-(j+1)}} \otimes \mathbf{P}_{2^{j+1}}) \\
&= (\mathbf{I}_{2^{r-(j+1)}} \otimes \mathbf{B}_{2^{j+1}})(\mathbf{I}_{2^{r-j}} \otimes \mathbf{F}_{2^j})(\mathbf{I}_{2^{r-(j+1)}} \otimes \mathbf{P}_{2^{j+1}}) \\
&= (\mathbf{I}_{2^{r-(j+1)}} \otimes \mathbf{B}_{2^{j+1}})\mathbf{L}_{2^j}\mathbf{R}_{2^j}(\mathbf{I}_{2^{r-(j+1)}} \otimes \mathbf{P}_{2^{j+1}}) \\
&= \mathbf{L}_{2^{j+1}}\mathbf{R}_{2^{j+1}}.
\end{aligned}
$$

Therefore, $\mathbf{I}_{2^{r-k}} \otimes \mathbf{F}_{2^k} = \mathbf{L}_{2^k}\mathbf{R}_{2^k}$ for $k = 1, 2, \ldots, r$, and when $k = r$ we have that $\mathbf{F}_n = \mathbf{L}_n\mathbf{R}_n$.

5.8.16. According to Exercise 5.8.10,

$$
\mathbf{F}_n\mathbf{a} = \mathbf{b}, \quad \text{where} \quad \mathbf{a} = \begin{pmatrix} \alpha_0 \\ \alpha_1 \\ \alpha_2 \\ \vdots \\ \alpha_{n-1} \end{pmatrix} \quad \text{and} \quad \mathbf{b} = \begin{pmatrix} p(1) \\ p(\xi) \\ p(\xi^2) \\ \vdots \\ p(\xi^{n-1}) \end{pmatrix}.
$$

By making use of the fact that $(1/\sqrt{n})\mathbf{F}_n$ is unitary we can write

$$
\sum_{k=0}^{n-1} \big|p(\xi^k)\big|^2 = \mathbf{b}^*\mathbf{b} = (\mathbf{F}_n\mathbf{a})^*(\mathbf{F}_n\mathbf{a}) = \mathbf{a}^*\mathbf{F}_n^*\mathbf{F}_n\mathbf{a} = \mathbf{a}^*(n\mathbf{I})\mathbf{a} = n \sum_{k=0}^{n-1} |\alpha_k|^2.
$$

5.8.17. Let $\mathbf{y} = (2/n)\mathbf{F}\mathbf{x}$, and use the result in (5.8.7) to write

$$
\begin{aligned}
\|\mathbf{y}\|^2 &= \left\| \sum_k \Big((\alpha_k - i\beta_k)\mathbf{e}_{f_k} + (\alpha_k + i\beta_k)\mathbf{e}_{n-f_k}\Big) \right\| \\
&= \sum_k \Big(|\alpha_k - i\beta_k|^2 + |\alpha_k + i\beta_k|^2\Big) \\
&= 2\sum_k \Big(\alpha_k^2 + \beta_k^2\Big).
\end{aligned}
$$

But because $\mathbf{F}^*\mathbf{F} = n\mathbf{I}$, it follows that

$$
\|\mathbf{y}\|^2 = \left\| \frac{2}{n}\mathbf{F}\mathbf{x} \right\|^2 = \frac{4}{n^2}\mathbf{x}^*\mathbf{F}^*\mathbf{F}\mathbf{x} = \frac{4}{n}\|\mathbf{x}\|^2,
$$

so combining these two statements produces the desired conclusion.

5.8.18. We know from (5.8.11) that if $p(x) = \sum_{k=0}^{n-1} \alpha_k x^k$, then

$$p^2(x) = \sum_{k=0}^{2n-2} [\mathbf{a} \odot \mathbf{a}]_k x^k.$$

The last component of $\mathbf{a} \odot \mathbf{a}$ is zero, so we can write

$$\mathbf{c}^T(\mathbf{a} \odot \mathbf{a}) = \sum_{k=0}^{2n-2} [\mathbf{a} \odot \mathbf{a}]_k \eta^k = p^2(\eta) = \left(\sum_{k=0}^{n-1} \alpha_k \eta^k \right)^2 = \left(\mathbf{c}^T \hat{\mathbf{a}} \right)^2.$$

5.8.19. Start with $\mathbf{X} \longleftarrow rev(\mathbf{x}) = (x_0\, x_4\, x_2\, x_6\, x_1\, x_5\, x_3\, x_7)$.

For $j = 0$:

$\mathbf{D} \longleftarrow (1)$

$\mathbf{X}^{(0)} \longleftarrow (\,x_0 \quad x_2 \quad x_1 \quad x_3\,)$

$\mathbf{X}^{(1)} \longleftarrow (\,x_4 \quad x_6 \quad x_5 \quad x_7\,)$

$\mathbf{X} \longleftarrow \begin{pmatrix} \mathbf{X}^{(0)} + \mathbf{D} \times \mathbf{X}^{(1)} \\ \mathbf{X}^{(0)} - \mathbf{D} \times \mathbf{X}^{(1)} \end{pmatrix}$

$= \begin{pmatrix} x_0 + x_4 & x_2 + x_6 & x_1 + x_5 & x_3 + x_7 \\ x_0 - x_4 & x_2 - x_6 & x_1 - x_5 & x_3 - x_7 \end{pmatrix}_{2 \times 8}$

For $j = 1$:

$\mathbf{D} \longleftarrow \begin{pmatrix} 1 \\ e^{-\pi i/2} \end{pmatrix} = \begin{pmatrix} 1 \\ \xi^2 \end{pmatrix}$

$\mathbf{X}^{(0)} \longleftarrow \begin{pmatrix} x_0 + x_4 & x_1 + x_5 \\ x_0 - x_4 & x_1 - x_5 \end{pmatrix}$

$\mathbf{X}^{(1)} \longleftarrow \begin{pmatrix} x_2 + x_6 & x_3 + x_7 \\ x_2 - x_6 & x_3 - x_7 \end{pmatrix}$

$\mathbf{X} \longleftarrow \begin{pmatrix} \mathbf{X}^{(0)} + \mathbf{D} \times \mathbf{X}^{(1)} \\ \mathbf{X}^{(0)} - \mathbf{D} \times \mathbf{X}^{(1)} \end{pmatrix}$

$= \left(\begin{array}{cccc|cccc} x_0 + x_4 + & x_2 + & x_6 & & x_1 + x_5 + & x_3 + & x_7 \\ x_0 - x_4 + \xi^2 x_2 & - \xi^2 x_6 & & & x_1 - x_5 + \xi^2 x_3 & - \xi^2 x_7 \\ x_0 + x_4 - & x_2 - & x_6 & & x_1 + x_5 - & x_3 - & x_7 \\ x_0 - x_4 - \xi^2 x_2 & + \xi^2 x_6 & & & x_1 - x_5 - \xi^2 x_3 & + \xi^2 x_7 \end{array} \right)_{4 \times 2}$

For $j = 2$:

$\mathbf{D} \longleftarrow \begin{pmatrix} 1 \\ e^{-\pi i/4} \\ e^{-2\pi i/4} \\ e^{-3\pi i/4} \end{pmatrix} = \begin{pmatrix} 1 \\ \xi \\ \xi^2 \\ \xi^3 \end{pmatrix}$

$$\mathbf{X}^{(0)} \longleftarrow \begin{pmatrix} x_0 + x_4 + & x_2 + & x_6 \\ x_0 - x_4 + \xi^2 x_2 - \xi^2 x_6 \\ x_0 + x_4 - & x_2 - & x_6 \\ x_0 - x_4 - \xi^2 x_2 + \xi^2 x_6 \end{pmatrix}$$

$$\mathbf{X}^{(1)} \longleftarrow \begin{pmatrix} x_1 + x_5 + & x_3 + & x_7 \\ x_1 - x_5 + \xi^2 x_3 - \xi^2 x_7 \\ x_1 + x_5 - & x_3 - & x_7 \\ x_1 - x_5 - \xi^2 x_3 + \xi^2 x_7 \end{pmatrix}$$

$$\mathbf{X} \longleftarrow \begin{pmatrix} \mathbf{X}^{(0)} + \mathbf{D} \times \mathbf{X}^{(1)} \\ \mathbf{X}^{(0)} - \mathbf{D} \times \mathbf{X}^{(1)} \end{pmatrix}$$

$$= \begin{pmatrix} x_0 + x_4 + & x_2 + & x_6 + & x_1 + & x_5 + & x_3 + & x_7 \\ x_0 - x_4 + \xi^2 x_2 - \xi^2 x_6 + \xi\ x_1 - & \xi x_5 + \xi^3 x_3 - \xi^3 x_7 \\ x_0 + x_4 - & x_2 - & x_6 + \xi^2 x_1 + \xi^2 x_5 - \xi^2 x_3 - \xi^2 x_7 \\ x_0 - x_4 - \xi^2 x_2 + \xi^2 x_6 + \xi^3 x_1 - \xi^3 x_5 - \xi^5 x_3 + \xi^5 x_7 \\ x_0 + x_4 + & x_2 + & x_6 - & x_1 - & x_5 - & x_3 - & x_7 \\ x_0 - x_4 + \xi^2 x_2 - \xi^2 x_6 - \xi\ x_1 + \xi\ x_5 - \xi^3 x_3 + \xi^3 x_7 \\ x_0 + x_4 - & x_2 - & x_6 - \xi^2 x_1 - \xi^2 x_5 + \xi^2 x_3 + \xi^2 x_7 \\ x_0 - x_4 - \xi^2 x_2 + \xi^2 x_6 - \xi^3 x_1 + \xi^3 x_5 + \xi^5 x_3 - \xi^5 x_7 \end{pmatrix}_{8 \times 1}$$

To verify that this is the same as $\mathbf{F_8 x_8}$, use the fact that $\xi = -\xi^5$, $\xi^2 = -\xi^6$, $\xi^3 = -\xi^7$, and $\xi^4 = -1$.

Solutions for exercises in section 5. 9

5.9.1. (a)   The fact that

$$rank\,(\mathbf{B}) = rank\left[\mathbf{X}\,|\,\mathbf{Y}\right] = rank \begin{pmatrix} 1 & 1 & 1 \\ 1 & 2 & 2 \\ 1 & 2 & 3 \end{pmatrix} = 3$$

implies $\mathcal{B_X} \cup \mathcal{B_Y}$ is a basis for $\Re^3$, so (5.9.4) guarantees that $\mathcal{X}$ and $\mathcal{Y}$ are complementary.

(b)   According to (5.9.12), the projector onto $\mathcal{X}$ along $\mathcal{Y}$ is

$$\mathbf{P} = \left[\mathbf{X}\,|\,\mathbf{0}\right]\left[\mathbf{X}\,|\,\mathbf{Y}\right]^{-1} = \begin{pmatrix} 1 & 1 & 0 \\ 1 & 2 & 0 \\ 1 & 2 & 0 \end{pmatrix} \begin{pmatrix} 1 & 1 & 1 \\ 1 & 2 & 2 \\ 1 & 2 & 3 \end{pmatrix}^{-1}$$

$$= \begin{pmatrix} 1 & 1 & 0 \\ 1 & 2 & 0 \\ 1 & 2 & 0 \end{pmatrix} \begin{pmatrix} 2 & -1 & 0 \\ -1 & 2 & -1 \\ 0 & -1 & 1 \end{pmatrix} = \begin{pmatrix} 1 & 1 & -1 \\ 0 & 3 & -2 \\ 0 & 3 & -2 \end{pmatrix},$$

and (5.9.9) insures that the complementary projector onto $\mathcal{Y}$ along $\mathcal{X}$ is

$$\mathbf{Q} = \mathbf{I} - \mathbf{P} = \begin{pmatrix} 0 & -1 & 1 \\ 0 & -2 & 2 \\ 0 & -3 & 3 \end{pmatrix}.$$

(c) $\mathbf{Qv} = \begin{pmatrix} 2 \\ 4 \\ 6 \end{pmatrix}$

(d)   Direct multiplication shows $\mathbf{P}^2 = \mathbf{P}$ and $\mathbf{Q}^2 = \mathbf{Q}$.

(e)   To verify that $R(\mathbf{P}) = \mathcal{X} = N(\mathbf{Q})$, you can use the technique of Example 4.2.2 to show that the basic columns of $\mathbf{P}$ (or the columns in a basis for $N(\mathbf{Q})$) span the same space generated by $\mathcal{B}_{\mathcal{X}}$. To verify that $N(\mathbf{P}) = \mathcal{Y}$, note that

$$\mathbf{P}\begin{pmatrix} 1 \\ 2 \\ 3 \end{pmatrix} = \begin{pmatrix} 0 \\ 0 \\ 0 \end{pmatrix} \text{ together with the fact that } \dim N(\mathbf{P}) = 3 - rank(\mathbf{P}) = 1.$$

5.9.2. There are many ways to do this. One way is to write down any basis for $\Re^5$—say $\mathcal{B} = \{\mathbf{x}_1, \mathbf{x}_2, \mathbf{x}_3, \mathbf{x}_4, \mathbf{x}_5\}$—and set

$$\mathcal{X} = span\{\mathbf{x}_1, \mathbf{x}_2\} \quad \text{and} \quad \mathcal{Y} = span\{\mathbf{x}_3, \mathbf{x}_4, \mathbf{x}_5\}.$$

Property (5.9.4) guarantees that $\mathcal{X}$ and $\mathcal{Y}$ are complementary.

5.9.3. Let $\mathcal{X} = \{(\alpha, \alpha) \,|\, \alpha \in \Re\}$ and $\mathcal{Y} = \Re^2$ so that $\Re^2 = \mathcal{X} + \mathcal{Y}$, but $\mathcal{X} \cap \mathcal{Y} \neq \mathbf{0}$. For each vector in $\Re^2$ we can write

$$(x, y) = (x, x) + (0, y - x) \quad \text{and} \quad (x, y) = (y, y) + (x - y, 0).$$

5.9.4. Exercise 3.2.6 says that each $\mathbf{A} \in \Re^{n \times n}$ can be uniquely written as the sum of a symmetric matrix and a skew-symmetric matrix according to the formula

$$\mathbf{A} = \frac{\mathbf{A} + \mathbf{A}^T}{2} + \frac{\mathbf{A} - \mathbf{A}^T}{2},$$

so (5.9.3) guarantees that $\Re^{n \times n} = \mathcal{S} \oplus \mathcal{K}$. By definition, the projection of $\mathbf{A}$ onto $\mathcal{S}$ along $\mathcal{K}$ is the $\mathcal{S}$-component of $\mathbf{A}$—namely $(\mathbf{A} + \mathbf{A}^T)/2$. For the given matrix, this is

$$\frac{\mathbf{A} + \mathbf{A}^T}{2} = \begin{pmatrix} 1 & 3 & 5 \\ 3 & 5 & 7 \\ 5 & 7 & 9 \end{pmatrix}.$$

5.9.5. (a)   Assume that $\mathcal{X} \cap \mathcal{Y} = \mathbf{0}$. To prove $\mathcal{B}_{\mathcal{X}} \cup \mathcal{B}_{\mathcal{Y}}$ is linearly independent, write

$$\sum_{i=1}^{m} \alpha_i \mathbf{x}_i + \sum_{j=1}^{n} \beta_j \mathbf{y}_j = \mathbf{0} \implies \sum_{i=1}^{m} \alpha_i \mathbf{x}_i = -\sum_{j=1}^{n} \beta_j \mathbf{y}_j$$

$$\implies \sum_{i=1}^{m} \alpha_i \mathbf{x}_i \in \mathcal{X} \cap \mathcal{Y} = \mathbf{0}$$

$$\implies \sum_{i=1}^{m} \alpha_i \mathbf{x}_i = \mathbf{0} \quad \text{and} \quad \sum_{j=1}^{n} \beta_j \mathbf{y}_j = \mathbf{0}$$

$$\implies \alpha_1 = \cdots = \alpha_m = \beta_1 = \cdots = \beta_n = 0$$

(because $\mathcal{B}_{\mathcal{X}}$ and $\mathcal{B}_{\mathcal{Y}}$ are both independent).

Conversely, if $\mathcal{B}_\mathcal{X} \cup \mathcal{B}_\mathcal{Y}$ is linearly independent, then

$$\mathbf{v} \in \mathcal{X} \cap \mathcal{Y} \implies \mathbf{v} = \sum_{i=1}^{m} \alpha_i \mathbf{x}_i \quad \text{and} \quad \mathbf{v} = \sum_{j=1}^{n} \beta_j \mathbf{y}_j$$

$$\implies \sum_{i=1}^{m} \alpha_i \mathbf{x}_i - \sum_{j=1}^{n} \beta_j \mathbf{y}_j = \mathbf{0}$$

$$\implies \alpha_1 = \cdots = \alpha_m = \beta_1 = \cdots = \beta_n = 0$$

$$\text{(because } \mathcal{B}_\mathcal{X} \cup \mathcal{B}_\mathcal{Y} \text{ is independent)}$$

$$\implies \mathbf{v} = \mathbf{0}.$$

(b) No. Take $\mathcal{X}$ to be the xy-plane and $\mathcal{Y}$ to be the yz-plane in $\Re^3$ with $\mathcal{B}_\mathcal{X} = \{\mathbf{e}_1, \mathbf{e}_2\}$ and $\mathcal{B}_\mathcal{Y} = \{\mathbf{e}_2, \mathbf{e}_3\}$. We have $\mathcal{B}_\mathcal{X} \cup \mathcal{B}_\mathcal{Y} = \{\mathbf{e}_1, \mathbf{e}_2, \mathbf{e}_3\}$, but $\mathcal{X} \cap \mathcal{Y} \neq \mathbf{0}$.

(c) No, the fact that $\mathcal{B}_\mathcal{X} \cup \mathcal{B}_\mathcal{Y}$ is linearly independent is no guarantee that $\mathcal{X} + \mathcal{Y}$ is the entire space—e.g., consider two distinct lines in $\Re^3$.

5.9.6. If $\mathbf{x}$ is a fixed point for $\mathbf{P}$, then $\mathbf{Px} = \mathbf{x}$ implies $\mathbf{x} \in R(\mathbf{P})$. Conversely, if $\mathbf{x} \in R(\mathbf{P})$, then $\mathbf{x} = \mathbf{Py}$ for some $\mathbf{y} \in \mathcal{V} \implies \mathbf{Px} = \mathbf{P}^2\mathbf{y} = \mathbf{Py} = \mathbf{x}$.

5.9.7. Use (5.9.10) (which you just validated in Exercise 5.9.6) in conjunction with the definition of a projector onto $\mathcal{X}$ to realize that

$$\mathbf{x} \in \mathcal{X} \iff \mathbf{Px} = \mathbf{x} \iff \mathbf{x} \in R(\mathbf{P}),$$

and

$$\mathbf{x} \in R(\mathbf{P}) \iff \mathbf{Px} = \mathbf{x} \iff (\mathbf{I} - \mathbf{P})\mathbf{x} = \mathbf{0} \iff \mathbf{x} \in N(\mathbf{I} - \mathbf{P}).$$

The statements concerning the complementary projector $\mathbf{I} - \mathbf{P}$ are proven in a similar manner.

5.9.8. If θ is the angle between $R(\mathbf{P})$ and $N(\mathbf{P})$, it follows from (5.9.18) that $\|\mathbf{P}\|_2 = (1/\sin\theta) \geq 1$. Furthermore, $\|\mathbf{P}\|_2 = 1$ if and only if $\sin\theta = 1$ (i.e., $\theta = \pi/2$), which is equivalent to saying $R(\mathbf{P}) \perp N(\mathbf{P})$.

5.9.9. Let θ be the angle between $R(\mathbf{P})$ and $N(\mathbf{P})$. We know from (5.9.11) that $R(\mathbf{I} - \mathbf{P}) = N(\mathbf{P})$ and $N(\mathbf{I} - \mathbf{P}) = R(\mathbf{P})$, so θ is also the angle between $R(\mathbf{I} - \mathbf{P})$ and $N(\mathbf{I} - \mathbf{P})$. Consequently, (5.9.18) says that

$$\|\mathbf{I} - \mathbf{P}\|_2 = \frac{1}{\sin\theta} = \|\mathbf{P}\|_2.$$

5.9.10. The trick is to observe that $\mathbf{P} = \mathbf{u}\mathbf{v}^T$ is a projector because $\mathbf{v}^T\mathbf{u} = 1$ implies $\mathbf{P}^2 = \mathbf{u}\mathbf{v}^T\mathbf{u}\mathbf{v}^T = \mathbf{u}\mathbf{v}^T = \mathbf{P}$, so the result of Exercise 5.9.9 insures that

$$\left\|\mathbf{I} - \mathbf{u}\mathbf{v}^T\right\|_2 = \left\|\mathbf{u}\mathbf{v}^T\right\|_2.$$

To prove that $\left\| \mathbf{u}\mathbf{v}^T \right\|_2 = \left\| \mathbf{u} \right\|_2 \left\| \mathbf{v} \right\|_2$, start with the definition of an induced matrix given in (5.2.4) on p. 280, and write $\left\| \mathbf{u}\mathbf{v}^T \right\|_2 = \max_{\|\mathbf{x}\|_2=1} \left\| \mathbf{u}\mathbf{v}^T\mathbf{x} \right\|_2$. If the maximum occurs at $\mathbf{x} = \mathbf{x}_0$ with $\|\mathbf{x}_0\|_2 = 1$, then

$$\left\| \mathbf{u}\mathbf{v}^T \right\|_2 = \left\| \mathbf{u}\mathbf{v}^T\mathbf{x}_0 \right\|_2 = \left\| \mathbf{u} \right\|_2 \left| \mathbf{v}^T\mathbf{x}_0 \right|$$
$$\leq \left\| \mathbf{u} \right\|_2 \left\| \mathbf{v} \right\|_2 \left\| \mathbf{x}_0 \right\|_2 \text{ by CBS inequality}$$
$$= \left\| \mathbf{u} \right\|_2 \left\| \mathbf{v} \right\|_2.$$

But we can also write

$$\left\| \mathbf{u} \right\|_2 \left\| \mathbf{v} \right\|_2 = \left\| \mathbf{u} \right\|_2 \frac{\left\| \mathbf{v} \right\|_2^2}{\left\| \mathbf{v} \right\|_2} = \left\| \mathbf{u} \right\|_2 \frac{(\mathbf{v}^T\mathbf{v})}{\left\| \mathbf{v} \right\|_2} = \frac{\left\| \mathbf{u}\mathbf{v}^T\mathbf{v} \right\|_2}{\left\| \mathbf{v} \right\|_2}$$
$$= \left\| \mathbf{u}\mathbf{v}^T \left(\frac{\mathbf{v}}{\left\| \mathbf{v} \right\|_2} \right) \right\|_2 \leq \max_{\|\mathbf{x}\|_2=1} \left\| \mathbf{u}\mathbf{v}^T\mathbf{x} \right\|_2$$
$$= \left\| \mathbf{u}\mathbf{v}^T \right\|_2,$$

so $\left\| \mathbf{u}\mathbf{v}^T \right\|_2 = \left\| \mathbf{u} \right\|_2 \left\| \mathbf{v} \right\|_2$. Finally, if $\mathbf{P} = \mathbf{u}\mathbf{v}^T$, use Example 3.6.5 to write

$$\left\| \mathbf{P} \right\|_F^2 = trace\left(\mathbf{P}^T\mathbf{P} \right) = trace(\mathbf{v}\mathbf{u}^T\mathbf{u}\mathbf{v}^T) = trace(\mathbf{u}^T\mathbf{u}\mathbf{v}^T\mathbf{v}) = \left\| \mathbf{u} \right\|_2^2 \left\| \mathbf{v} \right\|_2^2.$$

5.9.11. $\mathbf{p} = \mathbf{P}\mathbf{v} = [\mathbf{X}\,|\,\mathbf{0}][\mathbf{X}\,|\,\mathbf{Y}]^{-1}\mathbf{v} = [\mathbf{X}\,|\,\mathbf{0}]\mathbf{z} = \mathbf{X}\mathbf{z}_1$

5.9.12. (a)   Use (5.9.10) to conclude that

$$R\,(\mathbf{P}) = R\,(\mathbf{Q}) \implies \mathbf{P}\mathbf{Q}_{*j} = \mathbf{Q}_{*j} \quad \text{and} \quad \mathbf{Q}\mathbf{P}_{*j} = \mathbf{P}_{*j} \quad \forall\ j$$
$$\implies \mathbf{P}\mathbf{Q} = \mathbf{Q} \quad \text{and} \quad \mathbf{Q}\mathbf{P} = \mathbf{P}.$$

Conversely, use Exercise 4.2.12 to write

$$\left\{ \begin{array}{l} \mathbf{P}\mathbf{Q} = \mathbf{Q} \implies R\,(\mathbf{Q}) \subseteq R\,(\mathbf{P}) \\ \mathbf{Q}\mathbf{P} = \mathbf{P} \implies R\,(\mathbf{P}) \subseteq R\,(\mathbf{Q}) \end{array} \right\} \implies R\,(\mathbf{P}) = R\,(\mathbf{Q}).$$

(b)   Use $N\,(\mathbf{P}) = N\,(\mathbf{Q}) \Longleftrightarrow R\,(\mathbf{I} - \mathbf{P}) = R\,(\mathbf{I} - \mathbf{Q})$ together with part (a).

(c)   From part (a), $\mathbf{E}_i\mathbf{E}_j = \mathbf{E}_j$ so that

$$\left(\sum_j \alpha_j\mathbf{E}_j \right)^2 = \sum_i \sum_j \alpha_i\alpha_j\mathbf{E}_i\mathbf{E}_j = \sum_i \sum_j \alpha_i\alpha_j\mathbf{E}_j$$
$$= \left(\sum_i \alpha_i \right)\left(\sum_j \alpha_j\mathbf{E}_j \right) = \sum_j \alpha_j\mathbf{E}_j.$$

5.9.13. According to (5.9.12), the projector onto $\mathcal{X}$ along $\mathcal{Y}$ is $\mathbf{P} = \mathbf{B}\begin{pmatrix} \mathbf{I}_r & \mathbf{0} \\ \mathbf{0} & \mathbf{0} \end{pmatrix}\mathbf{B}^{-1}$, where $\mathbf{B} = [\mathbf{X}\,|\,\mathbf{Y}]$ in which the columns of $\mathbf{X}$ and $\mathbf{Y}$ form bases for $\mathcal{X}$

and $\mathcal{Y}$, respectively. Since multiplication by nonsingular matrices does not alter the rank, it follows that $rank\,(\mathbf{P}) = rank \begin{pmatrix} \mathbf{I}_r & \mathbf{0} \\ \mathbf{0} & \mathbf{0} \end{pmatrix} = r$. Using the result of Example 3.6.5 produces

$$trace\,(\mathbf{P}) = trace \left[\mathbf{B} \begin{pmatrix} \mathbf{I}_r & \mathbf{0} \\ \mathbf{0} & \mathbf{0} \end{pmatrix} \mathbf{B}^{-1} \right] = trace \left[\begin{pmatrix} \mathbf{I}_r & \mathbf{0} \\ \mathbf{0} & \mathbf{0} \end{pmatrix} \mathbf{B}^{-1} \mathbf{B} \right]$$

$$= trace \begin{pmatrix} \mathbf{I}_r & \mathbf{0} \\ \mathbf{0} & \mathbf{0} \end{pmatrix} = r = rank\,(\mathbf{P}).$$

5.9.14. **(i)** $\implies$ **(ii)** : If $\mathbf{v} = \mathbf{x}_1 + \cdots + \mathbf{x}_k$ and $\mathbf{v} = \mathbf{y}_1 + \cdots + \mathbf{y}_k$, where $\mathbf{x}_i, \mathbf{y}_i \in \mathcal{X}_i$, then

$$\sum_{i=1}^{k} (\mathbf{x}_i - \mathbf{y}_i) = \mathbf{0} \implies (\mathbf{x}_k - \mathbf{y}_k) = - \sum_{i=1}^{k-1} (\mathbf{x}_i - \mathbf{y}_i)$$

$$\implies (\mathbf{x}_k - \mathbf{y}_k) \in \mathcal{X}_k \cap (\mathcal{X}_1 + \cdots + \mathcal{X}_{k-1}) = \mathbf{0}$$

$$\implies \mathbf{x}_k = \mathbf{y}_k \quad \text{and} \quad \sum_{i=1}^{k-1} (\mathbf{x}_i - \mathbf{y}_i) = \mathbf{0}.$$

Now repeat the argument—to be formal, use induction.

(ii) $\implies$ **(iii)** : The proof is essentially the same argument as that used to establish (5.9.3) $\implies$ (5.9.4).

(iii) $\implies$ **(i)** : $\mathcal{B}$ always spans $\mathcal{X}_1 + \mathcal{X}_2 + \cdots + \mathcal{X}_k$, and since the hypothesis is that $\mathcal{B}$ is a basis for $\mathcal{V}$, it follows that $\mathcal{B}$ is a basis for both $\mathcal{V}$ and $\mathcal{X}_1 + \cdots + \mathcal{X}_k$. Consequently $\mathcal{V} = \mathcal{X}_1 + \mathcal{X}_2 + \cdots + \mathcal{X}_k$. Furthermore, the set $\mathcal{B}_1 \cup \cdots \cup \mathcal{B}_{k-1}$ is linearly independent (each subset of an independent set is independent), and it spans $\mathcal{V}_{k-1} = \mathcal{X}_1 + \cdots + \mathcal{X}_{k-1}$, so $\mathcal{B}_1 \cup \cdots \cup \mathcal{B}_{k-1}$ must be a basis for $\mathcal{V}_{k-1}$. Now, since $(\mathcal{B}_1 \cup \cdots \cup \mathcal{B}_{k-1}) \cup \mathcal{B}_k$ is a basis for $\mathcal{V} = (\mathcal{X}_1 + \cdots + \mathcal{X}_{k-1}) + \mathcal{X}_k$, it follows from (5.9.2)–(5.9.4) that $\mathcal{V} = (\mathcal{X}_1 + \cdots + \mathcal{X}_{k-1}) \oplus \mathcal{X}_k$, so $\mathcal{X}_k \cap (\mathcal{X}_1 + \cdots + \mathcal{X}_{k-1}) = \mathbf{0}$. The same argument can now be repeated on $\mathcal{V}_{k-1}$—to be formal, use induction.

5.9.15. We know from (5.9.12) that $\mathbf{P} = \mathbf{Q} \begin{pmatrix} \mathbf{I} & \mathbf{0} \\ \mathbf{0} & \mathbf{0} \end{pmatrix} \mathbf{Q}^{-1}$ and $\mathbf{I} - \mathbf{P} = \mathbf{Q} \begin{pmatrix} \mathbf{0} & \mathbf{0} \\ \mathbf{0} & \mathbf{I} \end{pmatrix} \mathbf{Q}^{-1}$, so

$$\mathbf{PAP} = \mathbf{Q} \begin{pmatrix} \mathbf{I} & \mathbf{0} \\ \mathbf{0} & \mathbf{0} \end{pmatrix} \mathbf{Q}^{-1} \mathbf{Q} \begin{pmatrix} \mathbf{A}_{11} & \mathbf{A}_{12} \\ \mathbf{A}_{21} & \mathbf{A}_{22} \end{pmatrix} \mathbf{Q}^{-1} \mathbf{Q} \begin{pmatrix} \mathbf{I} & \mathbf{0} \\ \mathbf{0} & \mathbf{0} \end{pmatrix} \mathbf{Q}^{-1}$$

$$= \mathbf{Q} \begin{pmatrix} \mathbf{A}_{11} & \mathbf{0} \\ \mathbf{0} & \mathbf{0} \end{pmatrix} \mathbf{Q}^{-1}.$$

The other three statements are derived in an analogous fashion.

5.9.16. According to (5.9.12), the projector onto $\mathcal{X}$ along $\mathcal{Y}$ is $\mathbf{P} = [\mathbf{X} \,|\, \mathbf{0}] [\mathbf{X} \,|\, \mathbf{Y}]^{-1}$, where the columns of $\mathbf{X}$ and $\mathbf{Y}$ are bases for $\mathcal{X}$ and $\mathcal{Y}$, respectively. If

$$\left[\mathbf{X}_{n\times r}\,|\,\mathbf{Y}\right]^{-1} = \begin{pmatrix} \mathbf{A}_{r\times n} \\ \mathbf{C} \end{pmatrix}, \text{ then}$$

$$\mathbf{P} = \left[\mathbf{X}_{n\times r}\,|\,\mathbf{0}\right]\begin{pmatrix} \mathbf{A}_{r\times n} \\ \mathbf{C} \end{pmatrix} = \mathbf{X}_{n\times r}\mathbf{A}_{r\times n}.$$

The nonsingularity of $\left[\mathbf{X}\,|\,\mathbf{Y}\right]$ and $\begin{pmatrix} \mathbf{A} \\ \mathbf{C} \end{pmatrix}$ insures that $\mathbf{X}$ has full column rank and $\mathbf{A}$ has full row rank. The fact that $\mathbf{AX} = \mathbf{I}_r$ is a consequence of

$$\begin{pmatrix} \mathbf{I}_r & \mathbf{0} \\ \mathbf{0} & \mathbf{I} \end{pmatrix} = \left[\mathbf{X}\,|\,\mathbf{Y}\right]^{-1}\left[\mathbf{X}\,|\,\mathbf{Y}\right] = \begin{pmatrix} \mathbf{A}_{r\times n} \\ \mathbf{C} \end{pmatrix}\left[\mathbf{X}_{n\times r}\,|\,\mathbf{Y}\right] = \begin{pmatrix} \mathbf{AX} & \mathbf{AY} \\ \mathbf{CX} & \mathbf{CY} \end{pmatrix}.$$

5.9.17. (a) Use the fact that a linear operator $\mathbf{P}$ is a projector if and only if $\mathbf{P}$ is idempotent. If $\mathbf{EF} = \mathbf{FE} = \mathbf{0}$, then $(\mathbf{E}+\mathbf{F})^2 = \mathbf{E}+\mathbf{F}$. Conversely, if $\mathbf{E}+\mathbf{F}$ is a projector, then

$$\begin{aligned}
(\mathbf{E}+\mathbf{F})^2 = \mathbf{E}+\mathbf{F} \implies\ & \mathbf{EF} + \mathbf{FE} = \mathbf{0} \\
\implies\ & \mathbf{E}(\mathbf{EF}+\mathbf{FE}) = \mathbf{0} \quad \text{and} \quad (\mathbf{EF}+\mathbf{FE})\mathbf{E} = \mathbf{0} \\
\implies\ & \mathbf{EF} = \mathbf{FE} \\
\implies\ & \mathbf{EF} = \mathbf{0} = \mathbf{FE} \quad (\text{because } \mathbf{EF}+\mathbf{FE} = \mathbf{0}).
\end{aligned}$$

Thus $\mathbf{P} = \mathbf{E}+\mathbf{F}$ is a projector if and only if $\mathbf{EF} = \mathbf{FE} = \mathbf{0}$. Now prove that under this condition $R\left(\mathbf{P}\right) = \mathcal{X}_1 \oplus \mathcal{X}_2$. Start with the fact that $\mathbf{z} \in R\left(\mathbf{P}\right)$ if and only if $\mathbf{Pz} = \mathbf{z}$, and write each such vector $\mathbf{z}$ as $\mathbf{z} = \mathbf{x}_1 + \mathbf{y}_1$ and $\mathbf{z} = \mathbf{x}_2 + \mathbf{y}_2$, where $\mathbf{x}_i \in \mathcal{X}_i$ and $\mathbf{y}_i \in \mathcal{Y}_i$ so that $\mathbf{Ex}_1 = \mathbf{x}_1$, $\mathbf{Ey}_1 = \mathbf{0}$, $\mathbf{Fx}_2 = \mathbf{x}_2$, and $\mathbf{Fy}_2 = \mathbf{0}$. To prove that $R\left(\mathbf{P}\right) = \mathcal{X}_1 + \mathcal{X}_2$, write

$$\begin{aligned}
\mathbf{z} \in R\left(\mathbf{P}\right) \implies\ & \mathbf{Pz} = \mathbf{z} \implies (\mathbf{E}+\mathbf{F})\mathbf{z} = \mathbf{z} \\
\implies\ & (\mathbf{E}+\mathbf{F})(\mathbf{x}_2+\mathbf{y}_2) = (\mathbf{x}_2+\mathbf{y}_2) \\
\implies\ & \mathbf{Ez} = \mathbf{y}_2 \implies \mathbf{x}_1 = \mathbf{y}_2 \\
\implies\ & \mathbf{z} = \mathbf{x}_1 + \mathbf{x}_2 \in \mathcal{X}_1 + \mathcal{X}_2 \implies R\left(\mathbf{P}\right) \subseteq \mathcal{X}_1 + \mathcal{X}_2.
\end{aligned}$$

Conversely, $\mathcal{X}_1 + \mathcal{X}_2 \subseteq R\left(\mathbf{P}\right)$ because

$$\begin{aligned}
\mathbf{z} \in \mathcal{X}_1 + \mathcal{X}_2 \implies\ & \mathbf{z} = \mathbf{x}_1 + \mathbf{x}_2, \quad \text{where} \quad \mathbf{x}_1 \in \mathcal{X}_1 \text{ and } \mathbf{x}_2 \in \mathcal{X}_2 \\
\implies\ & \mathbf{x}_1 = \mathbf{Ex}_1 \quad \text{and} \quad \mathbf{x}_2 = \mathbf{Fx}_2 \\
\implies\ & \mathbf{Fx}_1 = \mathbf{FEx}_1 = \mathbf{0} \text{ and } \mathbf{Ex}_2 = \mathbf{EFx}_2 = \mathbf{0} \\
\implies\ & \mathbf{Pz} = (\mathbf{E}+\mathbf{F})(\mathbf{x}_1+\mathbf{x}_2) = \mathbf{x}_1 + \mathbf{x}_2 = \mathbf{z} \\
\implies\ & \mathbf{z} \in R\left(\mathbf{P}\right).
\end{aligned}$$

The fact that $\mathcal{X}_1$ and $\mathcal{X}_2$ are disjoint follows by writing

$$\mathbf{z} \in \mathcal{X}_1 \cap \mathcal{X}_2 \implies \mathbf{Ez} = \mathbf{z} = \mathbf{Fz} \implies \mathbf{z} = \mathbf{EFz} = \mathbf{0},$$

and thus $R(\mathbf{P}) = \mathcal{X}_1 \oplus \mathcal{X}_2$ is established. To prove that $N(\mathbf{P}) = \mathcal{Y}_1 \cap \mathcal{Y}_2$, write

$$
\begin{aligned}
\mathbf{Pz} = 0 &\implies (\mathbf{E} + \mathbf{F})\mathbf{z} = 0 \implies \mathbf{Ez} = -\mathbf{Fz} \\
&\implies \mathbf{Ez} = -\mathbf{EFz} \quad \text{and} \quad \mathbf{FEz} = -\mathbf{Fz} \\
&\implies \mathbf{Ez} = 0 \quad \text{and} \quad 0 = \mathbf{Fz} \implies \mathbf{z} \in \mathcal{Y}_1 \cap \mathcal{Y}_2.
\end{aligned}
$$

5.9.18. Use the hint together with the result of Exercise 5.9.17 to write

$$
\begin{aligned}
\mathbf{E} - \mathbf{F} \text{ is a projector} &\iff \mathbf{I} - (\mathbf{E} - \mathbf{F}) \text{ is a projector} \\
&\iff (\mathbf{I} - \mathbf{E}) + \mathbf{F} \text{ is a projector} \\
&\iff (\mathbf{I} - \mathbf{E})\mathbf{F} = 0 = \mathbf{F}(\mathbf{I} - \mathbf{E}) \\
&\iff \mathbf{EF} = \mathbf{F} = \mathbf{FE}.
\end{aligned}
$$

Under this condition, Exercise 5.9.17 says that

$$
R(\mathbf{I} - \mathbf{E} + \mathbf{F}) = R(\mathbf{I} - \mathbf{E}) \oplus R(\mathbf{F}) \quad \text{and} \quad N(\mathbf{I} - \mathbf{E} + \mathbf{F}) = N(\mathbf{I} - \mathbf{E}) \cap N(\mathbf{F}),
$$

so (5.9.11) guarantees

$$
\begin{aligned}
R(\mathbf{E} - \mathbf{F}) &= N(\mathbf{I} - \mathbf{E} + \mathbf{F}) = N(\mathbf{I} - \mathbf{E}) \cap N(\mathbf{F}) = R(\mathbf{E}) \cap N(\mathbf{F}) = \mathcal{X}_1 \cap \mathcal{Y}_2 \\
N(\mathbf{E} - \mathbf{F}) &= R(\mathbf{I} - \mathbf{E} + \mathbf{F}) = R(\mathbf{I} - \mathbf{E}) \oplus R(\mathbf{F}) = N(\mathbf{E}) \oplus R(\mathbf{F}) = \mathcal{Y}_1 \oplus \mathcal{X}_2.
\end{aligned}
$$

5.9.19. If $\mathbf{EF} = \mathbf{P} = \mathbf{FE}$, then $\mathbf{P}$ is idempotent, and hence $\mathbf{P}$ is a projector. To prove that $R(\mathbf{P}) = \mathcal{X}_1 \cap \mathcal{X}_2$, write

$$
\begin{aligned}
\mathbf{z} \in R(\mathbf{P}) &\implies \mathbf{Pz} = \mathbf{z} \\
&\implies \mathbf{E}(\mathbf{Fz}) = \mathbf{z} \quad \text{and} \quad \mathbf{F}(\mathbf{Ez}) = \mathbf{z} \\
&\implies \mathbf{z} \in R(\mathbf{E}) \cap R(\mathbf{F}) = \mathcal{X}_1 \cap \mathcal{X}_2 \\
&\implies R(\mathbf{P}) \subseteq \mathcal{X}_1 \cap \mathcal{X}_2.
\end{aligned}
$$

Conversely,

$$
\mathbf{z} \in \mathcal{X}_1 \cap \mathcal{X}_2 \implies \mathbf{Ez} = \mathbf{z} = \mathbf{Fz} \implies \mathbf{Pz} = \mathbf{z} \implies \mathcal{X}_1 \cap \mathcal{X}_2 \subseteq R(\mathbf{P}),
$$

and hence $R(\mathbf{P}) = \mathcal{X}_1 \cap \mathcal{X}_2$. To prove that $N(\mathbf{P}) = \mathcal{Y}_1 + \mathcal{Y}_2$, first notice that

$$
\mathbf{z} \in N(\mathbf{P}) \implies \mathbf{FEz} = 0 \implies \mathbf{Ez} \in N(\mathbf{F}).
$$

This together with the fact that $(\mathbf{I} - \mathbf{E})\mathbf{z} \in N(\mathbf{E})$ allows us to conclude that

$$
\mathbf{z} = (\mathbf{I} - \mathbf{E})\mathbf{z} + \mathbf{Ez} \in N(\mathbf{E}) + N(\mathbf{F}) = \mathcal{Y}_1 + \mathcal{Y}_2 \implies N(\mathbf{P}) \subseteq \mathcal{Y}_1 + \mathcal{Y}_2.
$$

Conversely,

$$\mathbf{z} \in \mathcal{Y}_1 + \mathcal{Y}_2 \implies \mathbf{z} = \mathbf{y}_1 + \mathbf{y}_2, \text{ where } \mathbf{y}_i \in \mathcal{Y}_i \text{ for } i = 1, 2$$
$$\implies \mathbf{E}\mathbf{y}_1 = \mathbf{0} \quad \text{and} \quad \mathbf{F}\mathbf{y}_2 = \mathbf{0}$$
$$\implies \mathbf{P}\mathbf{z} = \mathbf{0} \implies \mathcal{Y}_1 + \mathcal{Y}_2 \subseteq N(\mathbf{P}).$$

Thus $N(\mathbf{P}) = \mathcal{Y}_1 + \mathcal{Y}_2$.

5.9.20. (a)   For every inner pseudoinverse, $\mathbf{A}\mathbf{A}^-$ is a projector onto $R(\mathbf{A})$, and $\mathbf{I} - \mathbf{A}^-\mathbf{A}$ is a projector onto $N(\mathbf{A})$. The system being consistent means that

$$\mathbf{b} \in R(\mathbf{A}) = R(\mathbf{A}\mathbf{A}^-) \implies \mathbf{A}\mathbf{A}^-\mathbf{b} = \mathbf{b},$$

so $\mathbf{A}^-\mathbf{b}$ is a particular solution. Therefore, the general solution of the system is

$$\mathbf{A}^-\mathbf{b} + N(\mathbf{A}) = \mathbf{A}^-\mathbf{b} + R(\mathbf{I} - \mathbf{A}^-\mathbf{A}).$$

(b) $\mathbf{A}^-\mathbf{A}$ is a projector along $N(\mathbf{A})$, so Exercise 5.9.12 insures $\mathbf{Q}(\mathbf{A}^-\mathbf{A}) = \mathbf{Q}$ and $(\mathbf{A}^-\mathbf{A})\mathbf{Q} = (\mathbf{A}^-\mathbf{A})$. This together with the fact that $\mathbf{P}\mathbf{A} = \mathbf{A}$ allows us to write

$$\mathbf{A}\mathbf{X}\mathbf{A} = \mathbf{A}\mathbf{Q}\mathbf{A}^-\mathbf{P}\mathbf{A} = \mathbf{A}\mathbf{Q}\mathbf{A}^-\mathbf{A} = \mathbf{A}\mathbf{Q} = \mathbf{A}\mathbf{A}^-\mathbf{A}\mathbf{Q} = \mathbf{A}\mathbf{A}^-\mathbf{A} = \mathbf{A}.$$

Similarly,

$$\mathbf{X}\mathbf{A}\mathbf{X} = (\mathbf{Q}\mathbf{A}^-\mathbf{P})\mathbf{A}(\mathbf{Q}\mathbf{A}^-\mathbf{P}) = \mathbf{Q}\mathbf{A}^-(\mathbf{P}\mathbf{A})\mathbf{Q}\mathbf{A}^-\mathbf{P} = \mathbf{Q}\mathbf{A}^-\mathbf{A}\mathbf{Q}\mathbf{A}^-\mathbf{P}$$
$$= \mathbf{Q}(\mathbf{A}^-\mathbf{A}\mathbf{Q})\mathbf{A}^-\mathbf{P} = \mathbf{Q}(\mathbf{A}^-\mathbf{A})\mathbf{A}^-\mathbf{P} = \mathbf{Q}\mathbf{A}^-\mathbf{P} = \mathbf{X},$$

so $\mathbf{X}$ is a reflexive pseudoinverse for $\mathbf{A}$. To show $\mathbf{X}$ has the prescribed range and nullspace, use the fact that $\mathbf{X}\mathbf{A}$ is a projector onto $R(\mathbf{X})$ and $\mathbf{A}\mathbf{X}$ is a projector along $N(\mathbf{X})$ to write

$$R(\mathbf{X}) = R(\mathbf{X}\mathbf{A}) = R(\mathbf{Q}\mathbf{A}^-\mathbf{P}\mathbf{A}) = R(\mathbf{Q}\mathbf{A}^-\mathbf{A}) = R(\mathbf{Q}) = \mathcal{L}$$

and

$$N(\mathbf{X}) = N(\mathbf{A}\mathbf{X}) = N(\mathbf{A}\mathbf{Q}\mathbf{A}^-\mathbf{P}) = N(\mathbf{A}\mathbf{A}^-\mathbf{A}\mathbf{Q}\mathbf{A}^-\mathbf{P})$$
$$= N(\mathbf{A}\mathbf{A}^-\mathbf{A}\mathbf{A}^-\mathbf{P}) = N(\mathbf{A}\mathbf{A}^-\mathbf{P}) = N(\mathbf{P}) = \mathcal{M}.$$

To prove uniqueness, suppose that $\mathbf{X}_1$ and $\mathbf{X}_2$ both satisfy the specified conditions. Then

$$N(\mathbf{X}_2) = \mathcal{M} = R(\mathbf{I} - \mathbf{A}\mathbf{X}_1) \implies \mathbf{X}_2(\mathbf{I} - \mathbf{A}\mathbf{X}_1) = \mathbf{0} \implies \mathbf{X}_2 = \mathbf{X}_2\mathbf{A}\mathbf{X}_1$$

and

$$R(\mathbf{X}_2\mathbf{A}) = R(\mathbf{X}_2) = \mathcal{L} = R(\mathbf{X}_1) \implies \mathbf{X}_2\mathbf{A}\mathbf{X}_1 = \mathbf{X}_1,$$

so $\mathbf{X}_2 = \mathbf{X}_1$.

Solutions for exercises in section 5. 10

5.10.1. Since $index(\mathbf{A}) = k$, we must have that

$$rank\left(\mathbf{A}^{k-1}\right) > rank\left(\mathbf{A}^k\right) = rank\left(\mathbf{A}^{k+1}\right) = \cdots = rank\left(\mathbf{A}^{2k}\right) = \cdots,$$

so $rank\left(\mathbf{A}^k\right) = rank(\mathbf{A}^k)^2$, and hence $index(\mathbf{A}^k) \le 1$. But $\mathbf{A}^k$ is singular (because $\mathbf{A}$ is singular) so that $index(\mathbf{A}^k) > 0$. Consequently, $index(\mathbf{A}^k) = 1$.

5.10.2. In this case, $R\left(\mathbf{A}^k\right) = \mathbf{0}$ and $N\left(\mathbf{A}^k\right) = \Re^n$. The nonsingular component $\mathbf{C}$ in (5.10.5) is missing, and you can take $\mathbf{Q} = \mathbf{I}$, thereby making $\mathbf{A}$ its own core-nilpotent decomposition.

5.10.3. If $\mathbf{A}$ is nonsingular, then $index(\mathbf{A}) = 0$, regardless of whether or not $\mathbf{A}$ is symmetric. If $\mathbf{A}$ is singular and symmetric, we want to prove $index(\mathbf{A}) = 1$. The strategy is to show that $R(\mathbf{A}) \cap N(\mathbf{A}) = \mathbf{0}$ because this implies that $R(\mathbf{A}) \oplus N(\mathbf{A}) = \Re^n$. To do so, start with

$$\mathbf{x} \in R(\mathbf{A}) \cap N(\mathbf{A}) \implies \mathbf{Ax} = \mathbf{0} \quad \text{and} \quad \mathbf{x} = \mathbf{Ay} \quad \text{for some} \quad \mathbf{y}.$$

Now combine this with the symmetry of $\mathbf{A}$ to obtain

$$\mathbf{x}^T = \mathbf{y}^T \mathbf{A}^T = \mathbf{y}^T \mathbf{A} \implies \mathbf{x}^T \mathbf{x} = \mathbf{y}^T \mathbf{Ax} = 0 \implies \|\mathbf{x}\|_2^2 = 0 \implies \mathbf{x} = \mathbf{0}.$$

5.10.4. $index(\mathbf{A}) = 0$ when $\mathbf{A}$ is nonsingular. If $\mathbf{A}$ is singular and normal we want to prove $index(\mathbf{A}) = 1$. The strategy is to show that $R(\mathbf{A}) \cap N(\mathbf{A}) = \mathbf{0}$ because this implies that $R(\mathbf{A}) \oplus N(\mathbf{A}) = \mathcal{C}^n$. Recall from (4.5.6) that $N(\mathbf{A}) = N(\mathbf{A}^*\mathbf{A})$ and $N(\mathbf{A}^*) = N(\mathbf{A}\mathbf{A}^*)$, so $N(\mathbf{A}) = N(\mathbf{A}^*)$. Start with

$$\mathbf{x} \in R(\mathbf{A}) \cap N(\mathbf{A}) \implies \mathbf{Ax} = \mathbf{0} \quad \text{and} \quad \mathbf{x} = \mathbf{Ay} \quad \text{for some} \quad \mathbf{y},$$

and combine this with $N(\mathbf{A}) = N(\mathbf{A}^*)$ to obtain

$$\mathbf{A}^*\mathbf{x} = \mathbf{0} \text{ and } \mathbf{x} = \mathbf{Ay} \implies \mathbf{x}^*\mathbf{x} = \mathbf{y}^*\mathbf{A}^*\mathbf{x} = 0 \implies \|\mathbf{x}\|_2^2 = 0 \implies \mathbf{x} = \mathbf{0}.$$

5.10.5. Compute $rank\left(\mathbf{A}^0\right) = 3$, $rank\left(\mathbf{A}\right) = 2$, $rank\left(\mathbf{A}^2\right) = 1$, and $rank\left(\mathbf{A}^3\right) = 1$, to see that $k = 2$ is the smallest integer such that $rank\left(\mathbf{A}^k\right) = rank\left(\mathbf{A}^{k+1}\right)$, so $index(\mathbf{A}) = 2$. The matrix $\mathbf{Q} = [\mathbf{X} \,|\, \mathbf{Y}]$ is a matrix in which the columns of $\mathbf{X}$ are a basis for $R\left(\mathbf{A}^2\right)$, and the columns of $\mathbf{Y}$ are a basis for $N\left(\mathbf{A}^2\right)$. Since

$$\mathbf{E}_{\mathbf{A}^2} = \begin{pmatrix} 1 & 1 & 0 \\ 0 & 0 & 0 \\ 0 & 0 & 0 \end{pmatrix},$$

we have

$$\mathbf{X} = \begin{pmatrix} -8 \\ 12 \\ 8 \end{pmatrix} \quad \text{and} \quad \mathbf{Y} = \begin{pmatrix} -1 & 0 \\ 1 & 0 \\ 0 & 1 \end{pmatrix}, \quad \text{so} \quad \mathbf{Q} = \begin{pmatrix} -8 & -1 & 0 \\ 12 & 1 & 0 \\ 8 & 0 & 1 \end{pmatrix}.$$

It can now be verified that

$$\mathbf{Q}^{-1}\mathbf{A}\mathbf{Q} = \begin{pmatrix} 1/4 & 1/4 & 0 \\ -3 & -2 & 0 \\ -2 & -2 & 1 \end{pmatrix} \begin{pmatrix} -2 & 0 & -4 \\ 4 & 2 & 4 \\ 3 & 2 & 2 \end{pmatrix} \begin{pmatrix} -8 & -1 & 0 \\ 12 & 1 & 0 \\ 8 & 0 & 1 \end{pmatrix} = \begin{pmatrix} 2 & 0 & 0 \\ 0 & -2 & 4 \\ 0 & -1 & 2 \end{pmatrix},$$

where

$$\mathbf{C} = [2] \quad \text{and} \quad \mathbf{N} = \begin{pmatrix} -2 & 4 \\ -1 & 2 \end{pmatrix},$$

and $\mathbf{N}^2 = \mathbf{0}$. Finally, $\mathbf{A}^D = \mathbf{Q}\begin{pmatrix} \mathbf{C}^{-1} & \mathbf{0} \\ \mathbf{0} & \mathbf{0} \end{pmatrix}\mathbf{Q}^{-1} = \begin{pmatrix} -1 & -1 & 0 \\ 3/2 & 3/2 & 0 \\ 1 & 1 & 0 \end{pmatrix}.$

5.10.6. (a)   Because

$$\mathbf{J} - \lambda\mathbf{I} = \begin{pmatrix} 1-\lambda & 0 & 0 & 0 & 0 \\ 0 & 1-\lambda & 0 & 0 & 0 \\ 0 & 0 & 1-\lambda & 0 & 0 \\ 0 & 0 & 0 & 2-\lambda & 0 \\ 0 & 0 & 0 & 0 & 2-\lambda \end{pmatrix},$$

and because a diagonal matrix is singular if and only if it has a zero-diagonal entry, it follows that $\mathbf{J} - \lambda\mathbf{I}$ is singular if and only if $\lambda = 1$ or $\lambda = 2$, so $\lambda_1 = 1$ and $\lambda_2 = 2$ are the two eigenvalues of $\mathbf{J}$. To find the index of λ_1, use block multiplication to observe that

$$\mathbf{J} - \mathbf{I} = \begin{pmatrix} \mathbf{0} & \mathbf{0} \\ \mathbf{0} & \mathbf{I}_{2\times 2} \end{pmatrix} \quad \Longrightarrow \quad rank\,(\mathbf{J} - \mathbf{I}) = 2 = rank\,(\mathbf{J} - \mathbf{I})^2.$$

Therefore, $index(\lambda_1) = 1$. Similarly,

$$\mathbf{J} - 2\mathbf{I} = \begin{pmatrix} -\mathbf{I}_{3\times 3} & \mathbf{0} \\ \mathbf{0} & \mathbf{0} \end{pmatrix} \quad \text{and} \quad rank\,(\mathbf{J} - 2\mathbf{I}) = 3 = rank\,(\mathbf{J} - 2\mathbf{I})^2,$$

so $index(\lambda_2) = 1$.

(b)   Since

$$\mathbf{J} - \lambda\mathbf{I} = \begin{pmatrix} 1-\lambda & 1 & 0 & 0 & 0 \\ 0 & 1-\lambda & 1 & 0 & 0 \\ 0 & 0 & 1-\lambda & 0 & 0 \\ 0 & 0 & 0 & 2-\lambda & 1 \\ 0 & 0 & 0 & 0 & 2-\lambda \end{pmatrix},$$

and since a triangular matrix is singular if and only if there exists a zero-diagonal entry (i.e., a zero pivot), it follows that $\mathbf{J} - \lambda\mathbf{I}$ is singular if and only if $\lambda = 1$

or $\lambda = 2$, so $\lambda_1 = 1$ and $\lambda_2 = 2$ are the two eigenvalues of $\mathbf{J}$. To find the index of λ_1, use block multiplication to compute

$$\mathbf{J} - \mathbf{I} = \begin{pmatrix} 0 & 1 & 0 & 0 & 0 \\ 0 & 0 & 1 & 0 & 0 \\ 0 & 0 & 0 & 0 & 0 \\ 0 & 0 & 0 & 1 & 1 \\ 0 & 0 & 0 & 0 & 1 \end{pmatrix}, \qquad (\mathbf{J} - \mathbf{I})^2 = \begin{pmatrix} 0 & 0 & 1 & 0 & 0 \\ 0 & 0 & 0 & 0 & 0 \\ 0 & 0 & 0 & 0 & 0 \\ 0 & 0 & 0 & 1 & 2 \\ 0 & 0 & 0 & 0 & 1 \end{pmatrix},$$

$$(\mathbf{J} - \mathbf{I})^3 = \begin{pmatrix} 0 & 0 & 0 & 0 & 0 \\ 0 & 0 & 0 & 0 & 0 \\ 0 & 0 & 0 & 0 & 0 \\ 0 & 0 & 0 & 1 & 3 \\ 0 & 0 & 0 & 0 & 1 \end{pmatrix}, \qquad (\mathbf{J} - \mathbf{I})^4 = \begin{pmatrix} 0 & 0 & 0 & 0 & 0 \\ 0 & 0 & 0 & 0 & 0 \\ 0 & 0 & 0 & 0 & 0 \\ 0 & 0 & 0 & 1 & 4 \\ 0 & 0 & 0 & 0 & 1 \end{pmatrix}.$$

Since

$$rank\,(\mathbf{J} - \mathbf{I}) > rank\,(\mathbf{J} - \mathbf{I})^2 > rank\,(\mathbf{J} - \mathbf{I})^3 = rank\,(\mathbf{J} - \mathbf{I})^4,$$

it follows that $index(\lambda_1) = 3$. A similar computation using λ_2 shows that

$$rank\,(\mathbf{J} - 2\mathbf{I}) > rank\,(\mathbf{J} - 2\mathbf{I})^2 = rank\,(\mathbf{J} - 2\mathbf{I})^3,$$

so $index(\lambda_2) = 2$. The fact that eigenvalues associated with diagonal matrices have index 1 while eigenvalues associated with triangular matrices can have higher indices is no accident. This will be discussed in detail in §7.8 (p. 587).

5.10.7. (a) If $\mathbf{P}$ is a projector, then, by (5.9.13), $\mathbf{P} = \mathbf{P}^2$, so $rank\,(\mathbf{P}) = rank\,(\mathbf{P}^2)$, and hence $index(\mathbf{P}) \le 1$. If $\mathbf{P} \neq \mathbf{I}$, then $\mathbf{P}$ is singular, and thus $index(\mathbf{P}) = 1$. If $\mathbf{P} = \mathbf{I}$, then $index(\mathbf{P}) = 0$. An alternate argument could be given on the basis of the observation that $\Re^n = R\,(\mathbf{P}) \oplus N\,(\mathbf{P})$.

(b) Recall from (5.9.12) that if the columns of $\mathbf{X}$ and $\mathbf{Y}$ constitute bases for $R\,(\mathbf{P})$ and $N\,(\mathbf{P})$, respectively, then for $\mathbf{Q} = [\mathbf{X}\,|\,\mathbf{Y}]$,

$$\mathbf{Q}^{-1}\mathbf{P}\mathbf{Q} = \begin{pmatrix} \mathbf{I} & \mathbf{0} \\ \mathbf{0} & \mathbf{0} \end{pmatrix},$$

and it follows that $\begin{pmatrix} \mathbf{I} & \mathbf{0} \\ \mathbf{0} & \mathbf{0} \end{pmatrix}$ is the core-nilpotent decomposition for $\mathbf{P}$.

5.10.8. Suppose that $\sum_{i=0}^{k-1} \alpha_i \mathbf{N}^i \mathbf{x} = \mathbf{0}$, and multiply both sides by $\mathbf{N}^{k-1}$ to obtain $\alpha_0 \mathbf{N}^{k-1} \mathbf{x} = \mathbf{0}$. By assumption, $\mathbf{N}^{k-1} \mathbf{x} \neq \mathbf{0}$, so $\alpha_0 = 0$, and hence $\sum_{i=1}^{k-1} \alpha_i \mathbf{N}^i \mathbf{x} = \mathbf{0}$. Now multiply both sides of this equation by $\mathbf{N}^{k-2}$ to produce $\alpha_1 \mathbf{N}^{k-1} \mathbf{x} = \mathbf{0}$, and conclude that $\alpha_1 = 0$. Continuing in this manner (or by making a formal induction argument) gives $\alpha_0 = \alpha_1 = \alpha_2 = \cdots = \alpha_{k-1} = 0$.

5.10.9. (a) $\mathbf{b} \in R\,(\mathbf{A}^k) \subseteq R\,(\mathbf{A}) \implies \mathbf{b} \in R\,(\mathbf{A}) \implies \mathbf{A}\mathbf{x} = \mathbf{b}$ is consistent.

(b)   We saw in Example 5.10.5 that when considered as linear operators restricted to $R\left(\mathbf{A}^k\right)$, both $\mathbf{A}$ and $\mathbf{A}^D$ are invertible, and in fact they are true inverses of each other. Consequently, $\mathbf{A}$ and $\mathbf{A}^D$ are one-to-one mappings on $R\left(\mathbf{A}^k\right)$ (recall Exercise 4.7.18), so for each $\mathbf{b} \in R\left(\mathbf{A}^k\right)$ there is a unique $\mathbf{x} \in R\left(\mathbf{A}^k\right)$ such that $\mathbf{Ax} = \mathbf{b}$, and this unique $\mathbf{x}$ is given by

$$\mathbf{x} = \left(\mathbf{A}_{/R(\mathbf{A}^k)}\right)^{-1} \mathbf{b} = \mathbf{A}^D \mathbf{b}.$$

(c)   Part (b) shows that $\mathbf{A}^D \mathbf{b}$ is a particular solution. The desired result follows because the general solution is any particular solution plus the general solution of the associated homogeneous equation.

5.10.10.   Notice that $\mathbf{A}\mathbf{A}^D = \mathbf{Q}\begin{pmatrix} \mathbf{I} & \mathbf{0} \\ \mathbf{0} & \mathbf{0} \end{pmatrix}\mathbf{Q}^{-1}$, and use the results from Example 5.10.3 (p. 398). $\mathbf{I} - \mathbf{A}\mathbf{A}^D$ is the complementary projector, so it projects onto $N\left(\mathbf{A}^k\right)$ along $R\left(\mathbf{A}^k\right)$.

5.10.11.   In each case verify that the axioms (A1), (A2), (A4), and (A5) in the definition of a vector space given on p. 160 hold for matrix multiplication (rather than +). In parts (a) and (b) the identity element is the ordinary identity matrix, and the inverse of each member is the ordinary inverse. In part (c), the identity element is $\mathbf{E} = \begin{pmatrix} 1/2 & 1/2 \\ 1/2 & 1/2 \end{pmatrix}$ because $\mathbf{AE} = \mathbf{A} = \mathbf{EA}$ for each $\mathbf{A} \in \mathcal{G}$, and $\begin{pmatrix} \alpha & \alpha \\ \alpha & \alpha \end{pmatrix}^{\#} = \frac{1}{4\alpha}\begin{pmatrix} 1 & 1 \\ 1 & 1 \end{pmatrix}$ because $\mathbf{AA}^{\#} = \mathbf{E} = \mathbf{A}^{\#}\mathbf{A}$.

5.10.12.   (a) $\Longrightarrow$ (b) : If $\mathbf{A}$ belongs to a matrix group $\mathcal{G}$ in which the identity element is $\mathbf{E}$, and if $\mathbf{A}^{\#}$ is the inverse of $\mathbf{A}$ in $\mathcal{G}$, then $\mathbf{A}^{\#}\mathbf{A}^2 = \mathbf{EA} = \mathbf{A}$, so

$$\begin{aligned} \mathbf{x} \in R\left(\mathbf{A}\right) \cap N\left(\mathbf{A}\right) \quad &\Longrightarrow \quad \mathbf{x} = \mathbf{Ay} \text{ for some } \mathbf{y} \text{ and } \mathbf{Ax} = \mathbf{0} \\ &\Longrightarrow \quad \mathbf{Ay} = \mathbf{A}^{\#}\mathbf{A}^2\mathbf{y} = \mathbf{A}^{\#}\mathbf{Ax} = \mathbf{0} \\ &\Longrightarrow \quad \mathbf{x} = \mathbf{0}. \end{aligned}$$

(b) $\Longrightarrow$ (c) : Suppose $\mathbf{A}$ is $n \times n$, and let $\mathcal{B}_R$ and $\mathcal{B}_N$ be bases for $R\left(\mathbf{A}\right)$ and $N\left(\mathbf{A}\right)$, respectively. Verify that $\mathcal{B} = R\left(\mathbf{A}\right) \cap N\left(\mathbf{A}\right) = \mathbf{0}$ implies $\mathcal{B}_R \cap \mathcal{B}_N$ is a linearly independent set, and use the fact that there are n vectors in $\mathcal{B}$ to conclude that $\mathcal{B}$ is a basis for $\Re^n$. Statement (c) now follows from (5.9.4).

(c) $\Longrightarrow$ (d) : Use the fact that $R\left(\mathbf{A}^k\right) \cap N\left(\mathbf{A}^k\right) = \mathbf{0}$.

(d) $\Longrightarrow$ (e) : Use the result of Example 5.10.5 together with the fact that the only nilpotent matrix of index 1 is the zero matrix.

(e) $\Longrightarrow$ (a) : It is straightforward to verify that the set

$$\mathcal{G} = \left\{ \mathbf{Q}\begin{pmatrix} \mathbf{X}_{r\times r} & \mathbf{0} \\ \mathbf{0} & \mathbf{0} \end{pmatrix}\mathbf{Q}^{-1} \,\middle|\, \mathbf{X} \text{ is nonsingular} \right\}$$

is a matrix group, and it's clear that $\mathbf{A} \in \mathcal{G}$.

5.10.13. (a) Use part (e) of Exercise 5.10.12 to write $\mathbf{A} = \mathbf{Q} \begin{pmatrix} \mathbf{C}_{r \times r} & \mathbf{0} \\ \mathbf{0} & \mathbf{0} \end{pmatrix} \mathbf{Q}^{-1}$. For the given $\mathbf{E}$, verify that $\mathbf{EA} = \mathbf{AE} = \mathbf{A}$ for all $\mathbf{A} \in \mathcal{G}$. The fact that $\mathbf{E}$ is the desired projector follows from (5.9.12).

(b) Simply verify that $\mathbf{AA}^{\#} = \mathbf{A}^{\#}\mathbf{A} = \mathbf{E}$. Notice that the group inverse agrees with the Drazin inverse of $\mathbf{A}$ described in Example 5.10.5. However, the Drazin inverse exists for all square matrices, but the concept of a group inverse makes sense only for group matrices—i.e., when $index(\mathbf{A}) = 1$.

Solutions for exercises in section 5. 11

5.11.1. Proceed as described on p. 199 to determine the following bases for each of the four fundamental subspaces.

$$R(\mathbf{A}) = span \left\{ \begin{pmatrix} 2 \\ -1 \\ -2 \end{pmatrix}, \begin{pmatrix} 1 \\ -1 \\ -1 \end{pmatrix} \right\} \quad N(\mathbf{A}^T) = span \left\{ \begin{pmatrix} 1 \\ 0 \\ 1 \end{pmatrix} \right\}$$

$$N(\mathbf{A}) = span \left\{ \begin{pmatrix} -1 \\ 1 \\ 1 \end{pmatrix} \right\} \quad R(\mathbf{A}^T) = span \left\{ \begin{pmatrix} 1 \\ 0 \\ 1 \end{pmatrix}, \begin{pmatrix} 0 \\ 1 \\ -1 \end{pmatrix} \right\}$$

Since each vector in a basis for $R(\mathbf{A})$ is orthogonal to each vector in a basis for $N(\mathbf{A}^T)$, it follows that $R(\mathbf{A}) \perp N(\mathbf{A}^T)$. The same logic also explains why $N(\mathbf{A}) \perp R(\mathbf{A}^T)$. Notice that $R(\mathbf{A})$ is a plane through the origin in $\Re^3$, and $N(\mathbf{A}^T)$ is the line through the origin perpendicular to this plane, so it is evident from the parallelogram law that $R(\mathbf{A}) \oplus N(\mathbf{A}^T) = \Re^3$. Similarly, $N(\mathbf{A})$ is the line through the origin normal to the plane defined by $R(\mathbf{A}^T)$, so $N(\mathbf{A}) \oplus R(\mathbf{A}^T) = \Re^3$.

5.11.2. $\mathcal{V}^{\perp} = \mathbf{0}$, and $\mathbf{0}^{\perp} = \mathcal{V}$.

5.11.3. If $\mathbf{A} = \begin{pmatrix} 1 & 2 \\ 2 & 4 \\ 0 & 1 \\ 3 & 6 \end{pmatrix}$, then $R(\mathbf{A}) = \mathcal{M}$, so (5.11.5) insures $\mathcal{M}^{\perp} = N(\mathbf{A}^T)$. Using

row operations, a basis for $N(\mathbf{A}^T)$ is computed to be $\left\{ \begin{pmatrix} -2 \\ 1 \\ 0 \\ 0 \end{pmatrix}, \begin{pmatrix} -3 \\ 0 \\ 0 \\ 1 \end{pmatrix} \right\}$.

5.11.4. Verify that $\mathcal{M}^{\perp}$ is closed with respect to vector addition and scalar multiplication. If $\mathbf{x}, \mathbf{y} \in \mathcal{M}^{\perp}$, then $\langle \mathbf{m} | \mathbf{x} \rangle = 0 = \langle \mathbf{m} | \mathbf{y} \rangle$ for each $\mathbf{m} \in \mathcal{M}$ so that $\langle \mathbf{m} | \mathbf{x} + \mathbf{y} \rangle = 0$ for each $\mathbf{m} \in \mathcal{M}$, and thus $\mathbf{x} + \mathbf{y} \in \mathcal{M}^{\perp}$. Similarly, for every scalar α we have $\langle \mathbf{m} | \alpha \mathbf{x} \rangle = \alpha \langle \mathbf{m} | \mathbf{x} \rangle = 0$ for each $\mathbf{m} \in \mathcal{M}$, so $\alpha \mathbf{x} \in \mathcal{M}^{\perp}$.

5.11.5. (a) $\mathbf{x} \in \mathcal{N}^{\perp} \Longrightarrow \mathbf{x} \perp \mathcal{N} \supseteq \mathcal{M} \Longrightarrow \mathbf{x} \perp \mathcal{M} \Longrightarrow \mathbf{x} \in \mathcal{M}^{\perp}$.

(b) Simply observe that

$$\mathbf{x} \in (\mathcal{M} + \mathcal{N})^{\perp} \Longleftrightarrow \mathbf{x} \perp (\mathcal{M} + \mathcal{N})$$
$$\Longleftrightarrow \mathbf{x} \perp \mathcal{M} \text{ and } \mathbf{x} \perp \mathcal{N}$$
$$\Longleftrightarrow \mathbf{x} \in (\mathcal{M}^{\perp} \cap \mathcal{N}^{\perp}).$$

(c) Use part (b) together with (5.11.4) to write

$$\left(\mathcal{M}^{\perp} + \mathcal{N}^{\perp}\right)^{\perp} = \mathcal{M}^{\perp\perp} \cap \mathcal{N}^{\perp\perp} = \mathcal{M} \cap \mathcal{N},$$

and then perp both sides.

5.11.6. Use the fact that $\dim R\left(\mathbf{A}^{T}\right) = rank\left(\mathbf{A}^{T}\right) = rank\left(\mathbf{A}\right) = \dim R\left(\mathbf{A}\right)$ together with (5.11.7) to conclude that

$$n = \dim N\left(\mathbf{A}\right) + \dim R\left(\mathbf{A}^{T}\right) = \dim N\left(\mathbf{A}\right) + \dim R\left(\mathbf{A}\right).$$

5.11.7. $\mathbf{U}$ is a unitary matrix in which the columns of $\mathbf{U}_1$ are an orthonormal basis for $R\left(\mathbf{A}\right)$ and the columns of $\mathbf{U}_2$ are an orthonormal basis for $N\left(\mathbf{A}^{T}\right)$, so setting $\mathbf{X} = \mathbf{U}_1$, $\mathbf{Y} = \mathbf{U}_2$, and $\left[\mathbf{X} \,|\, \mathbf{Y}\right]^{-1} = \mathbf{U}^{T}$ in (5.9.12) produces $\mathbf{P} = \mathbf{U}_1 \mathbf{U}_1^{T}$. According to (5.9.9), the projector onto $N\left(\mathbf{A}^{T}\right)$ along $R\left(\mathbf{A}\right)$ is $\mathbf{I} - \mathbf{P} = \mathbf{I} - \mathbf{U}_1 \mathbf{U}_1^{T} = \mathbf{U}_2 \mathbf{U}_2^{T}$.

5.11.8. Start with the first column of $\mathbf{A}$, and set $\mathbf{u} = \mathbf{A}_{*1} + 6\mathbf{e}_1 = \begin{pmatrix} 2 & 2 & -4 \end{pmatrix}^{T}$ to obtain

$$\mathbf{R}_1 = \mathbf{I} - \frac{2\mathbf{u}\mathbf{u}^{T}}{\mathbf{u}^{T}\mathbf{u}} = \frac{1}{3}\begin{pmatrix} 2 & -1 & 2 \\ -1 & 2 & 2 \\ 2 & 2 & -1 \end{pmatrix} \quad \text{and} \quad \mathbf{R}_1\mathbf{A} = \begin{pmatrix} -6 & 0 & -6 & -3 \\ 0 & 0 & 0 & 0 \\ 0 & -3 & 0 & 0 \end{pmatrix}.$$

Now set $\mathbf{u} = \begin{pmatrix} 0 \\ -3 \end{pmatrix} + 3\mathbf{e}_1 = \begin{pmatrix} 3 \\ -3 \end{pmatrix}$ to get

$$\hat{\mathbf{R}}_2 = \mathbf{I} - \frac{2\mathbf{u}\mathbf{u}^{T}}{\mathbf{u}^{T}\mathbf{u}} = \begin{pmatrix} 0 & 1 \\ 1 & 0 \end{pmatrix} \quad \text{and} \quad \mathbf{R}_2 = \begin{pmatrix} 1 & \mathbf{0} \\ \mathbf{0} & \hat{\mathbf{R}}_2 \end{pmatrix} = \begin{pmatrix} 1 & 0 & 0 \\ 0 & 0 & 1 \\ 0 & 1 & 0 \end{pmatrix},$$

so

$$\mathbf{P} = \mathbf{R}_2\mathbf{R}_1 = \frac{1}{3}\begin{pmatrix} 2 & -1 & 2 \\ 2 & 2 & -1 \\ -1 & 2 & 2 \end{pmatrix} \quad \text{and} \quad \mathbf{PA} = \begin{pmatrix} -6 & 0 & -6 & -3 \\ 0 & -3 & 0 & 0 \\ 0 & 0 & 0 & 0 \end{pmatrix} = \begin{pmatrix} \mathbf{B} \\ \mathbf{0} \end{pmatrix}.$$

Therefore, $rank\,(\mathbf{A}) = 2$, and orthonormal bases for $R\,(\mathbf{A})$ and $N\,(\mathbf{A}^T)$ are extracted from the columns of $\mathbf{U} = \mathbf{P}^T$ as shown below.

$$R\,(\mathbf{A}) = span\left\{\begin{pmatrix} 2/3 \\ -1/3 \\ 2/3 \end{pmatrix}, \begin{pmatrix} 2/3 \\ 2/3 \\ -1/3 \end{pmatrix}\right\} \quad \text{and} \quad N\,(\mathbf{A}^T) = span\left\{\begin{pmatrix} -1/3 \\ 2/3 \\ 2/3 \end{pmatrix}\right\}$$

Now work with $\mathbf{B}^T$, and set $\mathbf{u} = (\mathbf{B}_{1*})^T + 9\mathbf{e}_1 = (\,3 \quad 0 \quad -6 \quad -3\,)^T$ to get

$$\mathbf{Q} = \mathbf{I} - \frac{2\mathbf{u}\mathbf{u}^T}{\mathbf{u}^T\mathbf{u}} = \frac{1}{3}\begin{pmatrix} 2 & 0 & 2 & 1 \\ 0 & 3 & 0 & 0 \\ 2 & 0 & -1 & -2 \\ 1 & 0 & -2 & 2 \end{pmatrix} \quad \text{and} \quad \mathbf{Q}\mathbf{B}^T = \begin{pmatrix} -9 & 0 \\ 0 & -3 \\ 0 & 0 \\ 0 & 0 \end{pmatrix} = \begin{pmatrix} \mathbf{T} \\ \mathbf{0} \end{pmatrix}.$$

Orthonormal bases for $R\,(\mathbf{A}^T)$ and $N\,(\mathbf{A})$ are extracted from the columns of $\mathbf{V} = \mathbf{Q}^T = \mathbf{Q}$ as shown below.

$$R\,(\mathbf{A}^T) = span\left\{\begin{pmatrix} 2/3 \\ 0 \\ 2/3 \\ 1/3 \end{pmatrix}, \begin{pmatrix} 0 \\ 1 \\ 0 \\ 0 \end{pmatrix}\right\} \quad \text{and} \quad N\,(\mathbf{A}) = span\left\{\begin{pmatrix} 2/3 \\ 0 \\ -1/3 \\ -2/3 \end{pmatrix}, \begin{pmatrix} 1/3 \\ 0 \\ -2/3 \\ 2/3 \end{pmatrix}\right\}$$

A URV factorization is obtained by setting $\mathbf{U} = \mathbf{P}^T$, $\mathbf{V} = \mathbf{Q}^T$, and

$$\mathbf{R} = \begin{pmatrix} \mathbf{T}^T & \mathbf{0} \\ \mathbf{0} & \mathbf{0} \end{pmatrix} = \begin{pmatrix} -9 & 0 & 0 & 0 \\ 0 & -3 & 0 & 0 \\ 0 & 0 & 0 & 0 \end{pmatrix}.$$

5.11.9. Using $\mathbf{E}_{\mathbf{A}} = \begin{pmatrix} 1 & 0 & 1 & 1/2 \\ 0 & 1 & 0 & 0 \\ 0 & 0 & 0 & 0 \end{pmatrix}$ along with the standard methods of Chapter 4, we have

$$R\,(\mathbf{A}) = span\left\{\begin{pmatrix} -4 \\ 2 \\ -4 \end{pmatrix}, \begin{pmatrix} -2 \\ -2 \\ 1 \end{pmatrix}\right\} \quad \text{and} \quad N\,(\mathbf{A}^T) = span\left\{\begin{pmatrix} -1 \\ 2 \\ 2 \end{pmatrix}\right\},$$

$$R\,(\mathbf{A}^T) = span\left\{\begin{pmatrix} 1 \\ 0 \\ 1 \\ 1/2 \end{pmatrix}, \begin{pmatrix} 0 \\ 1 \\ 0 \\ 0 \end{pmatrix}\right\} \quad \text{and} \quad N\,(\mathbf{A}) = span\left\{\begin{pmatrix} -1 \\ 0 \\ 1 \\ 0 \end{pmatrix}, \begin{pmatrix} -1/2 \\ 0 \\ 0 \\ 1 \end{pmatrix}\right\}.$$

Applying the Gram–Schmidt procedure to each of these sets produces the following orthonormal bases for the four fundamental subspaces.

$$\mathcal{B}_{R(\mathbf{A})} = \left\{\frac{1}{3}\begin{pmatrix} -2 \\ 1 \\ -2 \end{pmatrix}, \frac{1}{3}\begin{pmatrix} -2 \\ -2 \\ 1 \end{pmatrix}\right\} \qquad \mathcal{B}_{N(\mathbf{A}^T)} = \left\{\frac{1}{3}\begin{pmatrix} -1 \\ 2 \\ 2 \end{pmatrix}\right\}$$

$$\mathcal{B}_{R(\mathbf{A}^T)} = \left\{ \frac{1}{3}\begin{pmatrix} 2 \\ 0 \\ 2 \\ 1 \end{pmatrix}, \begin{pmatrix} 0 \\ 1 \\ 0 \\ 0 \end{pmatrix} \right\} \qquad \mathcal{B}_{N(\mathbf{A})} = \left\{ \frac{1}{\sqrt{2}}\begin{pmatrix} -1 \\ 0 \\ 1 \\ 0 \end{pmatrix}, \frac{1}{3\sqrt{2}}\begin{pmatrix} -1 \\ 0 \\ -1 \\ 4 \end{pmatrix} \right\}$$

The matrices $\mathbf{U}$ and $\mathbf{V}$ were defined in (5.11.8) to be

$$\mathbf{U} = \left(\mathcal{B}_{R(\mathbf{A})} \cup \mathcal{B}_{N(\mathbf{A}^T)} \right) = \frac{1}{3}\begin{pmatrix} -2 & -2 & -1 \\ 1 & -2 & 2 \\ -2 & 1 & 2 \end{pmatrix}$$

and

$$\mathbf{V} = \left(\mathcal{B}_{R(\mathbf{A}^T)} \cup \mathcal{B}_{N(\mathbf{A})} \right) = \frac{1}{3}\begin{pmatrix} 2 & 0 & -3/\sqrt{2} & -1/\sqrt{2} \\ 0 & 3 & 0 & 0 \\ 2 & 0 & 3/\sqrt{2} & -1/\sqrt{2} \\ 1 & 0 & 0 & 4/\sqrt{2} \end{pmatrix}.$$

Direct multiplication now produces

$$\mathbf{R} = \mathbf{U}^T \mathbf{A} \mathbf{V} = \begin{pmatrix} 9 & 0 & 0 & 0 \\ 0 & 3 & 0 & 0 \\ 0 & 0 & 0 & 0 \end{pmatrix}.$$

5.11.10. According to the discussion of projectors on p. 386, the *unique* vectors satisfying $\mathbf{v} = \mathbf{x} + \mathbf{y}$, $\mathbf{x} \in R(\mathbf{A})$, and $\mathbf{y} \in N(\mathbf{A}^T)$ are given by $\mathbf{x} = \mathbf{Pv}$ and $\mathbf{y} = (\mathbf{I} - \mathbf{P})\mathbf{v}$, where $\mathbf{P}$ is the projector onto $R(\mathbf{A})$ along $N(\mathbf{A}^T)$. Use the results of Exercise 5.11.7 and Exercise 5.11.8 to compute

$$\mathbf{P} = \mathbf{U}_1 \mathbf{U}_1^T = \frac{1}{9}\begin{pmatrix} 8 & 2 & 2 \\ 2 & 5 & -4 \\ 2 & -4 & 5 \end{pmatrix}, \quad \mathbf{x} = \mathbf{Pv} = \begin{pmatrix} 4 \\ 1 \\ 1 \end{pmatrix}, \quad \mathbf{y} = (\mathbf{I} - \mathbf{P})\mathbf{v} = \begin{pmatrix} -1 \\ 2 \\ 2 \end{pmatrix}.$$

5.11.11. Observe that

$$\begin{aligned} R(\mathbf{A}) \cap N(\mathbf{A}) = \mathbf{0} &\implies index(\mathbf{A}) \leq 1, \\ R(\mathbf{A}) \not\perp N(\mathbf{A}) &\implies \mathbf{A} \text{ is singular}, \\ R(\mathbf{A}) \not\perp N(\mathbf{A}) &\implies R(\mathbf{A}^T) \neq R(\mathbf{A}). \end{aligned}$$

It is now trial and error to build a matrix that satisfies the three conditions on the right-hand side. One such matrix is $\mathbf{A} = \begin{pmatrix} 1 & 2 \\ 1 & 2 \end{pmatrix}$.

5.11.12. $R(\mathbf{A}) \perp N(\mathbf{A}) \implies R(\mathbf{A}) \cap N(\mathbf{A}) = \mathbf{0} \implies index(\mathbf{A}) = 1$ by using (5.10.4). The example in the solution to Exercise 5.11.11 shows that the converse is false.

5.11.13. The facts that real symmetric $\implies$ hermitian $\implies$ normal are direct consequences of the definitions. To show that normal $\implies$ RPN, use (4.5.5) to write $R(\mathbf{A}) = R(\mathbf{AA}^*) = R(\mathbf{A}^*\mathbf{A}) = R(\mathbf{A}^*)$. The matrix $\begin{pmatrix} 1 & i \\ -i & 2 \end{pmatrix}$ is hermitian but not symmetric. To construct a matrix that is normal but not hermitian or real symmetric, try to find an example with real numbers. If $\mathbf{A} = \begin{pmatrix} a & b \\ c & d \end{pmatrix}$, then

$$\mathbf{A}\mathbf{A}^T = \begin{pmatrix} a^2 + b^2 & ac + bd \\ ac + bd & c^2 + d^2 \end{pmatrix} \quad \text{and} \quad \mathbf{A}^T\mathbf{A} = \begin{pmatrix} a^2 + c^2 & ab + cd \\ ab + cd & b^2 + d^2 \end{pmatrix},$$

so we need to have $b^2 = c^2$. One such matrix is $\mathbf{A} = \begin{pmatrix} 1 & -1 \\ 1 & 1 \end{pmatrix}$. To construct a singular matrix that is RPN but not normal, try again to find an example with real numbers. For any orthogonal matrix $\mathbf{P}$ and nonsingular matrix $\mathbf{C}$, the matrix $\mathbf{A} = \mathbf{P} \begin{pmatrix} \mathbf{C} & \mathbf{0} \\ \mathbf{0} & \mathbf{0} \end{pmatrix} \mathbf{P}^T$ is RPN. To prevent $\mathbf{A}$ from being normal, simply choose $\mathbf{C}$ to be nonnormal. For example, let $\mathbf{C} = \begin{pmatrix} 1 & 2 \\ 3 & 4 \end{pmatrix}$ and $\mathbf{P} = \mathbf{I}$.

5.11.14. (a) $\mathbf{A}^*\mathbf{A} = \mathbf{A}\mathbf{A}^* \implies (\mathbf{A} - \lambda\mathbf{I})^* (\mathbf{A} - \lambda\mathbf{I}) = (\mathbf{A} - \lambda\mathbf{I}) (\mathbf{A} - \lambda\mathbf{I})^* \implies$
$(\mathbf{A} - \lambda\mathbf{I})$ is normal $\implies (\mathbf{A} - \lambda\mathbf{I})$ is RPN $\implies R(\mathbf{A} - \lambda\mathbf{I}) \perp N(\mathbf{A} - \lambda\mathbf{I})$.
(b) Suppose $\mathbf{x} \in N(\mathbf{A} - \lambda\mathbf{I})$ and $\mathbf{y} \in N(\mathbf{A} - \mu\mathbf{I})$, and use the fact that $N(\mathbf{A} - \lambda\mathbf{I}) = N(\mathbf{A} - \lambda\mathbf{I})^*$ to write

$$(\mathbf{A} - \lambda\mathbf{I})\mathbf{x} = \mathbf{0} \implies \mathbf{0} = \mathbf{x}^* (\mathbf{A} - \lambda\mathbf{I}) \implies \mathbf{0} = \mathbf{x}^* (\mathbf{A} - \lambda\mathbf{I})\mathbf{y}$$
$$= \mathbf{x}^*(\mu\mathbf{y} - \lambda\mathbf{y}) = \mathbf{x}^*\mathbf{y}(\mu - \lambda) \implies \mathbf{x}^*\mathbf{y} = 0.$$

Solutions for exercises in section 5. 12

5.12.1. Since $\mathbf{C}^T\mathbf{C} = \begin{pmatrix} 25 & 0 \\ 0 & 100 \end{pmatrix}$, $\sigma_1^2 = 100$, and it's clear that $\mathbf{x} = \mathbf{e}_2$ is a vector such that $(\mathbf{C}^T\mathbf{C} - 100\mathbf{I})\mathbf{x} = \mathbf{0}$ and $\|\mathbf{x}\|_2 = 1$. Let $\mathbf{y} = \mathbf{C}\mathbf{x}/\sigma_1 = \begin{pmatrix} -3/5 \\ -4/5 \end{pmatrix}$.
Following the procedure in Example 5.6.3, set $\mathbf{u}_x = \mathbf{x} - \mathbf{e}_1$ and $\mathbf{u}_y = \mathbf{y} - \mathbf{e}_1$, and construct

$$\mathbf{R}_x = \mathbf{I} - 2\frac{\mathbf{u}_x\mathbf{u}_x^T}{\mathbf{u}_x^T\mathbf{u}_x} = \begin{pmatrix} 0 & 1 \\ 1 & 0 \end{pmatrix} \quad \text{and} \quad \mathbf{R}_y = \mathbf{I} - 2\frac{\mathbf{u}_y\mathbf{u}_y^T}{\mathbf{u}_y^T\mathbf{u}_y} = \begin{pmatrix} -3/5 & -4/5 \\ -4/5 & 3/5 \end{pmatrix}.$$

Since $\mathbf{R}_y\mathbf{C}\mathbf{R}_x = \begin{pmatrix} 10 & 0 \\ 0 & 5 \end{pmatrix} = \mathbf{D}$, it follows that $\mathbf{C} = \mathbf{R}_y\mathbf{D}\mathbf{R}_x$ is a singular value decomposition of $\mathbf{C}$.

5.12.2. $\nu_1^2(\mathbf{A}) = \sigma_1^2 = \|\mathbf{A}\|_2^2$ needs no proof—it's just a restatement of (5.12.4). The fact that $\nu_r^2(\mathbf{A}) = \|\mathbf{A}\|_F^2$ amounts to observing that

$$\|\mathbf{A}\|_F^2 = trace\left(\mathbf{A}^T\mathbf{A}\right) = trace\mathbf{V}\begin{pmatrix} \mathbf{D}^2 & \mathbf{0} \\ \mathbf{0} & \mathbf{0} \end{pmatrix}\mathbf{V}^T = trace\left(\mathbf{D}^2\right) = \sigma_1^2 + \cdots + \sigma_r^2.$$

5.12.3. If $\sigma_1 \geq \cdots \geq \sigma_r$ are the nonzero singular values for $\mathbf{A}$, then it follows from Exercise 5.12.2 that $\|\mathbf{A}\|_2^2 = \sigma_1^2 \leq \sigma_1^2 + \sigma_2^2 + \cdots + \sigma_r^2 = \|\mathbf{A}\|_F^2 \leq n\sigma_1^2 = n\|\mathbf{A}\|_2^2$.

5.12.4. If $rank\,(\mathbf{A} + \mathbf{E}) = k < r$, then (5.12.10) implies that

$$\|\mathbf{E}\|_2 = \|\mathbf{A} - (\mathbf{A} + \mathbf{E})\|_2 \geq \min_{rank(\mathbf{B})=k} \|\mathbf{A} - \mathbf{B}\|_2 = \sigma_{k+1} \geq \sigma_r,$$

which is impossible. Hence $rank\,(\mathbf{A} + \mathbf{E}) \geq r = rank\,(\mathbf{A})$.

5.12.5. The argument is almost identical to that given for the nonsingular case except that $\mathbf{A}^\dagger$ replaces $\mathbf{A}^{-1}$. Start with SVDs

$$\mathbf{A} = \mathbf{U} \begin{pmatrix} \mathbf{D} & \mathbf{0} \\ \mathbf{0} & \mathbf{0} \end{pmatrix} \mathbf{V}^T \quad \text{and} \quad \mathbf{A}^\dagger = \mathbf{V} \begin{pmatrix} \mathbf{D}^{-1} & \mathbf{0} \\ \mathbf{0} & \mathbf{0} \end{pmatrix} \mathbf{U}^T,$$

where $\mathbf{D} = \mathrm{diag}\,(\sigma_1, \sigma_2, \ldots, \sigma_r)$, and note that $\|\mathbf{A}^\dagger \mathbf{A}\mathbf{x}\|_2 \leq \|\mathbf{A}^\dagger \mathbf{A}\|_2 \|\mathbf{x}\|_2 = 1$ with equality holding when $\mathbf{A}^\dagger \mathbf{A} = \mathbf{I}$ (i.e., when $r = n$). For each $\mathbf{y} \in \mathbf{A}(\mathcal{S}_2)$ there is an $\mathbf{x} \in \mathcal{S}_2$ such that $\mathbf{y} = \mathbf{A}\mathbf{x}$, so, with $\mathbf{w} = \mathbf{U}^T \mathbf{y}$,

$$1 \geq \|\mathbf{A}^\dagger \mathbf{A}\mathbf{x}\|_2^2 = \|\mathbf{A}^\dagger \mathbf{y}\|_2^2 = \|\mathbf{V}\mathbf{D}^{-1}\mathbf{U}^T \mathbf{y}\|_2^2 = \|\mathbf{D}^{-1}\mathbf{U}^T \mathbf{y}\|_2^2$$

$$= \|\mathbf{D}^{-1}\mathbf{w}\|_2^2 = \frac{w_1^2}{\sigma_1^2} + \frac{w_2^2}{\sigma_2^2} + \cdots + \frac{w_r^2}{\sigma_r^2}$$

with equality holding when $r = n$. In other words, the set $\mathbf{U}^T \mathbf{A}(\mathcal{S}_2)$ is an ellipsoid (degenerate if $r < n$) whose k^{th} semiaxis has length σ_k. To resolve the inequality with what it means for points to be *on* an ellipsoid, realize that the surface of a degenerate ellipsoid (one having some semiaxes with zero length) is actually the set of all points *in and on* a smaller dimension ellipsoid. For example, visualize an ellipsoid in $\Re^3$, and consider what happens as one of its semiaxes shrinks to zero. The skin of the three-dimensional ellipsoid degenerates to a solid planar ellipse. In other words, all points *on* a degenerate ellipsoid with semiaxes of length $\sigma_1 \neq 0$, $\sigma_2 \neq 0$, $\sigma_3 = 0$ are actually points *on and inside* a planar ellipse with semiaxes of length σ_1 and σ_2. Arguing that the kth semiaxis of $\mathbf{A}(\mathcal{S}_2)$ is $\sigma_k \mathbf{U}_{*k} = \mathbf{A}\mathbf{V}_{*k}$ is the same as the nonsingular case given in the text.

5.12.6. If $\mathbf{A} = \mathbf{U} \begin{pmatrix} \mathbf{D} & \mathbf{0} \\ \mathbf{0} & \mathbf{0} \end{pmatrix} \mathbf{V}^T$ and $\mathbf{A}_{n \times m}^\dagger = \mathbf{V} \begin{pmatrix} \mathbf{D}^{-1} & \mathbf{0} \\ \mathbf{0} & \mathbf{0} \end{pmatrix} \mathbf{U}^T$ are SVDs in which $\mathbf{V} = (\mathbf{V}_1 \,|\, \mathbf{V}_2)$, then the columns of $\mathbf{V}_1$ are an orthonormal basis for $R\,(\mathbf{A}^T)$, so $\mathbf{x} \in R\,(\mathbf{A}^T)$ and $\|\mathbf{x}\|_2 = 1$ if and only if $\mathbf{x} = \mathbf{V}_1 \mathbf{y}$ with $\|\mathbf{y}\|_2 = 1$. Since the 2-norm is unitarily invariant (Exercise 5.6.9),

$$\min_{\substack{\|\mathbf{x}\|_2=1 \\ \mathbf{x} \in R(\mathbf{A}^T)}} \|\mathbf{A}\mathbf{x}\|_2 = \min_{\|\mathbf{y}\|_2=1} \|\mathbf{A}\mathbf{V}_1\mathbf{y}\|_2 = \min_{\|\mathbf{y}\|_2=1} \|\mathbf{D}\mathbf{y}\|_2 = \frac{1}{\|\mathbf{D}^{-1}\|_2} = \sigma_r = \frac{1}{\|\mathbf{A}^\dagger\|_2}.$$

5.12.7. $\mathbf{x} = \mathbf{A}^{\dagger}\mathbf{b}$ and $\tilde{\mathbf{x}} = \mathbf{A}^{\dagger}(\mathbf{b} - \mathbf{e})$ are the respective solutions of minimal 2-norm of $\mathbf{Ax} = \mathbf{b}$ and $\mathbf{A\tilde{x}} = \tilde{\mathbf{b}} = \mathbf{b} - \mathbf{e}$. The development of the more general bound is the same as for (5.12.8).

$$\|\mathbf{x} - \tilde{\mathbf{x}}\| = \|\mathbf{A}^{\dagger}(\mathbf{b} - \tilde{\mathbf{b}})\| \leq \|\mathbf{A}^{\dagger}\| \, \|\mathbf{b} - \tilde{\mathbf{b}}\|,$$
$$\mathbf{b} = \mathbf{Ax} \implies \|\mathbf{b}\| \leq \|\mathbf{A}\| \, \|\mathbf{x}\| \implies 1/\|\mathbf{x}\| \leq \|\mathbf{A}\|/\|\mathbf{b}\|,$$

so

$$\frac{\|\mathbf{x} - \tilde{\mathbf{x}}\|}{\|\mathbf{x}\|} \leq \left(\|\mathbf{A}^{\dagger}\| \, \|\mathbf{b} - \tilde{\mathbf{b}}\|\right) \frac{\|\mathbf{A}\|}{\|\mathbf{b}\|} = \kappa \frac{\|\mathbf{e}\|}{\|\mathbf{b}\|}.$$

Similarly,

$$\|\mathbf{b} - \tilde{\mathbf{b}}\| = \|\mathbf{A}(\mathbf{x} - \tilde{\mathbf{x}})\| \leq \|\mathbf{A}\| \, \|\mathbf{x} - \tilde{\mathbf{x}}\|,$$
$$\mathbf{x} = \mathbf{A}^{\dagger}\mathbf{b} \implies \|\mathbf{x}\| \leq \|\mathbf{A}^{\dagger}\| \, \|\mathbf{b}\| \implies 1/\|\mathbf{b}\| \leq \|\mathbf{A}^{\dagger}\|/\|\mathbf{x}\|,$$

so

$$\frac{\|\mathbf{b} - \tilde{\mathbf{b}}\|}{\|\mathbf{b}\|} \leq (\|\mathbf{A}\| \, \|\mathbf{x} - \tilde{\mathbf{x}}\|) \frac{\|\mathbf{A}^{\dagger}\|}{\|\mathbf{x}\|} = \kappa \frac{\|\mathbf{x} - \tilde{\mathbf{x}}\|}{\|\mathbf{x}\|}.$$

Equality was attained in Example 5.12.1 by choosing $\mathbf{b}$ and $\mathbf{e}$ to point in special directions. But for these choices, $\mathbf{Ax} = \mathbf{b}$ and $\mathbf{A\tilde{x}} = \tilde{\mathbf{b}} = \mathbf{b} - \mathbf{e}$ cannot be guaranteed to be consistent for all singular or rectangular matrices $\mathbf{A}$, so the answer to the second part is "no." However, the argument of Example 5.12.1 proves equality for all $\mathbf{A}$ such that $\mathbf{AA}^{\dagger} = \mathbf{I}$ (i.e., when $rank\,(\mathbf{A}_{m \times n}) = m$).

5.12.8. If $\mathbf{A} = \mathbf{U} \begin{pmatrix} \mathbf{D} & \mathbf{0} \\ \mathbf{0} & \mathbf{0} \end{pmatrix} \mathbf{V}^{T}$ is an SVD, then $\mathbf{A}^{T}\mathbf{A} + \epsilon\mathbf{I} = \mathbf{U} \begin{pmatrix} \mathbf{D}^{2} + \epsilon\mathbf{I} & \mathbf{0} \\ \mathbf{0} & \epsilon\mathbf{I} \end{pmatrix} \mathbf{V}^{T}$ is an SVD with no zero singular values, so it's nonsingular. Furthermore,

$$(\mathbf{A}^{T}\mathbf{A} + \epsilon\mathbf{I})^{-1}\mathbf{A}^{T} = \mathbf{U} \begin{pmatrix} (\mathbf{D}^{2} + \epsilon\mathbf{I})^{-1}\mathbf{D} & \mathbf{0} \\ \mathbf{0} & \mathbf{0} \end{pmatrix} \mathbf{V}^{T} \to \mathbf{U} \begin{pmatrix} \mathbf{D}^{-1} & \mathbf{0} \\ \mathbf{0} & \mathbf{0} \end{pmatrix} \mathbf{V}^{T} = \mathbf{A}^{\dagger}.$$

5.12.9. Since $\mathbf{A}^{-1} = \begin{pmatrix} -266000 & 667000 \\ 333000 & -835000 \end{pmatrix}$, $\kappa_{\infty} = \|\mathbf{A}\|_{\infty} \, \|\mathbf{A}^{-1}\|_{\infty} = 1,754,336$. Similar to the 2-norm situation discussed in Example 5.12.1, the worst case is realized when $\mathbf{b}$ is in the direction of a maximal vector in $\mathbf{A}(\mathcal{S}_{\infty})$ while $\mathbf{e}$ is in the direction of a minimal vector in $\mathbf{A}(\mathcal{S}_{\infty})$. Sketch $\mathbf{A}(\mathcal{S}_{\infty})$ as shown below to see that $\mathbf{v} = (\,1.502 \quad .599\,)^{T}$ is a maximal vector in $\mathbf{A}(\mathcal{S}_{\infty})$.

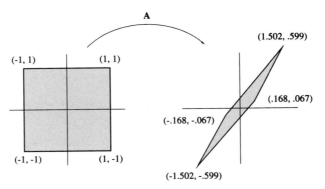

It's not clear which vector is minimal—don't assume $(.168 \quad .067)^T$ is. A minimal vector $\mathbf{y}$ in $\mathbf{A}(\mathcal{S}_\infty)$ satisfies $\|\mathbf{y}\|_\infty = \min_{\|\mathbf{x}\|_\infty = 1} \|\mathbf{Ax}\|_\infty = 1/\|\mathbf{A}^{-1}\|_\infty$ (see (5.2.6) on p. 280), so, for $\mathbf{y} = \mathbf{Ax}_0$ with $\|\mathbf{x}_0\|_\infty = 1$,

$$\left\| \mathbf{A}^{-1} \left(\frac{\mathbf{y}}{\|\mathbf{y}\|_\infty} \right) \right\|_\infty = \frac{\|\mathbf{x}_0\|_\infty}{\|\mathbf{y}\|_\infty} = \frac{1}{\|\mathbf{y}\|_\infty} = \|\mathbf{A}^{-1}\|_\infty = \max_{\|\mathbf{z}\|_\infty = 1} \|\mathbf{A}^{-1}\mathbf{z}\|_\infty .$$

In other words, $\hat{\mathbf{y}} = \mathbf{y}/\|\mathbf{y}\|_\infty$ must be a vector in $\mathcal{S}_\infty$ that receives maximal stretch under $\mathbf{A}^{-1}$. You don't have to look very hard to find such a vector because its components are ± 1—recall the proof of (5.2.15) on p. 283. Notice that $\hat{\mathbf{y}} = (1 \quad -1)^T \in \mathcal{S}_\infty$, and $\hat{\mathbf{y}}$ receives maximal stretch under $\mathbf{A}^{-1}$ because $\|\mathbf{A}^{-1}\mathbf{y}\|_\infty = 1,168,000 = \|\mathbf{A}^{-1}\|_\infty$, so setting

$$\mathbf{b} = \alpha \mathbf{v} = \alpha \begin{pmatrix} 1.502 \\ .599 \end{pmatrix} \quad \text{and} \quad \mathbf{e} = \beta \hat{\mathbf{y}} = \beta \begin{pmatrix} 1 \\ -1 \end{pmatrix}$$

produces equality in (5.12.8), regardless of α and β. You may wish to computationally verify that this is indeed the case.

5.12.10. (a) Consider $\mathbf{A} = \begin{pmatrix} \epsilon & -1 \\ 1 & 0 \end{pmatrix}$ or $\mathbf{A} = \begin{pmatrix} \epsilon & \epsilon^n \\ 0 & \epsilon \end{pmatrix}$ for small $\epsilon \neq 0$.

(b) For $\alpha > 1$, consider

$$\mathbf{A} = \begin{pmatrix} 1 & -\alpha & 0 & \cdots & 0 \\ 0 & 1 & -\alpha & \cdots & 0 \\ \vdots & \vdots & \ddots & \ddots & \vdots \\ 0 & 0 & \cdots & 1 & -\alpha \\ 0 & 0 & \cdots & 0 & 1 \end{pmatrix}_{n \times n} \quad \text{and} \quad \mathbf{A}^{-1} = \begin{pmatrix} 1 & \alpha & \cdots & \alpha^{n-2} & \alpha^{n-1} \\ 0 & 1 & \cdots & \alpha^{n-3} & \alpha^{n-2} \\ \vdots & \vdots & \ddots & \vdots & \vdots \\ 0 & 0 & \cdots & 1 & \alpha \\ 0 & 0 & \cdots & 0 & 1 \end{pmatrix} .$$

Regardless of which norm is used, $\|\mathbf{A}\| > \alpha$ and $\|\mathbf{A}^{-1}\| > \alpha^{n-1}$, so $\kappa > \alpha^n$ exhibits exponential growth. Even for moderate values of n and $\alpha > 1$, κ can be quite large.

5.12.11. For $\mathbf{B} = \mathbf{A}^{-1}\mathbf{E}$, write $(\mathbf{A} - \mathbf{E}) = \mathbf{A}(\mathbf{I} - \mathbf{B})$, and use the Neumann series expansion to obtain

$$\tilde{\mathbf{x}} = (\mathbf{A}-\mathbf{E})^{-1}\mathbf{b} = (\mathbf{I}-\mathbf{B})^{-1}\mathbf{A}^{-1}\mathbf{b} = (\mathbf{I}+\mathbf{B}+\mathbf{B}^2+\cdots)\mathbf{x} = \mathbf{x}+\mathbf{B}(\mathbf{I}+\mathbf{B}+\mathbf{B}^2+\cdots)\mathbf{x}.$$

Therefore, $\|\mathbf{x} - \tilde{\mathbf{x}}\| \le \|\mathbf{B}\| \sum_{n=0}^{\infty} \|\mathbf{B}\|^n \|\mathbf{x}\| \le \|\mathbf{A}^{-1}\| \|\mathbf{E}\| \|\mathbf{x}\| \sum_{n=0}^{\infty} \alpha^n$, so

$$\frac{\|\mathbf{x} - \tilde{\mathbf{x}}\|}{\|\mathbf{x}\|} \le \|\mathbf{A}^{-1}\| \|\mathbf{E}\| \frac{1}{1 - \alpha} = \|\mathbf{A}\| \|\mathbf{A}^{-1}\| \frac{\|\mathbf{E}\|}{\|\mathbf{A}\|} \frac{1}{1 - \alpha} = \frac{\kappa}{1 - \alpha} \frac{\|\mathbf{E}\|}{\|\mathbf{A}\|}.$$

5.12.12. Begin with

$$\mathbf{x} - \tilde{\mathbf{x}} = \mathbf{x} - (\mathbf{I} - \mathbf{B})^{-1}\mathbf{A}^{-1}(\mathbf{b} - \mathbf{e}) = \left(\mathbf{I} - (\mathbf{I} - \mathbf{B})^{-1}\right)\mathbf{x} + (\mathbf{I} - \mathbf{B})^{-1}\mathbf{A}^{-1}\mathbf{e}.$$

Use the triangle inequality with $\mathbf{b} = \mathbf{A}\mathbf{x} \Rightarrow 1/\|\mathbf{x}\| \le \|\mathbf{A}\|/\|\mathbf{b}\|$ to obtain

$$\frac{\|\mathbf{x} - \tilde{\mathbf{x}}\|}{\|\mathbf{x}\|} \le \|\mathbf{I} - (\mathbf{I} - \mathbf{B})^{-1}\| + \|(\mathbf{I} - \mathbf{B})^{-1}\| \kappa \frac{\|\mathbf{e}\|}{\|\mathbf{b}\|}.$$

Write $(\mathbf{I}-\mathbf{B})^{-1} = \sum_{i=0}^{\infty} \mathbf{B}^i$, and use the identity $\mathbf{I}-(\mathbf{I} - \mathbf{B})^{-1} = -\mathbf{B}(\mathbf{I} - \mathbf{B})^{-1}$ to produce

$$\|(\mathbf{I} - \mathbf{B})^{-1}\| \le \sum_{i=0}^{\infty} \|\mathbf{B}\|^i = \frac{1}{1 - \|\mathbf{B}\|} \quad \text{and} \quad \|\mathbf{I} - (\mathbf{I} - \mathbf{B})^{-1}\| \le \frac{\|\mathbf{B}\|}{1 - \|\mathbf{B}\|}.$$

Now combine everything above with $\|\mathbf{B}\| \le \|\mathbf{A}^{-1}\| \|\mathbf{E}\| = \kappa \|\mathbf{E}\|/\|\mathbf{A}\|$.

5.12.13. Even though the URV factors are not unique, $\mathbf{A}^\dagger$ is, so in each case you should arrive at the same matrix

$$\mathbf{A}^\dagger = \mathbf{V}\mathbf{R}^\dagger\mathbf{U}^T = \frac{1}{81}\begin{pmatrix} -4 & 2 & -4 \\ -18 & -18 & 9 \\ -4 & 2 & -4 \\ -2 & 1 & -2 \end{pmatrix}.$$

5.12.14. By (5.12.17), the minimum norm solution is $\mathbf{A}^\dagger\mathbf{b} = (1/9)\begin{pmatrix} 10 & 9 & 10 & 5 \end{pmatrix}^T$.

5.12.15. $\mathbf{U}$ is a unitary matrix in which the columns of $\mathbf{U}_1$ are an orthonormal basis for $R(\mathbf{A})$ and the columns of $\mathbf{U}_2$ are an orthonormal basis for $N(\mathbf{A}^T)$, so setting $\mathbf{X} = \mathbf{U}_1$, $\mathbf{Y} = \mathbf{U}_2$, and $[\mathbf{X}\,|\,\mathbf{Y}]^{-1} = \mathbf{U}^T$ in (5.9.12) produces $\mathbf{P} = \mathbf{U}_1\mathbf{U}_1^T$. Furthermore,

$$\mathbf{A}\mathbf{A}^\dagger = \mathbf{U}\begin{pmatrix} \mathbf{C} & \mathbf{0} \\ \mathbf{0} & \mathbf{0} \end{pmatrix}\mathbf{V}^T\mathbf{V}\begin{pmatrix} \mathbf{C}^{-1} & \mathbf{0} \\ \mathbf{0} & \mathbf{0} \end{pmatrix}\mathbf{U}^T = \mathbf{U}\begin{pmatrix} \mathbf{I} & \mathbf{0} \\ \mathbf{0} & \mathbf{0} \end{pmatrix}\mathbf{U}^T = \mathbf{U}_1\mathbf{U}_1^T.$$

According to (5.9.9), the projector onto $N(\mathbf{A}^T)$ along $R(\mathbf{A})$ is $\mathbf{I} - \mathbf{P} = \mathbf{I} - \mathbf{U}_1\mathbf{U}_1^T = \mathbf{U}_2\mathbf{U}_2^T = \mathbf{I} - \mathbf{A}\mathbf{A}^\dagger$.

5.12.16. (a)   When $\mathbf{A}$ is nonsingular, $\mathbf{U} = \mathbf{V} = \mathbf{I}$ and $\mathbf{R} = \mathbf{A}$, so $\mathbf{A}^\dagger = \mathbf{A}^{-1}$.

(b)   If $\mathbf{A} = \mathbf{U}\mathbf{R}\mathbf{V}^T$ is as given in (5.12.16), where $\mathbf{R} = \begin{pmatrix} \mathbf{C} & \mathbf{0} \\ \mathbf{0} & \mathbf{0} \end{pmatrix}$, it is clear that $(\mathbf{R}^\dagger)^\dagger = \mathbf{R}$, and hence $(\mathbf{A}^\dagger)^\dagger = (\mathbf{V}\mathbf{R}^\dagger\mathbf{U}^T)^\dagger = \mathbf{U}(\mathbf{R}^\dagger)^\dagger\mathbf{V}^T = \mathbf{U}\mathbf{R}\mathbf{V}^T = \mathbf{A}$.

(c)   For $\mathbf{R}$ as above, it is easy to see that $(\mathbf{R}^\dagger)^T = (\mathbf{R}^T)^\dagger$, so an argument similar to that used in part (b) leads to $(\mathbf{A}^\dagger)^T = (\mathbf{A}^T)^\dagger$.

(d)   When $rank\,(\mathbf{A}_{m\times n}) = n$, an SVD must have the form

$$\mathbf{A} = \mathbf{U}_{m\times m}\begin{pmatrix} \mathbf{D}_{n\times n} \\ \mathbf{0}_{m-n\times n} \end{pmatrix}\mathbf{I}_{n\times n}, \quad\text{so}\quad \mathbf{A}^\dagger = \mathbf{I}\,(\,\mathbf{D}^{-1} \quad \mathbf{0}\,)\,\mathbf{U}^T.$$

Furthermore, $\mathbf{A}^T\mathbf{A} = \mathbf{D}^2$, and $(\mathbf{A}^T\mathbf{A})^{-1}\mathbf{A}^T = \mathbf{I}\,(\,\mathbf{D}^{-1} \quad \mathbf{0}\,)\,\mathbf{U}^T = \mathbf{A}^\dagger$. The other part is similar.

(e) $\mathbf{A}^T\mathbf{A}\mathbf{A}^\dagger = \mathbf{V}\begin{pmatrix} \mathbf{C}^T & \mathbf{0} \\ \mathbf{0} & \mathbf{0} \end{pmatrix}\mathbf{U}^T\mathbf{U}\begin{pmatrix} \mathbf{C}_{r\times r} & \mathbf{0} \\ \mathbf{0} & \mathbf{0} \end{pmatrix}\mathbf{V}^T\mathbf{V}\begin{pmatrix} \mathbf{C}^{-1} & \mathbf{0} \\ \mathbf{0} & \mathbf{0} \end{pmatrix}\mathbf{U}^T = \mathbf{A}^T.$ The other part is similar.

(f)   Use an SVD to write

$$\mathbf{A}^T(\mathbf{A}\mathbf{A}^T)^\dagger = \mathbf{V}\begin{pmatrix} \mathbf{D}^T & \mathbf{0} \\ \mathbf{0} & \mathbf{0} \end{pmatrix}\mathbf{U}^T\mathbf{U}\begin{pmatrix} \mathbf{D}^{-2} & \mathbf{0} \\ \mathbf{0} & \mathbf{0} \end{pmatrix}\mathbf{U}^T = \mathbf{V}\begin{pmatrix} \mathbf{D}^{-1} & \mathbf{0} \\ \mathbf{0} & \mathbf{0} \end{pmatrix}\mathbf{U}^T = \mathbf{A}^\dagger.$$

The other part is similar.

(g)   The URV factorization insures that $rank\,(\mathbf{A}^\dagger) = rank\,(\mathbf{A}) = rank\,(\mathbf{A}^T)$, and part (f) implies $R\,(\mathbf{A}^\dagger) \subseteq R\,(\mathbf{A}^T)$, so $R\,(\mathbf{A}^\dagger) = R\,(\mathbf{A}^T)$. Argue that $R\,(\mathbf{A}^T) = R\,(\mathbf{A}^\dagger\mathbf{A})$ by using Exercise 5.12.15. The other parts are similar.

(h)   If $\mathbf{A} = \mathbf{U}\mathbf{R}\mathbf{V}^T$ is a URV factorization for $\mathbf{A}$, then $(\mathbf{P}\mathbf{U})\mathbf{R}(\mathbf{Q}^T\mathbf{V})^T$ is a URV factorization for $\mathbf{B} = \mathbf{P}\mathbf{A}\mathbf{Q}$. So, by (5.12.16), we have

$$\mathbf{B}^\dagger = \mathbf{Q}^T\mathbf{V}\begin{pmatrix} \mathbf{C}^{-1} & \mathbf{0} \\ \mathbf{0} & \mathbf{0} \end{pmatrix}\mathbf{U}^T\mathbf{P}^T = \mathbf{Q}^T\mathbf{A}^\dagger\mathbf{P}^T.$$

Almost any two singular or rectangular matrices can be used to build a counterexample to show that $(\mathbf{A}\mathbf{B})^\dagger$ is not always the same as $\mathbf{B}^\dagger\mathbf{A}^\dagger$.

(i)   If $\mathbf{A} = \mathbf{U}\mathbf{R}\mathbf{V}^T$, then $(\mathbf{A}^T\mathbf{A})^\dagger = (\mathbf{V}\mathbf{R}^T\mathbf{U}^T\mathbf{U}\mathbf{R}\mathbf{V})^\dagger = \mathbf{V}^T(\mathbf{R}^T\mathbf{R})^\dagger\mathbf{V}^T$. Similarly, $\mathbf{A}^\dagger(\mathbf{A}^T)^\dagger = \mathbf{V}\mathbf{R}^\dagger\mathbf{U}^T\mathbf{U}\mathbf{R}^{T\dagger}\mathbf{V}^T = \mathbf{V}\mathbf{R}^\dagger\mathbf{R}^{T\dagger}\mathbf{V}^T = \mathbf{V}^T(\mathbf{R}^T\mathbf{R})^\dagger\mathbf{V}^T$. The other part is argued in the same way.

5.12.17. If $\mathbf{A}$ is RPN, then $index(\mathbf{A}) = 1$, and the URV decomposition (5.11.15) is a similarity transformation of the kind (5.10.5). That is, $\mathbf{N} = \mathbf{0}$ and $\mathbf{Q} = \mathbf{U}$, so $\mathbf{A}^D$ as defined in (5.10.6) is the same as $\mathbf{A}^\dagger$ as defined by (5.12.16). Conversely, if $\mathbf{A}^\dagger = \mathbf{A}^D$, then

$$\mathbf{A}\mathbf{A}^D = \mathbf{A}^D\mathbf{A} \implies \mathbf{A}^\dagger\mathbf{A} = \mathbf{A}\mathbf{A}^\dagger \implies R\,(\mathbf{A}) = R\,(\mathbf{A}^T).$$

5.12.18. (a)   Recall that $\|\mathbf{B}\|_F^2 = trace\left(\mathbf{B}^T\mathbf{B}\right)$, and use the fact that $R\left(\mathbf{X}\right) \perp R\left(\mathbf{Y}\right)$ implies $\mathbf{X}^T\mathbf{Y} = \mathbf{0} = \mathbf{Y}^T\mathbf{X}$ to write

$$\|\mathbf{X} + \mathbf{Y}\|_F^2 = trace\left((\mathbf{X} + \mathbf{Y})^T(\mathbf{X} + \mathbf{Y})\right)$$
$$= trace\left(\mathbf{X}^T\mathbf{X} + \mathbf{X}^T\mathbf{Y} + \mathbf{Y}^T\mathbf{X} + \mathbf{Y}^T\mathbf{Y}\right)$$
$$= trace\left(\mathbf{X}^T\mathbf{X}\right) + trace\left(\mathbf{Y}^T\mathbf{Y}\right) = \|\mathbf{X}\|_F^2 + \|\mathbf{Y}\|_F^2\,.$$

(b)   Consider $\mathbf{X} = \begin{pmatrix} 2 & 0 \\ 0 & 0 \end{pmatrix}$ and $\mathbf{Y} = \begin{pmatrix} 0 & 0 \\ 0 & 3 \end{pmatrix}$.

(c)   Use the result of part (a) to write

$$\|\mathbf{I} - \mathbf{A}\mathbf{X}\|_F^2 = \left\|\mathbf{I} - \mathbf{A}\mathbf{A}^\dagger + \mathbf{A}\mathbf{A}^\dagger - \mathbf{A}\mathbf{X}\right\|_F^2$$
$$= \left\|\mathbf{I} - \mathbf{A}\mathbf{A}^\dagger\right\|_F^2 + \left\|\mathbf{A}\mathbf{A}^\dagger - \mathbf{A}\mathbf{X}\right\|_F^2$$
$$\geq \left\|\mathbf{I} - \mathbf{A}\mathbf{A}^\dagger\right\|_F^2,$$

with equality holding if and only if $\mathbf{A}\mathbf{X} = \mathbf{A}\mathbf{A}^\dagger$—i.e., if and only if $\mathbf{X} = \mathbf{A}^\dagger + \mathbf{Z}$, where $R\left(\mathbf{Z}\right) \subseteq N\left(\mathbf{A}\right) \perp R\left(\mathbf{A}^T\right) = R\left(\mathbf{A}^\dagger\right)$. Moreover, for any such $\mathbf{X}$,

$$\|\mathbf{X}\|_F^2 = \left\|\mathbf{A}^\dagger + \mathbf{Z}\right\|_F^2 = \left\|\mathbf{A}^\dagger\right\|_F^2 + \|\mathbf{Z}\|_F^2 \geq \left\|\mathbf{A}^\dagger\right\|_F^2$$

with equality holding if and only if $\mathbf{Z} = \mathbf{0}$.

Solutions for exercises in section 5. 13

5.13.1. $\mathbf{P}_\mathcal{M} = \mathbf{u}\mathbf{u}^T/(\mathbf{u}^T\mathbf{u}) = (1/10)\begin{pmatrix} 9 & 3 \\ 3 & 1 \end{pmatrix}$, and $\mathbf{P}_{\mathcal{M}^\perp} = \mathbf{I} - \mathbf{P}_\mathcal{M} = (1/10)\begin{pmatrix} 1 & -3 \\ -3 & 9 \end{pmatrix}$, so $\mathbf{P}_\mathcal{M}\mathbf{b} = \begin{pmatrix} 6 \\ 2 \end{pmatrix}$, and $\mathbf{P}_{\mathcal{M}^\perp}\mathbf{b} = \begin{pmatrix} -2 \\ 6 \end{pmatrix}$.

5.13.2. (a)   Use any of the techniques described in Example 5.13.3 to obtain the following.

$$\mathbf{P}_{R(\mathbf{A})} = \begin{pmatrix} .5 & 0 & .5 \\ 0 & 1 & 0 \\ .5 & 0 & .5 \end{pmatrix} \qquad \mathbf{P}_{N(\mathbf{A})} = \begin{pmatrix} .8 & -.4 & 0 \\ -.4 & .2 & 0 \\ 0 & 0 & 0 \end{pmatrix}$$

$$\mathbf{P}_{R(\mathbf{A}^T)} = \begin{pmatrix} .2 & .4 & 0 \\ .4 & .8 & 0 \\ 0 & 0 & 1 \end{pmatrix} \qquad \mathbf{P}_{N(\mathbf{A}^T)} = \begin{pmatrix} .5 & 0 & -.5 \\ 0 & 0 & 0 \\ -.5 & 0 & .5 \end{pmatrix}$$

(b)   The point in $N\left(\mathbf{A}\right)^\perp$ that is closest to $\mathbf{b}$ is

$$\mathbf{P}_{N(\mathbf{A})^\perp}\mathbf{b} = \mathbf{P}_{R(\mathbf{A}^T)}\mathbf{b} = \begin{pmatrix} .6 \\ 1.2 \\ 1 \end{pmatrix}.$$

5.13.3. If $\mathbf{x} \in R(\mathbf{P})$, then $\mathbf{Px} = \mathbf{x}$—recall (5.9.10)—so $\|\mathbf{Px}\|_2 = \|\mathbf{x}\|_2$. Conversely, suppose $\|\mathbf{Px}\|_2 = \|\mathbf{x}\|_2$, and let $\mathbf{x} = \mathbf{m} + \mathbf{n}$, where $\mathbf{m} \in R(\mathbf{P})$ and $\mathbf{n} \in N(\mathbf{P})$ so that $\mathbf{m} \perp \mathbf{n}$. The Pythagorean theorem (Exercise 5.4.14) guarantees that $\|\mathbf{x}\|_2^2 = \|\mathbf{m} + \mathbf{n}\|_2^2 = \|\mathbf{m}\|_2^2 + \|\mathbf{n}\|_2^2$. But we also have

$$\|\mathbf{x}\|_2^2 = \|\mathbf{Px}\|_2^2 = \|\mathbf{P}(\mathbf{m} + \mathbf{n})\|_2^2 = \|\mathbf{Pm}\|_2^2 = \|\mathbf{m}\|_2^2.$$

Therefore, $\mathbf{n} = \mathbf{0}$, and thus $\mathbf{x} = \mathbf{m} \in R(\mathbf{P})$.

5.13.4. $(\mathbf{A}^T \mathbf{P}_{R(\mathbf{A})})^T = \mathbf{P}_{R(\mathbf{A})}^T \mathbf{A} = \mathbf{P}_{R(\mathbf{A})}\mathbf{A} = \mathbf{A}$.

5.13.5. Equation (5.13.4) says that $\mathbf{P}_{\mathcal{M}} = \mathbf{U}\mathbf{U}^T = \sum_{i=1}^{r} \mathbf{u}_i \mathbf{u}_i{}^T$, where $\mathbf{U}$ contains the $\mathbf{u}_i$'s as columns.

5.13.6. The Householder (or Givens) reduction technique can be employed as described in Example 5.11.2 on p. 407 to compute orthogonal matrices $\mathbf{U} = (\mathbf{U}_1 \,|\, \mathbf{U}_2)$ and $\mathbf{V} = (\mathbf{V}_1 \,|\, \mathbf{V}_2)$, which are factors in a URV factorization of $\mathbf{A}$. Equation (5.13.12) insures that

$$\mathbf{P}_{R(\mathbf{A})} = \mathbf{U}_1 \mathbf{U}_1^T, \qquad \mathbf{P}_{N(\mathbf{A}^T)} = \mathbf{P}_{R(\mathbf{A})^\perp} = \mathbf{I} - \mathbf{U}_1 \mathbf{U}_1^T = \mathbf{U}_2 \mathbf{U}_2^T,$$
$$\mathbf{P}_{R(\mathbf{A}^T)} = \mathbf{V}_1 \mathbf{V}_1^T, \qquad \mathbf{P}_{N(\mathbf{A})} = \mathbf{P}_{R(\mathbf{A}^T)^\perp} = \mathbf{I} - \mathbf{V}_1 \mathbf{V}_1^T = \mathbf{V}_2 \mathbf{V}_2^T.$$

5.13.7. (a) The only nonsingular orthogonal projector (i.e., the only nonsingular symmetric idempotent matrix) is the identity matrix. Consequently, for all other orthogonal projectors $\mathbf{P}$, we must have $rank(\mathbf{P}) = 0$ or $rank(\mathbf{P}) = 1$, so $\mathbf{P} = \mathbf{0}$ or, by Example 5.13.1, $\mathbf{P} = (\mathbf{uu}^T)/\mathbf{u}^T\mathbf{u}$. In other words, the 2×2 orthogonal projectors are $\mathbf{P} = \mathbf{I}$, $\mathbf{P} = \mathbf{0}$, and, for a nonzero vector $\mathbf{u}^T = (\,\alpha \quad \beta\,)$,

$$\mathbf{P} = \frac{\mathbf{uu}^T}{\mathbf{u}^T\mathbf{u}} = \frac{1}{\alpha^2 + \beta^2} \begin{pmatrix} \alpha^2 & \alpha\beta \\ \alpha\beta & \beta^2 \end{pmatrix}.$$

(b) $\mathbf{P} = \mathbf{I}$, $\mathbf{P} = \mathbf{0}$, and, for nonzero vectors $\mathbf{u}, \mathbf{v} \in \Re^{2 \times 1}$, $\mathbf{P} = (\mathbf{uv}^T)/\mathbf{u}^T\mathbf{v}$.

5.13.8. If either $\mathbf{u}$ or $\mathbf{v}$ is the zero vector, then $\mathcal{L}$ is a one-dimensional subspace, and the solution is given in Example 5.13.1. Suppose that neither $\mathbf{u}$ nor $\mathbf{v}$ is the zero vector, and let $\mathbf{p}$ be the orthogonal projection of $\mathbf{b}$ onto $\mathcal{L}$. Since $\mathcal{L}$ is the translate of the subspace $span\{\mathbf{u} - \mathbf{v}\}$, subtracting $\mathbf{u}$ from everything moves the situation back to the origin—the following picture illustrates this in $\Re^2$.

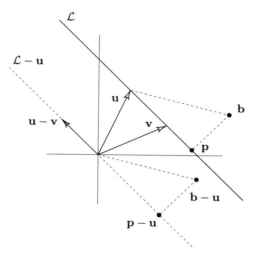

In other words, $\mathcal{L}$ is translated back down to $span\{\mathbf{u} - \mathbf{v}\}$, $\mathbf{b} \to \mathbf{b} - \mathbf{u}$, and $\mathbf{p} \to \mathbf{p} - \mathbf{u}$, so that $\mathbf{p} - \mathbf{u}$ must be the orthogonal projection of $\mathbf{b} - \mathbf{u}$ onto $span\{\mathbf{u} - \mathbf{v}\}$. Example 5.13.1 says that

$$\mathbf{p} - \mathbf{u} = \mathbf{P}_{span\{\mathbf{u}-\mathbf{v}\}}(\mathbf{b} - \mathbf{u}) = \frac{(\mathbf{u} - \mathbf{v})(\mathbf{u} - \mathbf{v})^T}{(\mathbf{u} - \mathbf{v})^T(\mathbf{u} - \mathbf{v})}(\mathbf{b} - \mathbf{u}),$$

and thus

$$\mathbf{p} = \mathbf{u} + \left[\frac{(\mathbf{u} - \mathbf{v})^T(\mathbf{b} - \mathbf{u})}{(\mathbf{u} - \mathbf{v})^T(\mathbf{u} - \mathbf{v})} \right](\mathbf{u} - \mathbf{v}).$$

5.13.9. $\left\| \mathbf{A}\hat{\mathbf{x}} - \mathbf{b} \right\|_2 = \left\| \mathbf{P}_{R(\mathbf{A})}\mathbf{b} - \mathbf{b} \right\|_2 = \left\| (\mathbf{I} - \mathbf{P}_{R(\mathbf{A})})\mathbf{b} \right\|_2 = \left\| \mathbf{P}_{N(\mathbf{A}^T)}\mathbf{b} \right\|_2$

5.13.10. Use (5.13.17) with $\mathbf{P}_{R(\mathbf{A})} = \mathbf{P}_{R(\mathbf{A})}^T = \mathbf{P}_{R(\mathbf{A})}^2$, to write

$$\begin{aligned}
\|\varepsilon\|_2^2 &= (\mathbf{b} - \mathbf{P}_{R(\mathbf{A})}\mathbf{b})^T(\mathbf{b} - \mathbf{P}_{R(\mathbf{A})}\mathbf{b}) \\
&= \mathbf{b}^T\mathbf{b} - \mathbf{b}^T\mathbf{P}_{R(\mathbf{A})}^T\mathbf{b} - \mathbf{b}^T\mathbf{P}_{R(\mathbf{A})}\mathbf{b} + \mathbf{b}^T\mathbf{P}_{R(\mathbf{A})}^T\mathbf{P}_{R(\mathbf{A})}\mathbf{b} \\
&= \mathbf{b}^T\mathbf{b} - \mathbf{b}^T\mathbf{P}_{R(\mathbf{A})}\mathbf{b} = \|\mathbf{b}\|_2^2 - \left\| \mathbf{P}_{R(\mathbf{A})}\mathbf{b} \right\|_2^2.
\end{aligned}$$

5.13.11. According to (5.13.13) we must show that $\sum_{i=1}^{r}(\mathbf{u}_i^T\mathbf{x})\mathbf{u}_i = \mathbf{P}_{\mathcal{M}}\mathbf{x}$. It follows from (5.13.4) that if $\mathbf{U}_{n\times r}$ is the matrix containing the vectors in $\mathcal{B}$ as columns, then

$$\mathbf{P}_{\mathcal{M}} = \mathbf{U}\mathbf{U}^T = \sum_{i=1}^{r}\mathbf{u}_i\mathbf{u}_i^T \implies \mathbf{P}_{\mathcal{M}}\mathbf{x} = \sum_{i=1}^{r}\mathbf{u}_i\mathbf{u}_i^T\mathbf{x} = \sum_{i=1}^{r}(\mathbf{u}_i^T\mathbf{x})\mathbf{u}_i.$$

5.13.12. Yes, the given spanning set $\{\mathbf{u}_1, \mathbf{u}_2, \mathbf{u}_3\}$ is an orthonormal basis for $\mathcal{M}$, so, by Exercise 5.13.11,

$$\mathbf{P}_{\mathcal{M}}\mathbf{b} = \sum_{i=1}^{3}(\mathbf{u}_i^T\mathbf{b})\mathbf{u}_i = \mathbf{u}_1 + 3\mathbf{u}_2 + 7\mathbf{u}_3 = \begin{pmatrix} 5 \\ 0 \\ 5 \\ 3 \end{pmatrix}.$$

5.13.13. (a)   Combine the fact that $\mathbf{P}_{\mathcal{M}}\mathbf{P}_{\mathcal{N}} = \mathbf{0}$ if and only if $R(\mathbf{P}_{\mathcal{N}}) \subseteq N(\mathbf{P}_{\mathcal{M}})$ with the facts $R(\mathbf{P}_{\mathcal{N}}) = \mathcal{N}$ and $N(\mathbf{P}_{\mathcal{M}}) = \mathcal{M}^{\perp}$ to write

$$\mathbf{P}_{\mathcal{M}}\mathbf{P}_{\mathcal{N}} = \mathbf{0} \iff \mathcal{N} \subseteq \mathcal{M}^{\perp} \iff \mathcal{N} \perp \mathcal{M}.$$

(b)   Yes—this is a direct consequence of part (a). Alternately, you could say

$$\mathbf{0} = \mathbf{P}_{\mathcal{M}}\mathbf{P}_{\mathcal{N}} \iff \mathbf{0} = (\mathbf{P}_{\mathcal{M}}\mathbf{P}_{\mathcal{N}})^{T} = \mathbf{P}_{\mathcal{N}}^{T}\mathbf{P}_{\mathcal{M}}^{T} = \mathbf{P}_{\mathcal{N}}\mathbf{P}_{\mathcal{M}}.$$

5.13.14. (a)   Use Exercise 4.2.9 along with (4.5.5) to write

$$R(\mathbf{P}_{\mathcal{M}}) + R(\mathbf{P}_{\mathcal{N}}) = R(\mathbf{P}_{\mathcal{M}} \mid \mathbf{P}_{\mathcal{N}}) = R\left((\mathbf{P}_{\mathcal{M}} \mid \mathbf{P}_{\mathcal{N}})\begin{pmatrix}\mathbf{P}_{\mathcal{M}}\\\mathbf{P}_{\mathcal{N}}\end{pmatrix}^{T}\right)$$

$$= R\left(\mathbf{P}_{\mathcal{M}}\mathbf{P}_{\mathcal{M}}^{T} + \mathbf{P}_{\mathcal{N}}\mathbf{P}_{\mathcal{N}}^{T}\right) = R\left(\mathbf{P}_{\mathcal{M}}^{2} + \mathbf{P}_{\mathcal{N}}^{2}\right)$$

$$= R(\mathbf{P}_{\mathcal{M}} + \mathbf{P}_{\mathcal{N}}).$$

(b) $\mathbf{P}_{\mathcal{M}}\mathbf{P}_{\mathcal{N}} = \mathbf{0} \iff R(\mathbf{P}_{\mathcal{N}}) \subseteq N(\mathbf{P}_{\mathcal{M}}) \iff \mathcal{N} \subseteq \mathcal{M}^{\perp} \iff \mathcal{M} \perp \mathcal{N}.$

(c)   Exercise 5.9.17 says $\mathbf{P}_{\mathcal{M}} + \mathbf{P}_{\mathcal{N}}$ is idempotent if and only if $\mathbf{P}_{\mathcal{M}}\mathbf{P}_{\mathcal{N}} = \mathbf{0} = \mathbf{P}_{\mathcal{N}}\mathbf{P}_{\mathcal{M}}$. Because $\mathbf{P}_{\mathcal{M}}$ and $\mathbf{P}_{\mathcal{N}}$ are symmetric, $\mathbf{P}_{\mathcal{M}}\mathbf{P}_{\mathcal{N}} = \mathbf{0}$ if and only if $\mathbf{P}_{\mathcal{M}}\mathbf{P}_{\mathcal{N}} = \mathbf{P}_{\mathcal{N}}\mathbf{P}_{\mathcal{M}} = \mathbf{0}$ (via the reverse order law for transposition). The fact that $R(\mathbf{P}_{\mathcal{M}} + \mathbf{P}_{\mathcal{N}}) = R(\mathbf{P}_{\mathcal{M}}) \oplus R(\mathbf{P}_{\mathcal{N}}) = \mathcal{M} \oplus \mathcal{N}$ was established in Exercise 5.9.17, and $\mathcal{M} \perp \mathcal{N}$ follows from part (b).

5.13.15.   First notice that $\mathbf{P}_{\mathcal{M}} + \mathbf{P}_{\mathcal{N}}$ is symmetric, so (5.13.12) and the result of Exercise 5.13.14, part (a), can be combined to conclude that

$$(\mathbf{P}_{\mathcal{M}} + \mathbf{P}_{\mathcal{N}})(\mathbf{P}_{\mathcal{M}} + \mathbf{P}_{\mathcal{N}})^{\dagger} = (\mathbf{P}_{\mathcal{M}} + \mathbf{P}_{\mathcal{N}})^{\dagger}(\mathbf{P}_{\mathcal{M}} + \mathbf{P}_{\mathcal{N}}) = \mathbf{P}_{R(\mathbf{P}_{\mathcal{M}}+\mathbf{P}_{\mathcal{N}})} = \mathbf{P}_{\mathcal{M}+\mathcal{N}}.$$

Now, $\mathcal{M} \subseteq \mathcal{M} + \mathcal{N}$ implies $\mathbf{P}_{\mathcal{M}+\mathcal{N}}\mathbf{P}_{\mathcal{M}} = \mathbf{P}_{\mathcal{M}}$, and the reverse order law for transposition yields $\mathbf{P}_{\mathcal{M}}\mathbf{P}_{\mathcal{M}+\mathcal{N}} = \mathbf{P}_{\mathcal{M}}$ so that $\mathbf{P}_{\mathcal{M}+\mathcal{N}}\mathbf{P}_{\mathcal{M}} = \mathbf{P}_{\mathcal{M}}\mathbf{P}_{\mathcal{M}+\mathcal{N}}$. In other words, $(\mathbf{P}_{\mathcal{M}} + \mathbf{P}_{\mathcal{N}})(\mathbf{P}_{\mathcal{M}} + \mathbf{P}_{\mathcal{N}})^{\dagger}\mathbf{P}_{\mathcal{M}} = \mathbf{P}_{\mathcal{M}}(\mathbf{P}_{\mathcal{M}} + \mathbf{P}_{\mathcal{N}})^{\dagger}(\mathbf{P}_{\mathcal{M}} + \mathbf{P}_{\mathcal{N}})$, or

$$\mathbf{P}_{\mathcal{M}}(\mathbf{P}_{\mathcal{M}} + \mathbf{P}_{\mathcal{N}})^{\dagger}\mathbf{P}_{\mathcal{M}} + \mathbf{P}_{\mathcal{N}}(\mathbf{P}_{\mathcal{M}} + \mathbf{P}_{\mathcal{N}})^{\dagger}\mathbf{P}_{\mathcal{M}}$$
$$= \mathbf{P}_{\mathcal{M}}(\mathbf{P}_{\mathcal{M}} + \mathbf{P}_{\mathcal{N}})^{\dagger}\mathbf{P}_{\mathcal{M}} + \mathbf{P}_{\mathcal{M}}(\mathbf{P}_{\mathcal{M}} + \mathbf{P}_{\mathcal{N}})^{\dagger}\mathbf{P}_{\mathcal{N}}.$$

Subtracting $\mathbf{P}_{\mathcal{M}}(\mathbf{P}_{\mathcal{M}} + \mathbf{P}_{\mathcal{N}})^{\dagger}\mathbf{P}_{\mathcal{M}}$ from both sides of this equation produces

$$\mathbf{P}_{\mathcal{M}}(\mathbf{P}_{\mathcal{M}} + \mathbf{P}_{\mathcal{N}})^{\dagger}\mathbf{P}_{\mathcal{N}} = \mathbf{P}_{\mathcal{N}}(\mathbf{P}_{\mathcal{M}} + \mathbf{P}_{\mathcal{N}})^{\dagger}\mathbf{P}_{\mathcal{M}}.$$

Let $\mathbf{Z} = 2\mathbf{P}_{\mathcal{M}}(\mathbf{P}_{\mathcal{M}}+\mathbf{P}_{\mathcal{N}})^{\dagger}\mathbf{P}_{\mathcal{N}} = 2\mathbf{P}_{\mathcal{N}}(\mathbf{P}_{\mathcal{M}}+\mathbf{P}_{\mathcal{N}})^{\dagger}\mathbf{P}_{\mathcal{M}}$, and notice that $R(\mathbf{Z}) \subseteq R(\mathbf{P}_{\mathcal{M}}) = \mathcal{M}$ and $R(\mathbf{Z}) \subseteq R(\mathbf{P}_{\mathcal{N}}) = \mathcal{N}$ implies $R(\mathbf{Z}) \subseteq \mathcal{M} \cap \mathcal{N}$. Furthermore, $\mathbf{P}_{\mathcal{M}}\mathbf{P}_{\mathcal{M}\cap\mathcal{N}} = \mathbf{P}_{\mathcal{M}\cap\mathcal{N}} = \mathbf{P}_{\mathcal{N}}\mathbf{P}_{\mathcal{M}\cap\mathcal{N}}$, and $\mathbf{P}_{\mathcal{M}+\mathcal{N}}\mathbf{P}_{\mathcal{M}\cap\mathcal{N}} = \mathbf{P}_{\mathcal{M}\cap\mathcal{N}}$, so, by the reverse order law for transposition, $\mathbf{P}_{\mathcal{M} \cap \mathcal{N}} \mathbf{P}_{\mathcal{M}} = \mathbf{P}_{\mathcal{M} \cap \mathcal{N}} = \mathbf{P}_{\mathcal{M} \cap \mathcal{N}} \mathbf{P}_{\mathcal{N}}$ and $\mathbf{P}_{\mathcal{M} \cap \mathcal{N}} \mathbf{P}_{\mathcal{M} + \mathcal{N}} = \mathbf{P}_{\mathcal{M} \cap \mathcal{N}}$. Consequently,

$$
\begin{aligned}
\mathbf{Z} = \mathbf{P}_{\mathcal{M} \cap \mathcal{N}} \mathbf{Z} &= \mathbf{P}_{\mathcal{M} \cap \mathcal{N}} \left[\mathbf{P}_{\mathcal{M}} (\mathbf{P}_{\mathcal{M}} + \mathbf{P}_{\mathcal{N}})^{\dagger} \mathbf{P}_{\mathcal{N}} + \mathbf{P}_{\mathcal{N}} (\mathbf{P}_{\mathcal{M}} + \mathbf{P}_{\mathcal{N}})^{\dagger} \mathbf{P}_{\mathcal{M}} \right] \\
&= \mathbf{P}_{\mathcal{M} \cap \mathcal{N}} (\mathbf{P}_{\mathcal{M}} + \mathbf{P}_{\mathcal{N}})^{\dagger} (\mathbf{P}_{\mathcal{M}} + \mathbf{P}_{\mathcal{N}}) = \mathbf{P}_{\mathcal{M} \cap \mathcal{N}} \mathbf{P}_{\mathcal{M} + \mathcal{N}} = \mathbf{P}_{\mathcal{M} \cap \mathcal{N}}.
\end{aligned}
$$

5.13.16. (a) Use the fact that $\mathbf{A}^{T} = \mathbf{A}^{T} \mathbf{P}_{R(\mathbf{A})} = \mathbf{A}^{T} \mathbf{A} \mathbf{A}^{\dagger}$ (see Exercise 5.13.4) to write

$$
\begin{aligned}
\int_{0}^{\infty} e^{-\mathbf{A}^{T} \mathbf{A} t} \mathbf{A}^{T} dt = \int_{0}^{\infty} e^{-\mathbf{A}^{T} \mathbf{A} t} \mathbf{A}^{T} \mathbf{A} \mathbf{A}^{\dagger} dt &= \left(\int_{0}^{\infty} e^{-\mathbf{A}^{T} \mathbf{A} t} \mathbf{A}^{T} \mathbf{A} dt \right) \mathbf{A}^{\dagger} \\
&= \left[-e^{-\mathbf{A}^{T} \mathbf{A} t} \right]_{0}^{\infty} \mathbf{A}^{\dagger} = [\mathbf{0} - (-\mathbf{I})] \mathbf{A}^{\dagger} = \mathbf{A}^{\dagger}.
\end{aligned}
$$

(b) Recall from Example 5.10.5 that $\mathbf{A}^{k} = \mathbf{A}^{k+1} \mathbf{A}^{D} = \mathbf{A}^{k} \mathbf{A} \mathbf{A}^{D}$, and write

$$
\begin{aligned}
\int_{0}^{\infty} e^{-\mathbf{A}^{k+1} t} \mathbf{A}^{k} dt = \int_{0}^{\infty} e^{-\mathbf{A}^{k+1} t} \mathbf{A}^{k} \mathbf{A} \mathbf{A}^{D} dt &= \left(\int_{0}^{\infty} e^{-\mathbf{A}^{k+1} t} \mathbf{A}^{k+1} dt \right) \mathbf{A}^{D} \\
&= \left[-e^{-\mathbf{A}^{k+1} t} \right]_{0}^{\infty} \mathbf{A}^{D} = [\mathbf{0} - (-\mathbf{I})] \mathbf{A}^{D} = \mathbf{A}^{D}.
\end{aligned}
$$

(c) This is just a special case of the formula in part (b) with $k = 0$. However, it is easy to derive the formula directly by writing

$$
\begin{aligned}
\int_{0}^{\infty} e^{-\mathbf{A} t} dt = \int_{0}^{\infty} e^{-\mathbf{A} t} \mathbf{A} \mathbf{A}^{-1} dt &= \left(\int_{0}^{\infty} e^{-\mathbf{A} t} \mathbf{A} dt \right) \mathbf{A}^{-1} \\
&= \left[e^{-\mathbf{A} t} \right]_{0}^{\infty} \mathbf{A}^{-1} = [\mathbf{0} - (-\mathbf{I})] \mathbf{A}^{-1} = \mathbf{A}^{-1}.
\end{aligned}
$$

5.13.17. (a) The points in $\mathcal{H}$ are just solutions to a linear system $\mathbf{u}^{T} \mathbf{x} = \beta$. Using the fact that the general solution of any linear system is a particular solution plus the general solution of the associated homogeneous equation produces

$$
\mathcal{H} = \frac{\beta \mathbf{u}}{\mathbf{u}^{T} \mathbf{u}} + N(\mathbf{u}^{T}) = \frac{\beta \mathbf{u}}{\mathbf{u}^{T} \mathbf{u}} + [R(\mathbf{u})]^{\perp} = \frac{\beta \mathbf{u}}{\mathbf{u}^{T} \mathbf{u}} + \mathbf{u}^{\perp},
$$

where $\mathbf{u}^{\perp}$ denotes the orthogonal complement of the one-dimensional space spanned by the vector $\mathbf{u}$. Thus $\mathcal{H} = \mathbf{v} + \mathcal{M}$, where $\mathbf{v} = \beta \mathbf{u} / \mathbf{u}^{T} \mathbf{u}$ and $\mathcal{M} = \mathbf{u}^{\perp}$. The fact that $\dim (\mathbf{u}^{\perp}) = n - 1$ follows directly from (5.11.3).

(b) Use (5.13.14) with part (a) and the fact that $\mathbf{P}_{\mathbf{u}^{\perp}} = \mathbf{I} - \mathbf{u} \mathbf{u}^{T} / \mathbf{u}^{T} \mathbf{u}$ to write

$$
\mathbf{p} = \frac{\beta \mathbf{u}}{\mathbf{u}^{T} \mathbf{u}} + \left(\mathbf{I} - \frac{\mathbf{u} \mathbf{u}^{T}}{\mathbf{u}^{T} \mathbf{u}} \right) \left(\mathbf{b} - \frac{\beta \mathbf{u}}{\mathbf{u}^{T} \mathbf{u}} \right) = \frac{\beta \mathbf{u}}{\mathbf{u}^{T} \mathbf{u}} + \mathbf{b} - \frac{\mathbf{u} \mathbf{u}^{T} \mathbf{b}}{\mathbf{u}^{T} \mathbf{u}} = \mathbf{b} - \left(\frac{\mathbf{u}^{T} \mathbf{b} - \beta}{\mathbf{u}^{T} \mathbf{u}} \right) \mathbf{u}.
$$

5.13.18. (a) $\mathbf{u}^T\mathbf{w} \neq 0$ implies $\mathcal{M} \cap \mathcal{W} = \mathbf{0}$ so that

$$\dim(\mathcal{M} + \mathcal{W}) = \dim\mathcal{M} + \dim\mathcal{W} = (n-1) + 1 = n.$$

Therefore, $\mathcal{M} + \mathcal{W} = \Re^n$. This together with $\mathcal{M} \cap \mathcal{W} = \mathbf{0}$ means $\Re^n = \mathcal{M} \oplus \mathcal{W}$.

(b)   Write

$$\mathbf{b} = \mathbf{b} - \frac{\mathbf{u}^T\mathbf{b}}{\mathbf{u}^T\mathbf{w}}\mathbf{w} + \frac{\mathbf{u}^T\mathbf{b}}{\mathbf{u}^T\mathbf{w}}\mathbf{w} = \mathbf{p} + \frac{\mathbf{u}^T\mathbf{b}}{\mathbf{u}^T\mathbf{w}}\mathbf{w},$$

and observe that $\mathbf{p} \in \mathcal{M}$ (because $\mathbf{u}^T\mathbf{p} = 0$) and $(\mathbf{u}^T\mathbf{b}/\mathbf{u}^T\mathbf{w})\mathbf{w} \in \mathcal{W}$. By definition, $\mathbf{p}$ is the projection of $\mathbf{b}$ onto $\mathcal{M}$ along $\mathcal{W}$.

(c)   We know from Exercise 5.13.17, part (a), that $\mathcal{H} = \mathbf{v} + \mathcal{M}$, where $\mathbf{v} = \beta\mathbf{u}/\mathbf{u}^T\mathbf{u}$ and $\mathcal{M} = \mathbf{u}^\perp$, so subtracting $\mathbf{v} = \beta\mathbf{u}/\mathbf{u}^T\mathbf{u}$ from everything in $\mathcal{H}$ as well as from $\mathbf{b}$ translates the situation back to the origin. Sketch a picture similar to that of Figure 5.13.5 to see that this moves $\mathcal{H}$ back to $\mathcal{M}$, it translates $\mathbf{b}$ to $\mathbf{b} - \mathbf{v}$, and it translates $\mathbf{p}$ to $\mathbf{p} - \mathbf{v}$. Now, $\mathbf{p} - \mathbf{v}$ should be the projection of $\mathbf{b} - \mathbf{v}$ onto $\mathcal{M}$ along $\mathcal{W}$, so by the result of part (b),

$$\mathbf{p} - \mathbf{v} = \mathbf{b} - \mathbf{v} - \frac{\mathbf{u}^T(\mathbf{b} - \mathbf{v})}{\mathbf{u}^T\mathbf{w}}\mathbf{w} \implies \mathbf{p} = \mathbf{b} - \frac{\mathbf{u}^T(\mathbf{b} - \mathbf{v})}{\mathbf{u}^T\mathbf{w}}\mathbf{w} = \mathbf{b} - \left(\frac{\mathbf{u}^T\mathbf{b} - \beta}{\mathbf{u}^T\mathbf{w}}\right)\mathbf{w}.$$

5.13.19. For convenience, set $\beta = \mathbf{A}_{i*}\mathbf{p}_{kn+i-1} - b_i$ so that $\mathbf{p}_{kn+i} = \mathbf{p}_{kn+i-1} - \beta(\mathbf{A}_{i*})^T$. Use the fact that

$$\mathbf{A}_{i*}\left(\mathbf{p}_{kn+i-1} - \mathbf{x}\right) = \mathbf{A}_{i*}\mathbf{p}_{kn+i-1} - b_i = \beta$$

together with $\|\mathbf{A}_{i*}\|_2 = 1$ to write

$$\begin{aligned}
\|\mathbf{p}_{kn+i} - \mathbf{x}\|_2^2 &= \left\|\mathbf{p}_{kn+i-1} - \beta(\mathbf{A}_{i*})^T - \mathbf{x}\right\|_2^2 \\
&= \left\|(\mathbf{p}_{kn+i-1} - \mathbf{x}) - \beta(\mathbf{A}_{i*})^T\right\|_2^2 \\
&= (\mathbf{p}_{kn+i-1} - \mathbf{x})^T(\mathbf{p}_{kn+i-1} - \mathbf{x}) \\
&\quad - 2\beta\mathbf{A}_{i*}(\mathbf{p}_{kn+i-1} - \mathbf{x}) + \beta^2\mathbf{A}_{i*}(\mathbf{A}_{i*})^T \\
&= \|\mathbf{p}_{kn+i-1} - \mathbf{x}\|_2^2 - \beta^2.
\end{aligned}$$

Consequently, $\|\mathbf{p}_{kn+i} - \mathbf{x}\|_2 \leq \|\mathbf{p}_{kn+i-1} - \mathbf{x}\|_2$, with equality holding if and only if $\beta = 0$ or, equivalently, if and only if $\mathbf{p}_{kn+i-1} \in \mathcal{H}_{i-1} \cap \mathcal{H}_i$. Therefore, the sequence of norms $\|\mathbf{p}_{kn+i} - \mathbf{x}\|_2$ is monotonically decreasing, and hence it must have a limiting value. This implies that the sequence of the β's defined above must approach 0, and thus the sequence of the $\mathbf{p}_{kn+i}$'s converges to $\mathbf{x}$.

5.13.20. Refer to Figure 5.13.8, and notice that the line passing from $\mathbf{p}_1^{(1)}$ to $\mathbf{p}_2^{(1)}$ is parallel to $\mathcal{V} = span\left(\mathbf{p}_1^{(1)} - \mathbf{p}_2^{(1)}\right)$, so projecting $\mathbf{p}_1^{(1)}$ through $\mathbf{p}_2^{(1)}$ onto $\mathcal{H}_2$

is exactly the same as projecting $\mathbf{p}_1^{(1)}$ onto $\mathcal{H}_2$ along (i.e., parallel to) $\mathcal{V}$. According to part (c) of Exercise 5.13.18, this projection is given by

$$\mathbf{p}_2^{(2)} = \mathbf{p}_1^{(1)} - \frac{\mathbf{A}_{2*}\left(\mathbf{p}_1^{(1)} - b_1\mathbf{A}_{2*}^T\right)}{\mathbf{A}_{2*}\left(\mathbf{p}_1^{(1)} - \mathbf{p}_2^{(1)}\right)}\left(\mathbf{p}_1^{(1)} - \mathbf{p}_2^{(1)}\right)$$

$$= \mathbf{p}_1^{(1)} - \frac{\left(\mathbf{A}_{2*}\mathbf{p}_1^{(1)} - b_1\right)}{\mathbf{A}_{2*}\left(\mathbf{p}_1^{(1)} - \mathbf{p}_2^{(1)}\right)}\left(\mathbf{p}_1^{(1)} - \mathbf{p}_2^{(1)}\right).$$

All other projections are similarly derived. It is now straightforward to verify that the points created by the algorithm are exactly the same points described in Steps $1, 2, \ldots, n - 1$.

Note: The condition that $\left\{\left(\mathbf{p}_1^{(1)} - \mathbf{p}_2^{(1)}\right), \left(\mathbf{p}_1^{(1)} - \mathbf{p}_3^{(1)}\right), \ldots, \left(\mathbf{p}_1^{(1)} - \mathbf{p}_n^{(1)}\right)\right\}$ is independent insures that $\left\{\left(\mathbf{p}_2^{(2)} - \mathbf{p}_3^{(2)}\right), \left(\mathbf{p}_2^{(2)} - \mathbf{p}_4^{(2)}\right), \ldots, \left(\mathbf{p}_2^{(2)} - \mathbf{p}_n^{(2)}\right)\right\}$ is also independent. The same holds at each subsequent step. Furthermore, $\mathbf{A}_{2*}\left(\mathbf{p}_1^{(1)} - \mathbf{p}_k^{(1)}\right) \neq 0$ for $k > 1$ implies that $\mathcal{V}_k = span\left(\mathbf{p}_1^{(1)} - \mathbf{p}_k^{(1)}\right)$ is not parallel to $\mathcal{H}_2$, so all projections onto $\mathcal{H}_2$ along $\mathcal{V}_k$ are well defined. It can be argued that the analogous situation holds at each step of the process—i.e., the initial conditions insure $\mathbf{A}_{i+1*}\left(\mathbf{p}_i^{(i)} - \mathbf{p}_k^{(i)}\right) \neq 0$ for $k > i$.

5.13.21. Equation (5.13.13) says that the orthogonal distance between $\mathbf{x}$ and $\mathcal{M}^\perp$ is

$$dist\left(\mathbf{x}, \mathcal{M}^\perp\right) = \|\mathbf{x} - \mathbf{P}_{\mathcal{M}^\perp}\mathbf{x}\|_2 = \|(\mathbf{I} - \mathbf{P}_{\mathcal{M}^\perp})\mathbf{x}\|_2 = \|\mathbf{P}_{\mathcal{M}}\mathbf{x}\|_2.$$

Similarly,

$$dist\left(\mathbf{R}\mathbf{x}, \mathcal{M}^\perp\right) = \|\mathbf{P}_{\mathcal{M}}\mathbf{R}\mathbf{x}\|_2 = \|-\mathbf{P}_{\mathcal{M}}\mathbf{x}\|_2 = \|\mathbf{P}_{\mathcal{M}}\mathbf{x}\|_2.$$

5.13.22. (a) We know from Exercise 5.13.17 that $\mathcal{H} = \mathbf{v} + \mathbf{u}^\perp$, where $\mathbf{v} = \beta\mathbf{u}$, so subtracting $\mathbf{v}$ from everything in $\mathcal{H}$ as well as from $\mathbf{b}$ translates the situation back to the origin. As depicted in the diagram below, this moves $\mathcal{H}$ down to $\mathbf{u}^\perp$, and it translates $\mathbf{b}$ to $\mathbf{b} - \mathbf{v}$ and $\mathbf{r}$ to $\mathbf{r} - \mathbf{v}$.

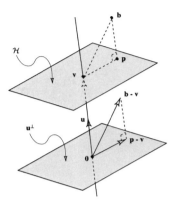

Now, we know from (5.6.8) that the reflection of $\mathbf{b} - \mathbf{v}$ about $\mathbf{u}^\perp$ is

$$\mathbf{r} - \mathbf{v} = \mathbf{R}(\mathbf{b} - \mathbf{v}) = (\mathbf{I} - 2\mathbf{u}\mathbf{u}^T)(\mathbf{b} - \mathbf{v}) = \mathbf{b} + (\beta - 2\mathbf{u}^T\mathbf{b})\mathbf{u},$$

and therefore the reflection of $\mathbf{b}$ about $\mathcal{H}$ is

$$\mathbf{r} = \mathbf{R}(\mathbf{b} - \mathbf{v}) + \mathbf{v} = \mathbf{b} - 2(\mathbf{u}^T\mathbf{b} - \beta)\mathbf{u}.$$

(b)   From part (a), the reflection of $\mathbf{r}_0$ about $\mathcal{H}_i$ is

$$\mathbf{r}_i = \mathbf{r}_0 - 2(\mathbf{A}_{i*}\mathbf{r}_0 - b_i)\left(\mathbf{A}_{i*}\right)^T,$$

and therefore the mean value of all of the reflections $\{\mathbf{r}_1, \mathbf{r}_2, \ldots, \mathbf{r}_n\}$ is

$$\mathbf{m} = \frac{1}{n}\sum_{i=1}^{n}\mathbf{r}_i = \frac{1}{n}\sum_{i=1}^{n}\left(\mathbf{r}_0 - 2(\mathbf{A}_{i*}\mathbf{r}_0 - b_i)\left(\mathbf{A}_{i*}\right)^T\right)$$

$$= \mathbf{r}_0 - \frac{2}{n}\sum_{i=1}^{n}(\mathbf{A}_{i*}\mathbf{r}_0 - b_i)(\mathbf{A}_{i*})^T$$

$$= \mathbf{r}_0 - \frac{2}{n}\mathbf{A}^T\left(\mathbf{A}\mathbf{r}_0 - \mathbf{b}\right) = \mathbf{r}_0 - \frac{2}{n}\mathbf{A}^T\boldsymbol{\varepsilon}.$$

Note: If weights $w_i > 0$ such that $\sum w_i = 1$ are used, then the weighted mean is

$$\mathbf{m} = \sum_{i=1}^{n}w_i\mathbf{r}_i = \sum_{i=1}^{n}w_i\left(\mathbf{r}_0 - 2(\mathbf{A}_{i*}\mathbf{r}_0 - b_i)\left(\mathbf{A}_{i*}\right)^T\right)$$

$$= \mathbf{r}_0 - 2\sum_{i=1}^{n}w_i(\mathbf{A}_{i*}\mathbf{r}_0 - b_i)(\mathbf{A}_{i*})^T$$

$$= \mathbf{r}_0 - \frac{2}{n}\mathbf{A}^T\mathbf{W}\left(\mathbf{A}\mathbf{r}_0 - \mathbf{b}\right) = \mathbf{r}_0 - \frac{2}{n}\mathbf{A}^T\mathbf{W}\boldsymbol{\varepsilon},$$

where $\mathbf{W} = diag\{w_1, w_2, \ldots, w_n\}$.

(c) First observe that

$$\mathbf{x} - \mathbf{m}_k = \mathbf{x} - \mathbf{m}_{k-1} + \frac{2}{n}\mathbf{A}^T \boldsymbol{\varepsilon}_{k-1}$$

$$= \mathbf{x} - \mathbf{m}_{k-1} + \frac{2}{n}\mathbf{A}^T(\mathbf{A}\mathbf{m}_{k-1} - \mathbf{b})$$

$$= \mathbf{x} - \mathbf{m}_{k-1} + \frac{2}{n}\mathbf{A}^T(\mathbf{A}\mathbf{m}_{k-1} - \mathbf{A}\mathbf{x})$$

$$= \mathbf{x} - \mathbf{m}_{k-1} + \frac{2}{n}\mathbf{A}^T\mathbf{A}(\mathbf{m}_{k-1} - \mathbf{x})$$

$$= \left(\mathbf{I} - \frac{2}{n}\mathbf{A}^T\mathbf{A}\right)(\mathbf{x} - \mathbf{m}_{k-1}),$$

and then use successive substitution to conclude that

$$\mathbf{x} - \mathbf{m}_k = \left(\mathbf{I} - \frac{2}{n}\mathbf{A}^T\mathbf{A}\right)^k (\mathbf{x} - \mathbf{m}_0).$$

Solutions for exercises in section 5. 14

5.14.1. Use (5.14.5) to observe that

$$E[y_i y_j] = \text{Cov}[y_i, y_j] + \mu_{y_i}\mu_{y_j} = \begin{cases} \sigma^2 + (\mathbf{X}_{i*}\boldsymbol{\beta})^2 & \text{if } i = j, \\ (\mathbf{X}_{i*}\boldsymbol{\beta})(\mathbf{X}_{j*}\boldsymbol{\beta}) & \text{if } i \neq j, \end{cases}$$

so that

$$E[\mathbf{y}\mathbf{y}^T] = \sigma^2\mathbf{I} + (\mathbf{X}\boldsymbol{\beta})(\mathbf{X}\boldsymbol{\beta})^T = \sigma^2\mathbf{I} + \mathbf{X}\boldsymbol{\beta}\boldsymbol{\beta}^T\mathbf{X}^T.$$

Write $\hat{\mathbf{e}} = \mathbf{y} - \mathbf{X}\hat{\boldsymbol{\beta}} = (\mathbf{I} - \mathbf{X}\mathbf{X}^\dagger)\mathbf{y}$, and use the fact that $\mathbf{I} - \mathbf{X}\mathbf{X}^\dagger$ is idempotent to obtain

$$\hat{\mathbf{e}}^T\hat{\mathbf{e}} = \mathbf{y}^T(\mathbf{I} - \mathbf{X}\mathbf{X}^\dagger)\mathbf{y} = trace\left((\mathbf{I} - \mathbf{X}\mathbf{X}^\dagger)\mathbf{y}\mathbf{y}^T\right).$$

Now use the linearity of trace and expectation together with the result of Exercise 5.9.13 and the fact that $(\mathbf{I} - \mathbf{X}\mathbf{X}^\dagger)\mathbf{X} = \mathbf{0}$ to write

$$E[\hat{\mathbf{e}}^T\hat{\mathbf{e}}] = E\left[trace\left((\mathbf{I} - \mathbf{X}\mathbf{X}^\dagger)\mathbf{y}\mathbf{y}^T\right)\right] = trace\left(E[(\mathbf{I} - \mathbf{X}\mathbf{X}^\dagger)\mathbf{y}\mathbf{y}^T]\right)$$

$$= trace\left((\mathbf{I} - \mathbf{X}\mathbf{X}^\dagger)E[\mathbf{y}\mathbf{y}^T]\right) = trace\left((\mathbf{I} - \mathbf{X}\mathbf{X}^\dagger)(\sigma^2\mathbf{I} + \mathbf{X}\boldsymbol{\beta}\boldsymbol{\beta}^T\mathbf{X}^T)\right)$$

$$= \sigma^2 trace\left(\mathbf{I} - \mathbf{X}\mathbf{X}^\dagger\right) = \sigma^2\left(m - trace\left(\mathbf{X}\mathbf{X}^\dagger\right)\right)$$

$$= \sigma^2\left(m - rank\left(\mathbf{X}\mathbf{X}^\dagger\right)\right) = \sigma^2(m - n).$$

Solutions for exercises in section 5. 15

5.15.1. (a) $\theta_{min} = 0$, and $\theta_{max} = \theta = \phi = \pi/4$.

(b) $\theta_{min} = \theta = \phi = \pi/4$, and $\theta_{max} = 1$.

5.15.2. (a)   The first principal angle is $\theta_1 = \theta_{min} = 0$, and we can take $\mathbf{u}_1 = \mathbf{v}_1 = \mathbf{e}_1$. This means that

$$\mathcal{M}_2 = \mathbf{u}_1^\perp \cap \mathcal{M} = span\,\{\mathbf{e}_2\} \quad \text{and} \quad \mathcal{N}_2 = \mathbf{v}_1^\perp \cap \mathcal{N} = span\,\{(0,\,1,\,1)\}.$$

The second principal angle is the minimal angle between $\mathcal{M}_2$ and $\mathcal{N}_2$, and this is just the angle between $\mathbf{e}_2$ and $(0,\,1,\,1)$, so $\theta_2 = \pi/4$.

(b)   This time the first principal angle is $\theta_1 = \theta_{min} = \pi/4$, and we can take $\mathbf{u}_1 = \mathbf{e}_1$ and $\mathbf{v}_1 = (0,\,1/\sqrt{2},\,1/\sqrt{2})$. There are no more principal angles because $\mathcal{N}_2 = \mathbf{v}_1^\perp \cap \mathcal{N} = \mathbf{0}$.

5.15.3. (a)   This follows from (5.15.16) because $\mathbf{P}_\mathcal{M} = \mathbf{P}_\mathcal{N}$ if and only if $\mathcal{M} = \mathcal{N}$.

(b)   If $\mathbf{0} \neq \mathbf{x} \in \mathcal{M} \cap \mathcal{N}$, then (5.15.1) evaluates to 1 with the maximum being attained at $\mathbf{u} = \mathbf{v} = \mathbf{x}/\|\mathbf{x}\|_2$. Conversely, $\cos\theta_{min} = 1 \implies \mathbf{v}^T\mathbf{u} = 1$ for some $\mathbf{u} \in \mathcal{M}$ and $\mathbf{v} \in \mathcal{N}$ such that $\|\mathbf{u}\|_2 = 1 = \|\mathbf{v}\|_2$. But $\mathbf{v}^T\mathbf{u} = 1 = \|\mathbf{u}\|_2\|\mathbf{v}\|_2$ represents equality in the CBS inequality (5.1.3), and we know this occurs if and only if $\mathbf{v} = \alpha\mathbf{u}$ for $\alpha = \mathbf{v}^T\mathbf{u}/\mathbf{u}^*\mathbf{u} = 1/1 = 1$. Thus $\mathbf{u} = \mathbf{v} \in \mathcal{M} \cap \mathcal{N}$.

(c) $\max_{\substack{\mathbf{u}\in\mathcal{M},\,\mathbf{v}\in\mathcal{N}\\ \|\mathbf{u}\|_2 = \|\mathbf{v}\|_2 = 1}} \mathbf{v}^T\mathbf{u} = 0 \iff \mathbf{v}^T\mathbf{u} = 0 \,\,\forall\,\mathbf{u}\in\mathcal{M},\,\mathbf{v}\in\mathcal{V} \iff \mathcal{M}\perp\mathcal{N}$.

5.15.4. You can use either (5.15.3) or (5.15.4) to arrive at the result. The latter is used by observing

$$\left\|(\mathbf{P}_{\mathcal{M}^\perp} - \mathbf{P}_{\mathcal{N}^\perp})^{-1}\right\|_2 = \left\|\left((\mathbf{I} - \mathbf{P}_\mathcal{M}) - (\mathbf{I} - \mathbf{P}_\mathcal{N})\right)^{-1}\right\|_2$$
$$= \left\|(\mathbf{P}_\mathcal{N} - \mathbf{P}_\mathcal{M})^{-1}\right\|_2 = \left\|(\mathbf{P}_\mathcal{M} - \mathbf{P}_\mathcal{N})^{-1}\right\|_2.$$

5.15.5. $\mathcal{M} \oplus \mathcal{N}^\perp = \Re^n \implies \dim\mathcal{M} = \dim\mathcal{N} \implies \sin\theta_{max} = \delta(\mathcal{M},\mathcal{N}) = \delta(\mathcal{N},\mathcal{M})$, so $\cos\tilde\theta_{min} = \|\mathbf{P}_\mathcal{M}\mathbf{P}_{\mathcal{N}^\perp}\|_2 = \|\mathbf{P}_\mathcal{M}(\mathbf{I} - \mathbf{P}_\mathcal{N})\|_2 = \delta(\mathcal{M},\mathcal{N}) = \sin\theta_{max}$.

5.15.6. It was argued in the proof of (5.15.4) that $\mathbf{P}_\mathcal{M} - \mathbf{P}_\mathcal{N}$ is nonsingular whenever $\mathcal{M}$ and $\mathcal{N}$ are complementary, so we need only prove the converse. Suppose $\dim\mathcal{M} = r > 0$ and $\dim\mathcal{N} = k > 0$ (the problem is trivial if $r = 0$ or $k = 0$) so that $\mathbf{U}_1^T\mathbf{V}_1$ is $r \times n - k$ and $\mathbf{U}_2^T\mathbf{V}_2$ is $n - r \times k$. If $\mathbf{P}_\mathcal{M} - \mathbf{P}_\mathcal{N}$ is nonsingular, then (5.15.7) insures that the rows as well as the columns in each of these products must be linearly independent. That is, $\mathbf{U}_1^T\mathbf{V}_1$ and $\mathbf{U}_2^T\mathbf{V}_2$ must both be square and nonsingular, so $r + k = n$. Combine this with the formula for the rank of a product (4.5.1) to conclude

$$k = rank\,(\mathbf{U}_2^T\mathbf{V}_2) = rank\,(\mathbf{U}_2^T) - \dim N\,(\mathbf{U}_2^T) \cap R\,(\mathbf{V}_2)$$
$$= n - r - \dim\mathcal{M}\cap\mathcal{N} = k - \dim\mathcal{M}\cap\mathcal{N}.$$

It follows that $\mathcal{M}\cap\mathcal{N} = \mathbf{0}$, and hence $\mathcal{M}\oplus\mathcal{N} = \Re^n$.

5.15.7. (a)   This can be derived from (5.15.7), or it can be verified by direct multiplication by using $\mathbf{P}_{\mathcal{N}}(\mathbf{I} - \mathbf{P}) = \mathbf{I} - \mathbf{P} \implies \mathbf{P} - \mathbf{P}_{\mathcal{N}}\mathbf{P} = \mathbf{I} - \mathbf{P}_{\mathcal{N}}$ to write

$$(\mathbf{P}_{\mathcal{M}} - \mathbf{P}_{\mathcal{N}})(\mathbf{P} - \mathbf{Q}) = \mathbf{P}_{\mathcal{M}}\mathbf{P} - \mathbf{P}_{\mathcal{M}}\mathbf{Q} - \mathbf{P}_{\mathcal{N}}\mathbf{P} + \mathbf{P}_{\mathcal{N}}\mathbf{Q}$$
$$= \mathbf{P} - \mathbf{0} - \mathbf{P}_{\mathcal{N}}\mathbf{P} + \mathbf{P}_{\mathcal{N}}\mathbf{Q} = \mathbf{I} - \mathbf{P}_{\mathcal{N}} + \mathbf{P}_{\mathcal{N}}\mathbf{Q}$$
$$= \mathbf{I} - \mathbf{P}_{\mathcal{N}}(\mathbf{I} - \mathbf{Q}) = \mathbf{I}.$$

(b) and (c) follow from (a) in conjunction with (5.15.3) and (5.15.4).

5.15.8. Since we are maximizing over a larger set, $\max_{\|\mathbf{x}\|=1} f(\mathbf{x}) \leq \max_{\|\mathbf{x}\|\leq 1} f(\mathbf{x})$. A strict inequality here implies the existence of a nonzero vector $\mathbf{x}_0$ such that $\|\mathbf{x}_0\| < 1$ and $f(\mathbf{x}) < f(\mathbf{x}_0)$ for all vectors such that $\|\mathbf{x}\| = 1$. But then

$$f(\mathbf{x}_0) > f(\mathbf{x}_0/\|\mathbf{x}_0\|) = f(\mathbf{x}_0)/\|\mathbf{x}_0\| \implies \|\mathbf{x}_0\| f(\mathbf{x}_0) > f(\mathbf{x}_0),$$

which is impossible because $\|\mathbf{x}_0\| < 1$.

5.15.9. (a)   We know from equation (5.15.6) that $\mathbf{P}_{\mathcal{MN}} = \mathbf{U} \begin{pmatrix} \mathbf{C} & \mathbf{0} \\ \mathbf{0} & \mathbf{0} \end{pmatrix} \mathbf{V}^T$ in which $\mathbf{C}$ is nonsingular and $\mathbf{C}^{-1} = \mathbf{V}_1^T \mathbf{U}_1$. Consequently,

$$\mathbf{P}_{\mathcal{MN}}^{\dagger} = \mathbf{V} \begin{pmatrix} \mathbf{C}^{-1} & \mathbf{0} \\ \mathbf{0} & \mathbf{0} \end{pmatrix} \mathbf{U}^T = \mathbf{V}_1 \mathbf{C}^{-1} \mathbf{U}_1^T = \mathbf{V}_1 \mathbf{V}_1^T \mathbf{U}_1 \mathbf{U}_1^T = \mathbf{P}_{\mathcal{N}^{\perp}} \mathbf{P}_{\mathcal{M}}.$$

(b)   Use the fact

$$\left\|(\mathbf{U}_1^T \mathbf{V}_1)^{-1}\right\|_2 = \left\|(\mathbf{V}_1^T \mathbf{U}_1)^{-1}\right\|_2 = \left\|\mathbf{U}_1(\mathbf{V}_1^T \mathbf{U}_1)^{-1}\mathbf{V}_1^T\right\|_2$$
$$= \left\|(\mathbf{V}_1 \mathbf{V}_1^T \mathbf{U}_1 \mathbf{U}_1^T)^{\dagger}\right\|_2 = \left\|\left[(\mathbf{I} - \mathbf{P}_{\mathcal{N}})\mathbf{P}_{\mathcal{M}}\right]^{\dagger}\right\|_2$$

(and similarly for the other term) to show that

$$\left\|\left[(\mathbf{I} - \mathbf{P}_{\mathcal{N}})\mathbf{P}_{\mathcal{M}}\right]^{\dagger}\right\|_2 = \left\|(\mathbf{U}_1^T \mathbf{V}_1)^{-1}\right\|_2 = \left\|\left[\mathbf{P}_{\mathcal{M}}(\mathbf{I} - \mathbf{P}_{\mathcal{N}})\right]^{\dagger}\right\|_2,$$

and

$$\left\|\left[(\mathbf{I} - \mathbf{P}_{\mathcal{M}})\mathbf{P}_{\mathcal{N}}\right]^{\dagger}\right\|_2 = \left\|(\mathbf{U}_2^T \mathbf{V}_2)^{-1}\right\|_2 = \left\|\left[\mathbf{P}_{\mathcal{N}}(\mathbf{I} - \mathbf{P}_{\mathcal{M}})\right]^{\dagger}\right\|_2.$$

It was established in the proof of (5.15.4) that $\left\|(\mathbf{U}_1^T \mathbf{V}_1)^{-1}\right\|_2 = \left\|(\mathbf{U}_2^T \mathbf{V}_2)^{-1}\right\|_2$, so combining this with the result of part (a) and (5.15.3) produces the desired conclusion.

5.15.10. (a)   We know from (5.15.2) that $\cos \bar{\theta}_{min} = \|\mathbf{P}_{\mathcal{N}^{\perp}} \mathbf{P}_{\mathcal{M}}\|_2 = \|(\mathbf{I} - \mathbf{P}_{\mathcal{N}})\mathbf{P}_{\mathcal{M}}\|_2$, and we know from Exercise 5.15.9 that $\mathbf{P}_{\mathcal{MN}} = \left[(\mathbf{I} - \mathbf{P}_{\mathcal{N}})\mathbf{P}_{\mathcal{M}}\right]^{\dagger}$, so taking the pseudoinverse of both sides of this yields the desired result.

(b)   Use (5.15.3) together with part (a), (5.13.10), and (5.13.12) to write

$$1 = \left\| \mathbf{P}_{\mathcal{M}\mathcal{N}} \mathbf{P}_{\mathcal{M}\mathcal{N}}^{\dagger} \right\|_2 \le \left\| \mathbf{P}_{\mathcal{M}\mathcal{N}} \right\|_2 \left\| \mathbf{P}_{\mathcal{M}\mathcal{N}}^{\dagger} \right\|_2 = \frac{\cos \bar{\theta}_{min}}{\sin \theta_{min}}.$$

5.15.11. (a)   Use the facts that $\|\mathbf{A}\|_2 = \|\mathbf{A}^T\|_2$ and $(\mathbf{A}^T)^{-1} = (\mathbf{A}^{-1})^T$ to write

$$\frac{1}{\left\| (\mathbf{U}_2^T \mathbf{V}_2)^{-1} \right\|_2^2} = \frac{1}{\left\| (\mathbf{V}_2^T \mathbf{U}_2)^{-1} \right\|_2^2} = \min_{\|\mathbf{x}\|_2 = 1} \left\| \mathbf{V}_2^T \mathbf{U}_2 \mathbf{x} \right\|_2^2$$

$$= \min_{\|\mathbf{x}\|_2 = 1} \mathbf{x}^T \mathbf{U}_2^T \mathbf{V}_2 \mathbf{V}_2^T \mathbf{U}_2 \mathbf{x}$$

$$= \min_{\|\mathbf{x}\|_2 = 1} \mathbf{x}^T \mathbf{U}_2^T (\mathbf{I} - \mathbf{V}_1 \mathbf{V}_1^T) \mathbf{U}_2 \mathbf{x} = \min_{\|\mathbf{x}\|_2 = 1} \left(1 - \left\| \mathbf{V}_1^T \mathbf{U}_2 \mathbf{x} \right\|_2^2 \right)$$

$$= 1 - \max_{\|\mathbf{x}\|_2 = 1} \left\| \mathbf{V}_1^T \mathbf{U}_2 \mathbf{x} \right\|_2^2 = 1 - \left\| \mathbf{V}_1^T \mathbf{U}_2 \right\|_2^2 = 1 - \left\| \mathbf{U}_2^T \mathbf{V}_1 \right\|_2^2.$$

(b)   Use a similar technique to write

$$\left\| \mathbf{U}_2^T \mathbf{V}_2 \right\|_2^2 = \left\| \mathbf{U}_2^T \mathbf{V}_2 \mathbf{V}_2^T \right\|_2^2 = \left\| \mathbf{U}_2^T (\mathbf{I} - \mathbf{V}_1 \mathbf{V}_1^T) \right\|_2^2$$

$$= \left\| (\mathbf{I} - \mathbf{V}_1 \mathbf{V}_1^T) \mathbf{U}_2 \right\|_2^2 = \max_{\|\mathbf{x}\|_2 = 1} \mathbf{x}^T \mathbf{U}_2^T (\mathbf{I} - \mathbf{V}_1 \mathbf{V}_1^T) \mathbf{U}_2 \mathbf{x}$$

$$= 1 - \min_{\|\mathbf{x}\|_2 = 1} \left\| \mathbf{V}_1^T \mathbf{U}_2 \mathbf{x} \right\|_2^2 = 1 - \frac{1}{\left\| (\mathbf{V}_1^T \mathbf{U}_2)^{-1} \right\|_2^2}$$

$$= 1 - \frac{1}{\left\| (\mathbf{U}_2^T \mathbf{V}_1)^{-1} \right\|_2^2}.$$

Solutions for Chapter 6

Solutions for exercises in section 6. 1

6.1.1. (a) -1 (b) 8 (c) $-\alpha\beta\gamma$

(d) $a_{11}a_{22}a_{33} + a_{12}a_{23}a_{31} + a_{13}a_{21}a_{32} - (a_{11}a_{23}a_{32} + a_{12}a_{21}a_{33} + a_{13}a_{22}a_{31})$

(This is where the "diagonal rule" you learned in high school comes from.)

6.1.2. If $\mathbf{A} = [\mathbf{x}_1 \,|\, \mathbf{x}_2 \,|\, \mathbf{x}_3]$, then $V_3 = \left[\det\left(\mathbf{A}^T\mathbf{A}\right)\right]^{1/2} = 20$ (recall Example 6.1.4). But you could also realize that the $\mathbf{x}_i$'s are mutually orthogonal to conclude that $V_3 = \|\mathbf{x}_1\|_2 \|\mathbf{x}_2\|_2 \|\mathbf{x}_3\|_2 = 20$.

6.1.3. (a) 10 (b) 0 (c) 120 (d) 39 (e) 1 (f) $(n-1)!$

6.1.4. $rank(\mathbf{A}) = 2$

6.1.5. A square system has a unique solution if and only if its coefficient matrix is nonsingular—recall the discussion in §2.5. Consequently, (6.1.13) guarantees that a square system has a unique solution if and only if the determinant of the coefficient matrix is nonzero. Since

$$\begin{vmatrix} 1 & \alpha & 0 \\ 0 & 1 & -1 \\ \alpha & 0 & 1 \end{vmatrix} = 1 - \alpha^2,$$

it follows that there is a unique solution if and only if $\alpha \neq \pm 1$.

6.1.6. $\mathbf{I} = \mathbf{A}^{-1}\mathbf{A} \implies \det(\mathbf{I}) = \det\left(\mathbf{A}^{-1}\mathbf{A}\right) = \det\left(\mathbf{A}^{-1}\right)\det(\mathbf{A})$

$\implies 1 = \det\left(\mathbf{A}^{-1}\right)\det(\mathbf{A}) \implies \det\left(\mathbf{A}^{-1}\right) = 1/\det(\mathbf{A})$.

6.1.7. Use the product rule (6.1.15) to write

$$\det\left(\mathbf{P}^{-1}\mathbf{A}\mathbf{P}\right) = \det\left(\mathbf{P}^{-1}\right)\det(\mathbf{A})\det(\mathbf{P}) = \det\left(\mathbf{P}^{-1}\right)\det(\mathbf{P})\det(\mathbf{A})$$
$$= \det\left(\mathbf{P}^{-1}\mathbf{P}\right)\det(\mathbf{A}) = \det(\mathbf{I})\det(\mathbf{A}) = \det(\mathbf{A}).$$

6.1.8. Use (6.1.4) together with the fact that $\overline{z_1 z_2} = \bar{z}_1 \bar{z}_2$ and $\overline{z_1 + z_2} = \bar{z}_1 + \bar{z}_2$ for all complex numbers to write

$$\det(\mathbf{A}^*) = \det\left(\bar{\mathbf{A}}^T\right) = \det\left(\bar{\mathbf{A}}\right) = \sum_p \sigma(p)\overline{a_{1p_1}}\cdots\overline{a_{np_n}}$$

$$= \sum_p \sigma(p)\overline{a_{1p_1}\cdots a_{np_n}} = \overline{\sum_p \sigma(p)a_{1p_1}\cdots a_{np_n}} = \overline{\det(\mathbf{A})}.$$

6.1.9. (a) $\mathbf{I} = \mathbf{Q}^*\mathbf{Q} \implies 1 = \det(\mathbf{Q}^*\mathbf{Q}) = \det(\mathbf{Q}^*)\det(\mathbf{Q}) = [\det(\mathbf{Q})]^2$ by Exercise 6.1.8.

(b)   If $\mathbf{A} = \mathbf{UDV}^*$ is an SVD, then, by part (a),

$$|\det(\mathbf{A})| = |\det(\mathbf{UDV}^*)| = |\det(\mathbf{U})|\,|\det(\mathbf{D})|\,|\det(\mathbf{V}^*)|$$
$$= \det(\mathbf{D}) = \sigma_1 \sigma_2 \cdots \sigma_n.$$

6.1.10. Let $r = rank(\mathbf{A})$, and let $\sigma_1 \geq \cdots \geq \sigma_r$ be the nonzero singular values of $\mathbf{A}$. If $\mathbf{A} = \mathbf{U}_{m \times m} \begin{pmatrix} \mathbf{D}_{r \times r} & \mathbf{0} \\ \mathbf{0} & \mathbf{0} \end{pmatrix}_{m \times n} (\mathbf{V}^*)_{n \times n}$ is an SVD, then, by Exercises 6.1.9 and 6.1.8, $\det(\mathbf{V})\det(\mathbf{V}^*) = |\det(\mathbf{V})|^2 = 1$, so

$$\det(\mathbf{A}^*\mathbf{A}) = \det(\mathbf{VD}^*\mathbf{DV}^*) = \det(\mathbf{V}) \begin{vmatrix} (\mathbf{D}^*\mathbf{D})_{r \times r} & \mathbf{0} \\ \mathbf{0} & \mathbf{0} \end{vmatrix}_{n \times n} \det(\mathbf{V}^*)$$
$$= \sigma_1^2 \sigma_2^2 \cdots \sigma_r^2 \underbrace{0 \cdots 0}_{n-r}, \quad \text{and this is} \quad \begin{cases} = 0 & \text{when } r < n, \\ > 0 & \text{when } r = n. \end{cases}$$

Note: You can't say $\det(\mathbf{A}^*\mathbf{A}) = \overline{\det(\mathbf{A})}\det(\mathbf{A}) = |\det(\mathbf{A})|^2 \geq 0$ because $\mathbf{A}$ need not be square.

6.1.11. $\alpha\mathbf{A} = (\alpha\mathbf{I})\mathbf{A} \implies \det(\alpha\mathbf{A}) = \det(\alpha\mathbf{I})\det(\mathbf{A}) = \alpha^n\det(\mathbf{A})$.

6.1.12. $\mathbf{A} = -\mathbf{A}^T \implies \det(\mathbf{A}) = \det(-\mathbf{A}^T) = \det(-\mathbf{A}) = (-1)^n\det(\mathbf{A})$ (by Exercise 6.1.11) $\implies \det(\mathbf{A}) = -\det(\mathbf{A})$ when n is odd $\implies \det(\mathbf{A}) = 0$.

6.1.13. If $\mathbf{A} = \mathbf{LU}$, where $\mathbf{L}$ is lower triangular and $\mathbf{U}$ is upper triangular where each has 1's on its diagonal and random integers in the remaining nonzero positions, then $\det(\mathbf{A}) = \det(\mathbf{L})\det(\mathbf{U}) = 1 \times 1 = 1$, and the entries of $\mathbf{A}$ are rather random integers.

6.1.14. According to the definition,

$$\det(\mathbf{A}) = \sum_p \sigma(p) a_{1p_1} \cdots a_{kp_k} \cdots a_{np_n}$$
$$= \sum_p \sigma(p) a_{1p_1} \cdots (x_{p_k} + y_{p_k} + \cdots + z_{p_k}) \cdots a_{np_n}$$
$$= \sum_p \sigma(p) a_{1p_1} \cdots x_{p_k} \cdots a_{np_n} + \sum_p \sigma(p) a_{1p_1} \cdots y_{p_k} \cdots a_{np_n}$$
$$+ \cdots + \sum_p \sigma(p) a_{1p_1} \cdots z_{p_k} \cdots a_{np_n}$$
$$= \det\begin{pmatrix} \mathbf{A}_{1*} \\ \vdots \\ \mathbf{x}^T \\ \vdots \\ \mathbf{A}_{n*} \end{pmatrix} + \det\begin{pmatrix} \mathbf{A}_{1*} \\ \vdots \\ \mathbf{y}^T \\ \vdots \\ \mathbf{A}_{n*} \end{pmatrix} + \cdots + \det\begin{pmatrix} \mathbf{A}_{1*} \\ \vdots \\ \mathbf{z}^T \\ \vdots \\ \mathbf{A}_{n*} \end{pmatrix}.$$

6.1.15. If $\mathbf{A}_{n \times 2} = [\mathbf{x} \,|\, \mathbf{y}]$, then the result of Exercise 6.1.10 implies

$$0 \le \det(\mathbf{A}^* \mathbf{A}) = \begin{vmatrix} \mathbf{x}^* \mathbf{x} & \mathbf{x}^* \mathbf{y} \\ \mathbf{y}^* \mathbf{x} & \mathbf{y}^* \mathbf{y} \end{vmatrix} = (\mathbf{x}^* \mathbf{x})(\mathbf{y}^* \mathbf{y}) - (\mathbf{x}^* \mathbf{y})(\mathbf{y}^* \mathbf{x})$$

$$= \|\mathbf{x}\|_2^2 \|\mathbf{y}\|_2^2 - (\mathbf{x}^* \mathbf{y}) \overline{(\mathbf{x}^* \mathbf{y})}$$

$$= \|\mathbf{x}\|_2^2 \|\mathbf{y}\|_2^2 - |\mathbf{x}^* \mathbf{y}|^2,$$

with equality holding if and only if $rank(\mathbf{A}) < 2$ —i.e., if and only if $\mathbf{y}$ **is a** scalar multiple of $\mathbf{x}$.

6.1.16. Partition $\mathbf{A}$ as

$$\mathbf{A} = \mathbf{LU} = \begin{pmatrix} \mathbf{L}_k & \mathbf{0} \\ \mathbf{L}_{21} & \mathbf{L}_{22} \end{pmatrix} \begin{pmatrix} \mathbf{U}_k & \mathbf{U}_{12} \\ \mathbf{0} & \mathbf{U}_{22} \end{pmatrix} = \begin{pmatrix} \mathbf{L}_k \mathbf{U}_k & * \\ * & * \end{pmatrix}$$

to deduce that $\mathbf{A}_k$ can be written in the form

$$\mathbf{A}_k = \mathbf{L}_k \mathbf{U}_k = \begin{pmatrix} \mathbf{L}_{k-1} & \mathbf{0} \\ \mathbf{d}^T & 1 \end{pmatrix} \begin{pmatrix} \mathbf{U}_{k-1} & \mathbf{c} \\ \mathbf{0} & u_{kk} \end{pmatrix} \quad \text{and} \quad \mathbf{A}_{k-1} = \mathbf{L}_{k-1} \mathbf{U}_{k-1}.$$

The product rule (6.1.15) shows that

$$\det(\mathbf{A}_k) = \det(\mathbf{U}_{k-1}) \times u_{kk} = \det(\mathbf{A}_{k-1}) \times u_{kk},$$

and the desired conclusion follows.

6.1.17. According to (3.10.12), a matrix has an LU factorization if and only if **each** leading principal submatrix is nonsingular. The leading $k \times k$ principal submatrix of $\mathbf{A}^T \mathbf{A}$ is given by $\mathbf{P}_k = \mathbf{A}_k^T \mathbf{A}_k$, where $\mathbf{A}_k = [\mathbf{A}_{*1} \,|\, \mathbf{A}_{*2} \,|\, \cdots \,|\, \mathbf{A}_{*k}]$. If $\mathbf{A}$ has full column rank, then any nonempty subset of columns is linearly independent, so $rank(\mathbf{A}_k) = k$. Therefore, the results of Exercise 6.1.10 insure that $\det(\mathbf{P}_k) = \det(\mathbf{A}_k^T \mathbf{A}_k) > 0$ for each k, and hence $\mathbf{A}^T \mathbf{A}$ has an LU factorization. The fact that each pivot is positive follows from Exercise 6.1.16.

6.1.18. (a) To evaluate $\det(\mathbf{A})$, use Gaussian elimination as shown below.

$$\begin{pmatrix} 2-x & 3 & 4 \\ 0 & 4-x & -5 \\ 1 & -1 & 3-x \end{pmatrix} \longrightarrow \begin{pmatrix} 1 & -1 & 3-x \\ 0 & 4-x & -5 \\ 2-x & 3 & 4 \end{pmatrix}$$

$$\longrightarrow \begin{pmatrix} 1 & -1 & 3-x \\ 0 & 4-x & -5 \\ 0 & 5-x & -x^2+5x-2 \end{pmatrix} \longrightarrow \begin{pmatrix} 1 & -1 & 3-x \\ 0 & 4-x & -5 \\ 0 & 0 & \frac{x^3-9x^2+17x+17}{4-x} \end{pmatrix} = \mathbf{U}.$$

Since one interchange was used, $\det(\mathbf{A})$ is (-1) times the product of the diagonal entries of $\mathbf{U}$, so

$$\det(\mathbf{A}) = -x^3 + 9x^2 - 17x - 17 \quad \text{and} \quad \frac{d\big(\det(\mathbf{A})\big)}{dx} = -3x^2 + 18x - 17.$$

(b)   Using formula (6.1.19) produces

$$\frac{d\big(\det(\mathbf{A})\big)}{dx} = \begin{vmatrix} -1 & 0 & 0 \\ 0 & 4-x & -5 \\ 1 & -1 & 3-x \end{vmatrix} + \begin{vmatrix} 2-x & 3 & 4 \\ 0 & -1 & 0 \\ 1 & -1 & 3-x \end{vmatrix} + \begin{vmatrix} 2-x & 3 & 4 \\ 0 & 4-x & -5 \\ 0 & 0 & -1 \end{vmatrix}$$

$$= (-x^2 + 7x - 7) + (-x^2 + 5x - 2) + (-x^2 + 6x - 8)$$
$$= -3x^2 + 18x - 17.$$

6.1.19. No—almost any 2×2 example will show that this cannot hold in general.

6.1.20. It was argued in Example 4.3.6 that if there is at least one value of x for which the Wronski matrix

$$\mathbf{W}(x) = \begin{pmatrix} f_1(x) & f_2(x) & \cdots & f_n(x) \\ f_1'(x) & f_2'(x) & \cdots & f_n'(x) \\ \vdots & \vdots & \ddots & \vdots \\ f_1^{(n-1)}(x) & f_2^{(n-1)}(x) & \cdots & f_n^{(n-1)}(x) \end{pmatrix}$$

is nonsingular, then $\mathcal{S}$ is a linearly independent set. This is equivalent to saying that if $\mathcal{S}$ is a linearly *dependent* set, then the Wronski matrix $\mathbf{W}(x)$ is singular for all values of x. But (6.1.14) insures that a matrix is singular if and only if its determinant is zero, so, if $\mathcal{S}$ is linearly dependent, then the Wronskian $w(x)$ must vanish for every value of x. The converse of this statement is false (Exercise 4.3.14).

6.1.21. (a) $(n!)(n-1)$ (b) 11×11 (c)   About 9.24×10^{153} sec $\approx 3 \times 10^{146}$ years   (d)   About 3×10^{150} mult/sec. (Now this would truly be a "super computer.")

Solutions for exercises in section 6. 2

6.2.1. (a)   8   (b)   39   (c) -3

6.2.2. (a) $\mathbf{A}^{-1} = \dfrac{\text{adj}(\mathbf{A})}{\det(\mathbf{A})} = \dfrac{1}{8} \begin{pmatrix} 0 & 1 & -1 \\ -8 & 4 & 4 \\ 16 & -6 & -2 \end{pmatrix}$

(b) $\mathbf{A}^{-1} = \dfrac{\text{adj}(\mathbf{A})}{\det(\mathbf{A})} = \dfrac{1}{39} \begin{pmatrix} -12 & 25 & -14 & 7 \\ -9 & 9 & 9 & 15 \\ -6 & 6 & 6 & -3 \\ 9 & 4 & 4 & -2 \end{pmatrix}$

6.2.3. (a) $x_1 = 1 - \beta, \quad x_2 = \alpha + \beta - 1, \quad x_3 = 1 - \alpha$

(b)   Cramer's rule yields

$$x_2(t) = \frac{\begin{vmatrix} 1 & t^4 & t^2 \\ t^2 & t^3 & t \\ t & 0 & 1 \end{vmatrix}}{\begin{vmatrix} 1 & t & t^2 \\ t^2 & 1 & t \\ t & t^2 & 1 \end{vmatrix}} = \frac{t\begin{vmatrix} t^4 & t^2 \\ t^3 & t \end{vmatrix} + \begin{vmatrix} 1 & t^4 \\ t^2 & t^3 \end{vmatrix}}{\begin{vmatrix} 1 & t \\ t^2 & 1 \end{vmatrix} - t\begin{vmatrix} t^2 & t \\ t & 1 \end{vmatrix} + t^2\begin{vmatrix} t^2 & 1 \\ t & t^2 \end{vmatrix}}$$

$$= \frac{t^3 - t^6}{(t^3 - 1)(t^3 - 1)} = \frac{-t^3}{(t^3 - 1)},$$

and hence

$$\lim_{t \to \infty} x_2(t) = \lim_{t \to \infty} \frac{-1}{1 - 1/t^3} = -1.$$

6.2.4. Yes.

6.2.5. (a)   Almost any two matrices will do the job. One example is $\mathbf{A} = \mathbf{I}$ and $\mathbf{B} = -\mathbf{I}$.

(b)   Again, almost anything you write down will serve the purpose. One example is $\mathbf{A} = \mathbf{D} = \mathbf{0}_{2\times 2}, \mathbf{B} = \mathbf{C} = \mathbf{I}_{2\times 2}$.

6.2.6. Recall from Example 5.13.3 that $\mathbf{Q} = \mathbf{I} - \mathbf{B}\mathbf{B}^T\mathbf{B}^{-1}\mathbf{B}^T$. According to (6.2.1),

$$\det\left(\mathbf{A}^T\mathbf{A}\right) = \det\begin{pmatrix} \mathbf{B}^T\mathbf{B} & \mathbf{B}^T\mathbf{c} \\ \mathbf{c}^T\mathbf{B} & \mathbf{c}^T\mathbf{c} \end{pmatrix} = \det\left(\mathbf{B}^T\mathbf{B}\right)\left(\mathbf{c}^T\mathbf{Q}\mathbf{c}\right).$$

Since $\det\left(\mathbf{B}^T\mathbf{B}\right) > 0$ (by Exercise 6.1.10), $\mathbf{c}^T\mathbf{Q}\mathbf{c} = \det\left(\mathbf{A}^T\mathbf{A}\right)/\det\left(\mathbf{B}^T\mathbf{B}\right)$.

6.2.7. Expand $\begin{vmatrix} \mathbf{A} & -\mathbf{C} \\ \mathbf{D}^T & \mathbf{I}_k \end{vmatrix}$ both of the ways indicated in (6.2.1).

6.2.8. The result follows from Example 6.2.8, which says $\mathbf{A}[\text{adj}(\mathbf{A})] = \det(\mathbf{A})\mathbf{I}$, together with the fact that $\mathbf{A}$ is singular if and only if $\det(\mathbf{A}) = 0$.

6.2.9. The solution is $\mathbf{x} = \mathbf{A}^{-1}\mathbf{b}$, and Example 6.2.7 says that the entries in $\mathbf{A}^{-1}$ are continuous functions of the entries in $\mathbf{A}$. Since $x_i = \sum_k [\mathbf{A}^{-1}]_{ik}b_k$, and since the sum of continuous functions is again continuous, it follows that each x_i is a continuous function of the a_{ij}'s.

6.2.10. If $\mathbf{B} = \alpha\mathbf{A}$, then Exercise 6.1.11 implies $\mathring{B}_{ij} = \alpha^{n-1}\mathring{A}_{ij}$, so $\mathring{\mathbf{B}} = \alpha^{n-1}\mathring{\mathbf{A}}$, and hence $\text{adj}(\mathbf{B}) = \alpha^{n-1}\text{adj}(\mathbf{A})$.

6.2.11. (a)   We saw in §6.1 that $rank(\mathbf{A})$ is the order of the largest nonzero minor of $\mathbf{A}$. If $rank(\mathbf{A}) < n - 1$, then every minor of order $n - 1$ (as well as $\det(\mathbf{A})$ itself) must be zero. Consequently, $\mathring{\mathbf{A}} = \mathbf{0}$, and thus $\text{adj}(\mathbf{A}) = \mathring{\mathbf{A}}^T = \mathbf{0}$.

(b) $rank(\mathbf{A}) = n - 1 \implies$ at least one minor of order $n - 1$ is nonzero

$\implies$ some $\mathring{A}_{ij} \neq 0 \implies \text{adj}(\mathbf{A}) \neq \mathbf{0}$

$\implies rank(\text{adj}(\mathbf{A})) \geq 1$.

Also, $rank\,(\mathbf{A}) = n - 1 \implies \det(\mathbf{A}) = 0$

$\implies \mathbf{A}[\mathrm{adj}\,(\mathbf{A})] = \mathbf{0}$ (by Exercise 6.2.8)

$\implies R\,(\mathrm{adj}\,(\mathbf{A})) \subseteq N\,(\mathbf{A})$

$\implies \dim R\,(\mathrm{adj}\,(\mathbf{A})) \le \dim N\,(\mathbf{A})$

$\implies rank\,(\mathrm{adj}\,(\mathbf{A})) \le n - rank\,(\mathbf{A}) = 1.$

(c) $rank\,(\mathbf{A}) = n \implies \det(\mathbf{A}) \ne 0 \implies \mathrm{adj}\,(\mathbf{A}) = \det(\mathbf{A})\,\mathbf{A}^{-1}$

$\implies rank\,(\mathrm{adj}\,(\mathbf{A})) = n$

6.2.12. If $\det(\mathbf{A}) = 0$, then Exercise 6.2.11 insures that $rank\,(\mathrm{adj}\,(\mathbf{A})) \le 1$. Consequently, $\det(\mathrm{adj}\,(\mathbf{A})) = 0$, and the result is trivially true because both sides are zero. If $\det(\mathbf{A}) \ne 0$, apply the product rule (6.1.15) to $\mathbf{A}[\mathrm{adj}\,(\mathbf{A})] = \det(\mathbf{A})\,\mathbf{I}$ (from Example 6.2.8) to obtain $\det(\mathbf{A})\det(\mathrm{adj}\,(\mathbf{A})) = [\det(\mathbf{A})]^n$, so that $\det(\mathrm{adj}\,(\mathbf{A})) = [\det(\mathbf{A})]^{n-1}$.

6.2.13. Expanding in terms of cofactors of the first row produces $D_n = 2\mathring{A}_{11} - \mathring{A}_{12}$. But $\mathring{A}_{11} = D_{n-1}$ and expansion using the first column yields

$$\mathring{A}_{12} = (-1)\begin{vmatrix} -1 & -1 & 0 & \cdots & 0 \\ 0 & 2 & -1 & \cdots & 0 \\ 0 & -1 & 2 & \cdots & 0 \\ \vdots & \vdots & \vdots & \ddots & \vdots \\ 0 & 0 & 0 & \cdots & 2 \end{vmatrix} = (-1)(-1)D_{n-2},$$

so $D_n = 2D_{n-1} - D_{n-2}$. By recursion (or by direct substitution), it is easy to see that the solution of this equation is $D_n = n + 1$.

6.2.14. (a) Use the results of Example 6.2.1 with $\lambda_i = 1/\alpha_i$.

(b) Recognize that the matrix $\mathbf{A}$ is a rank-one updated matrix in the sense that

$$\mathbf{A} = (\alpha - \beta)\mathbf{I} + \beta\mathbf{e}\mathbf{e}^T, \quad \text{where} \quad \mathbf{e} = \begin{pmatrix} 1 \\ \vdots \\ 1 \end{pmatrix}.$$

If $\alpha = \beta$, then $\mathbf{A}$ is singular, so $\det(\mathbf{A}) = 0$. If $\alpha \ne \beta$, then (6.2.3) may be applied to obtain

$$\det(\mathbf{A}) = \det\big((\alpha - \beta)\mathbf{I}\big)\left(1 + \frac{\beta\mathbf{e}^T\mathbf{e}}{\alpha - \beta}\right) = (\alpha - \beta)^n\left(1 + \frac{n\beta}{\alpha - \beta}\right).$$

(c) Recognize that the matrix is $\mathbf{I} + \mathbf{e}\mathbf{d}^T$, where

$$\mathbf{e} = \begin{pmatrix} 1 \\ 1 \\ \vdots \\ 1 \end{pmatrix} \quad \text{and} \quad \mathbf{d} = \begin{pmatrix} \alpha_1 \\ \alpha_2 \\ \vdots \\ \alpha_n \end{pmatrix}.$$

Apply (6.2.2) to produce the desired formula.

6.2.15. (a)   Use the second formula in (6.2.1).

(b)   Apply the first formula in (6.2.1) along with (6.2.7).

6.2.16. If $\lambda = 0$, then the result is trivially true because both sides are zero. If $\lambda \neq 0$, then expand $\begin{vmatrix} \lambda \mathbf{I}_m & \lambda \mathbf{B} \\ \mathbf{C} & \lambda \mathbf{I}_n \end{vmatrix}$ both of the ways indicated in (6.2.1).

6.2.17. (a)   Use the product rule (6.1.15) together with (6.2.2) to write

$$\mathbf{A} + \mathbf{cd}^T = \mathbf{A} + \mathbf{Axd}^T = \mathbf{A}\left(\mathbf{I} + \mathbf{xd}^T\right).$$

(b)   Apply the same technique used in part (a) to obtain

$$\mathbf{A} + \mathbf{cd}^T = \mathbf{A} + \mathbf{cy}^T\mathbf{A} = \left(\mathbf{I} + \mathbf{cy}^T\right)\mathbf{A}.$$

6.2.18. For an elementary reflector $\mathbf{R} = \mathbf{I} - 2\mathbf{uu}^T/\mathbf{u}^T\mathbf{u}$, (6.2.2) insures $\det(\mathbf{R}) = -1$. If $\mathbf{A}_{n \times n}$ is reduced to upper-triangular form (say $\mathbf{PA} = \mathbf{T}$) by Householder reduction as explained on p. 341, then $\det(\mathbf{P})\det(\mathbf{A}) = \det(\mathbf{T}) = t_{11} \cdots t_{nn}$. Since $\mathbf{P}$ is the product of elementary reflectors, $\det(\mathbf{A}) = (-1)^k t_{11} \cdots t_{nn}$, where k is the number of reflections used in the reduction process. In general, one reflection is required to annihilate entries below a diagonal position, so, if no reduction steps can be skipped, then $\det(\mathbf{A}) = (-1)^{n-1} t_{11} \cdots t_{nn}$. If $\mathbf{P}_{ij}$ is a plane rotation, then there is a permutation matrix (a product of interchange matrices) $\mathbf{B}$ such that $\mathbf{P}_{ij} = \mathbf{B}^T \begin{pmatrix} \mathbf{Q} & \mathbf{0} \\ \mathbf{0} & \mathbf{I} \end{pmatrix} \mathbf{B}$, where $\mathbf{Q} = \begin{pmatrix} c & s \\ -s & c \end{pmatrix}$ with $c^2 + s^2 = 1$. Consequently, $\det(\mathbf{P}_{ij}) = \det\left(\mathbf{B}^T\right) \begin{vmatrix} \mathbf{Q} & \mathbf{0} \\ \mathbf{0} & \mathbf{I} \end{vmatrix} \det(\mathbf{B}) = \det(\mathbf{Q}) = 1$ because $\det(\mathbf{B})\det\left(\mathbf{B}^T\right) = \det(\mathbf{B})^2 = 1$ by (6.1.9). Since Givens reduction produces $\mathbf{PA} = \mathbf{T}$, where $\mathbf{P}$ is a product of plane rotations and $\mathbf{T}$ is upper triangular, the product rule (6.1.15) insures $\det(\mathbf{P}) = 1$, so $\det(\mathbf{A}) = \det(\mathbf{T}) = t_{11} \cdots t_{nn}$.

6.2.19. If $\det(\mathbf{A}) = \pm 1$, then (6.2.7) implies $\mathbf{A}^{-1} = \pm\text{adj}(\mathbf{A})$, and thus $\mathbf{A}^{-1}$ is an integer matrix because the cofactors are integers. Conversely, if $\mathbf{A}^{-1}$ is an integer matrix, then $\det\left(\mathbf{A}^{-1}\right)$ and $\det(\mathbf{A})$ are both integers. Since

$$\mathbf{AA}^{-1} = \mathbf{I} \implies \det(\mathbf{A})\det\left(\mathbf{A}^{-1}\right) = 1,$$

it follows that $\det(\mathbf{A}) = \pm 1$.

6.2.20. (a)   Exercise 6.2.19 guarantees that $\mathbf{A}^{-1}$ has integer entries if and only if $\det(\mathbf{A}) = \pm 1$, and (6.2.2) says that $\det(\mathbf{A}) = 1 - 2\mathbf{v}^T\mathbf{u}$, so $\mathbf{A}^{-1}$ has integer entries if and only if $\mathbf{v}^T\mathbf{u}$ is either 0 or 1.

(b)   According to (3.9.1),

$$\mathbf{A}^{-1} = \left(\mathbf{I} - 2\mathbf{uv}^T\right)^{-1} = \mathbf{I} - \frac{2\mathbf{uv}^T}{2\mathbf{v}^T\mathbf{u} - 1},$$

and thus $\mathbf{A}^{-1} = \mathbf{A}$ when $\mathbf{v}^T\mathbf{u} = 1$.

6.2.21. For $n = 2$, two multiplications are required, and $c(2) = 2$. Assume $c(k)$ multiplications are required to evaluate any $k \times k$ determinant by cofactors. For a $k+1 \times k+1$ matrix, the cofactor expansion in terms of the ith row is

$$\det(\mathbf{A}) = a_{i1}\mathring{A}_{i1} + \cdots + a_{ik}\mathring{A}_{ik} + a_{ik+1}\mathring{A}_{ik+1}.$$

Each $\mathring{A}_{ij}$ requires $c(k)$ multiplications, so the above expansion contains

$$(k+1) + (k+1)c(k) = (k+1) + (k+1)k!\left(1 + \frac{1}{2!} + \frac{1}{3!} + \cdots + \frac{1}{(k-1)!}\right)$$
$$= (k+1)!\left(\frac{1}{k!} + \left(1 + \frac{1}{2!} + \frac{1}{3!} + \cdots + \frac{1}{(k-1)!}\right)\right)$$
$$= c(k+1)$$

multiplications. Remember that $e^x = 1 + x + x^2/2! + x^3/3! + \cdots$, so for $n = 100$,

$$1 + \frac{1}{2!} + \frac{1}{3!} + \cdots + \frac{1}{99!} \approx e - 1,$$

and $c(100) \approx 100!(e-1)$. Consequently, approximately 1.6×10^{152} seconds (i.e., 5.1×10^{144} years) are required.

6.2.22. $\mathbf{A} - \lambda\mathbf{I}$ is singular if and only if $\det(\mathbf{A} - \lambda\mathbf{I}) = 0$. The cofactor expansion in terms of the first row yields

$$\det(\mathbf{A} - \lambda\mathbf{I}) = -\lambda\begin{vmatrix} 5-\lambda & 2 \\ -3 & -\lambda \end{vmatrix} + 3\begin{vmatrix} 2 & 2 \\ -2 & -\lambda \end{vmatrix} - 2\begin{vmatrix} 2 & 5-\lambda \\ -2 & -3 \end{vmatrix}$$
$$= -\lambda^3 + 5\lambda^2 - 8\lambda + 4,$$

so $\mathbf{A} - \lambda\mathbf{I}$ is singular if and only if $\lambda^3 - 5\lambda^2 + 8\lambda - 4 = 0$. According to the hint, the integer roots of $p(\lambda) = \lambda^3 - 5\lambda^2 + 8\lambda - 4$ are a subset of $\{\pm 4, \pm 2, \pm 1\}$. Evaluating $p(\lambda)$ at these points reveals that $\lambda = 2$ is a root, and either ordinary or synthetic division produces

$$\frac{p(\lambda)}{\lambda - 2} = \lambda^2 - 3\lambda + 2 = (\lambda - 2)(\lambda - 1).$$

Therefore, $p(\lambda) = (\lambda - 2)^2(\lambda - 1)$, so $\lambda = 2$ and $\lambda = 1$ are the roots of $p(\lambda)$, and these are the values for which $\mathbf{A} - \lambda\mathbf{I}$ is singular.

6.2.23. The indicated substitutions produce the system

$$\begin{pmatrix} x_1' \\ x_2' \\ \vdots \\ x_{n-1}' \\ x_n' \end{pmatrix} = \begin{pmatrix} 0 & 1 & 0 & \cdots & 0 \\ 0 & 0 & 1 & \cdots & 0 \\ \vdots & \vdots & \vdots & \ddots & \vdots \\ 0 & 0 & 0 & \cdots & 1 \\ -p_n & -p_{n-1} & -p_{n-2} & \cdots & -p_1 \end{pmatrix} \begin{pmatrix} x_1 \\ x_2 \\ \vdots \\ x_{n-1} \\ x_n \end{pmatrix}.$$

Each of the n vectors $\mathbf{w}_i = \begin{pmatrix} f_i(t) & f_i'(t) & \cdots & f_i^{(n-1)} \end{pmatrix}^T$ for $i = 1, 2, \ldots, n$ satisfies this system, so (6.2.8) may be applied to produce the desired conclusion.

6.2.24. The result is clearly true for $n = 2$. Assume the formula holds for $n = k - 1$, and prove that it must also hold for $n = k$. According to the cofactor expansion in terms of the first row, $\deg p(\lambda) = k - 1$, and it's clear that

$$p(x_2) = p(x_3) = \cdots = p(x_k) = 0,$$

so $x_2, x_3, \ldots, x_k$ are the $k - 1$ roots of $p(\lambda)$. Consequently,

$$p(\lambda) = \alpha(\lambda - x_2)(\lambda - x_3) \cdots (\lambda - x_k),$$

where α is the coefficient of λ^{k-1}. But the coefficient of λ^{k-1} is the **cofactor** associated with the $(1, k)$-entry, so the induction hypothesis yields

$$\alpha = (-1)^{k-1} \begin{vmatrix} 1 & x_2 & x_2^2 & \cdots & x_2^{k-2} \\ 1 & x_3 & x_3^2 & \cdots & x_3^{k-2} \\ \vdots & \vdots & \vdots & \cdots & \vdots \\ 1 & x_k & x_k^2 & \cdots & x_k^{k-2} \end{vmatrix}_{k-1 \times k-1} = (-1)^{k-1} \prod_{j > i \geq 2} (x_j - x_i).$$

Therefore,

$$\begin{aligned} \det(\mathbf{V}_k) = p(x_1) &= (x_1 - x_2)(x_1 - x_3) \cdots (x_1 - x_k)\alpha \\ &= (x_1 - x_2)(x_1 - x_3) \cdots (x_1 - x_k)\left((-1)^{k-1} \prod_{j > i \geq 2} (x_j - x_i)\right) \\ &= (x_2 - x_1)(x_3 - x_1) \cdots (x_k - x_1) \prod_{j > i \geq 2} (x_j - x_i) \\ &= \prod_{j > i} (x_j - x_i), \end{aligned}$$

and the formula is proven. The determinant is nonzero if and only if the x_i's are distinct numbers, and this agrees with the conclusion in Example 4.3.4.

6.2.25. According to (6.1.19),

$$\frac{d\left(\det(\mathbf{A})\right)}{dx} = \det(\mathbf{D}_1) + \det(\mathbf{D}_2) + \cdots + \det(\mathbf{D}_n),$$

where $\mathbf{D}_i$ is the matrix

$$\mathbf{D}_i = \begin{pmatrix} a_{11} & a_{12} & \cdots & a_{1n} \\ \vdots & \vdots & \cdots & \vdots \\ a_{i1}' & a_{i2}' & \cdots & a_{in}' \\ \vdots & \vdots & \cdots & \vdots \\ a_{n1} & a_{n2} & \cdots & a_{nn} \end{pmatrix}.$$

Expanding $\det(\mathbf{D}_i)$ in terms of cofactors of the ith row yields

$$\det(\mathbf{A}_i) = a'_{i1}\mathring{A}_{i1} + a'_{i2}\mathring{A}_{i2} + \cdots + a'_{in}\mathring{A}_{in},$$

so the desired conclusion is obtained.

6.2.26. According to (6.1.19),

$$\frac{\partial \det(\mathbf{A})}{\partial a_{ij}} = \det(\mathbf{D}_i) = \begin{vmatrix} a_{11} & \cdots & a_{1j} & \cdots & a_{1n} \\ \vdots & \cdots & \vdots & \cdots & \vdots \\ 0 & \cdots & 1 & \cdots & 0 \\ \vdots & \cdots & \vdots & \cdots & \vdots \\ a_{n1} & \cdots & a_{nj} & \cdots & a_{nn} \end{vmatrix} \leftarrow \text{row } i = \mathring{A}_{ij}.$$

6.2.27. The $\binom{4}{2} = 6$ ways to choose pairs of column indices are

$$\begin{array}{ccc} (1,2) & (1,3) & (1,4) \\ & (2,3) & (2,4) \\ & & (3,4) \end{array}$$

so that the Laplace expansion using $i_1 = 1$ and $i_2 = 3$ is

$$\begin{aligned} \det(\mathbf{A}) = {} & \det \mathbf{A}(1,3 \,|\, 1,2)\,\mathring{A}(1,3 \,|\, 1,2) + \det \mathbf{A}(1,3 \,|\, 1,3)\,\mathring{A}(1,3 \,|\, 1,3) \\ & + \det \mathbf{A}(1,3 \,|\, 1,4)\,\mathring{A}(1,3 \,|\, 1,4) + \det \mathbf{A}(1,3 \,|\, 2,3)\,\mathring{A}(1,3 \,|\, 2,3) \\ & + \det \mathbf{A}(1,3 \,|\, 2,4)\,\mathring{A}(1,3 \,|\, 2,4) + \det \mathbf{A}(1,3 \,|\, 3,4)\,\mathring{A}(1,3 \,|\, 3,4) \\ = {} & 0 + (-2)(-4) + (-1)(3)(-2) + 0 + (-3)(-3) + (-1)(-8)(2) \\ = {} & 39. \end{aligned}$$

Solutions for Chapter 7

Solutions for exercises in section 7. 1

7.1.1. $\sigma(\mathbf{A}) = \{-3, 4\}$

$$N(\mathbf{A} + 3\mathbf{I}) = span\left\{\begin{pmatrix} -1 \\ 1 \end{pmatrix}\right\} \quad \text{and} \quad N(\mathbf{A} - 4\mathbf{I}) = span\left\{\begin{pmatrix} -1/2 \\ 1 \end{pmatrix}\right\}$$

$\sigma(\mathbf{B}) = \{-2, 2\}$ in which the algebraic multiplicity of $\lambda = -2$ is two.

$$N(\mathbf{B} + 2\mathbf{I}) = span\left\{\begin{pmatrix} -4 \\ 1 \\ 0 \end{pmatrix}, \begin{pmatrix} -2 \\ 0 \\ 1 \end{pmatrix}\right\} \text{ and } N(\mathbf{B} - 2\mathbf{I}) = span\left\{\begin{pmatrix} -1/2 \\ -1/2 \\ 1 \end{pmatrix}\right\}$$

$\sigma(\mathbf{C}) = \{3\}$ in which the algebraic multiplicity of $\lambda = 3$ is three.

$$N(\mathbf{C} - 3\mathbf{I}) = span\left\{\begin{pmatrix} 1 \\ 0 \\ 0 \end{pmatrix}\right\}$$

$\sigma(\mathbf{D}) = \{3\}$ in which the algebraic multiplicity of $\lambda = 3$ is three.

$$N(\mathbf{D} - 3\mathbf{I}) = span\left\{\begin{pmatrix} 2 \\ 1 \\ 0 \end{pmatrix}, \begin{pmatrix} 1 \\ 0 \\ 1 \end{pmatrix}\right\}$$

$\sigma(\mathbf{E}) = \{3\}$ in which the algebraic multiplicity of $\lambda = 3$ is three.

$$N(\mathbf{E} - 3\mathbf{I}) = span\left\{\begin{pmatrix} 1 \\ 0 \\ 0 \end{pmatrix}, \begin{pmatrix} 0 \\ 1 \\ 0 \end{pmatrix}, \begin{pmatrix} 0 \\ 0 \\ 1 \end{pmatrix}\right\}$$

Matrices $\mathbf{C}$ and $\mathbf{D}$ are deficient in eigenvectors.

7.1.2. Form the product $\mathbf{Ax}$, and answer the question, "Is $\mathbf{Ax}$ some multiple of $\mathbf{x}$?" When the answer is *yes,* then $\mathbf{x}$ is an eigenvector for $\mathbf{A}$, and the multiplier is the associated eigenvalue. For this matrix, (a), (c), and (d) are eigenvectors associated with eigenvalues 1, 3, and 3, respectively.

7.1.3. The characteristic polynomial for $\mathbf{T}$ is

$$\det\left(\mathbf{T} - \lambda\mathbf{I}\right) = \left(t_{11} - \lambda\right)\left(t_{22} - \lambda\right)\cdots\left(t_{nn} - \lambda\right),$$

so the roots are the t_{ii}'s.

7.1.4. This follows directly from (6.1.16) because

$$\det\left(\mathbf{T} - \lambda\mathbf{I}\right) = \begin{vmatrix} \mathbf{A} - \lambda\mathbf{I} & \mathbf{B} \\ \mathbf{0} & \mathbf{C} - \lambda\mathbf{I} \end{vmatrix} = \det\left(\mathbf{A} - \lambda\mathbf{I}\right)\det\left(\mathbf{C} - \lambda\mathbf{I}\right).$$

7.1.5. If λ_i is not repeated, then $N\left(\mathbf{A} - \lambda_i\mathbf{I}\right) = span\left\{\mathbf{e}_i\right\}$. If the algebraic multiplicity of λ_i is k, and if λ_i occupies positions $i_1, i_2, \ldots, i_k$ in $\mathbf{D}$, then

$$N\left(\mathbf{A} - \lambda_i\mathbf{I}\right) = span\left\{\mathbf{e}_{i_1}, \mathbf{e}_{i_2}, \ldots, \mathbf{e}_{i_k}\right\}.$$

7.1.6. $\mathbf{A}$ singular $\iff \det\left(\mathbf{A}\right) = 0 \iff 0$ solves $\det\left(\mathbf{A} - \lambda\mathbf{I}\right) = 0 \iff 0 \in \sigma\left(\mathbf{A}\right)$.

7.1.7. Zero is not in or on any Gerschgorin circle. You could also say that $\mathbf{A}$ is non-singular because it is diagonally dominant—see Example 7.1.6 on p. 499.

7.1.8. If $(\lambda, \mathbf{x})$ is an eigenpair for $\mathbf{A}^*\mathbf{A}$, then $\|\mathbf{A}\mathbf{x}\|_2^2 / \|\mathbf{x}\|_2^2 = \mathbf{x}^*\mathbf{A}^*\mathbf{A}\mathbf{x}/\mathbf{x}^*\mathbf{x} = \lambda$ is real and nonnegative. Furthermore, $\lambda > 0$ if and only if $\mathbf{A}^*\mathbf{A}$ is nonsingular or, equivalently, $n = rank\left(\mathbf{A}^*\mathbf{A}\right) = rank\left(\mathbf{A}\right)$. Similar arguments apply to $\mathbf{A}\mathbf{A}^*$.

7.1.9. (a) $\mathbf{A}\mathbf{x} = \lambda\mathbf{x} \implies \mathbf{x} = \lambda\mathbf{A}^{-1}\mathbf{x} \implies (1/\lambda)\mathbf{x} = \mathbf{A}^{-1}\mathbf{x}$.

     (b) $\mathbf{A}\mathbf{x} = \lambda\mathbf{x} \iff (\mathbf{A} - \alpha\mathbf{I})\mathbf{x} = (\lambda - \alpha)\mathbf{x} \iff (\lambda - \alpha)^{-1}\mathbf{x} = (\mathbf{A} - \alpha\mathbf{I})^{-1}\mathbf{x}$.

7.1.10. (a) Successively use $\mathbf{A}$ as a left-hand multiplier to produce

$$\mathbf{A}\mathbf{x} = \lambda\mathbf{x} \implies \mathbf{A}^2\mathbf{x} = \lambda\mathbf{A}\mathbf{x} = \lambda^2\mathbf{x}$$
$$\implies \mathbf{A}^3\mathbf{x} = \lambda^2\mathbf{A}\mathbf{x} = \lambda^3\mathbf{x}$$
$$\implies \mathbf{A}^4\mathbf{x} = \lambda^3\mathbf{A}\mathbf{x} = \lambda^4\mathbf{x}$$

$$\text{etc.}$$

     (b) Use part (a) to write

$$p(\mathbf{A})\mathbf{x} = \left(\sum_i \alpha_i \mathbf{A}^i\right)\mathbf{x} = \sum_i \alpha_i \mathbf{A}^i \mathbf{x} = \sum_i \alpha_i \lambda^i \mathbf{x} = \left(\sum_i \alpha_i \lambda^i\right)\mathbf{x} = p(\lambda)\mathbf{x}.$$

7.1.11. Since one Geschgorin circle (derived from row sums and shown below) is isolated

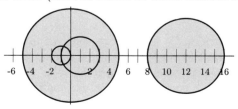

from the union of the other three circles, statement (7.1.14) on p. 498 insures that there is one eigenvalue in the isolated circle and three eigenvalues in the union of the other three. But, as discussed on p. 492, the eigenvalues of real matrices occur in conjugate pairs. So, the root in the isolated circle must be real and there must be at least one real root in the union of the other three circles. Computation reveals that $\sigma(\mathbf{A}) = \{\pm i, 2, 10\}$.

7.1.12. Use Exercise 7.1.10 to deduce that

$$\lambda \in \sigma(\mathbf{A}) \implies \lambda^k \in \sigma(\mathbf{A}^k) \implies \lambda^k = 0 \implies \lambda = 0.$$

Therefore, (7.1.7) insures that $trace(\mathbf{A}) = \sum_i \lambda_i = 0$.

7.1.13. This is true because $N(\mathbf{A} - \lambda\mathbf{I})$ is a subspace—recall that subspaces are closed under vector addition and scalar multiplication.

7.1.14. If there exists a nonzero vector $\mathbf{x}$ that satisfies $\mathbf{A}\mathbf{x} = \lambda_1\mathbf{x}$ and $\mathbf{A}\mathbf{x} = \lambda_2\mathbf{x}$, where $\lambda_1 \neq \lambda_2$, then

$$\mathbf{0} = \mathbf{A}\mathbf{x} - \mathbf{A}\mathbf{x} = \lambda_1\mathbf{x} - \lambda_2\mathbf{x} = (\lambda_1 - \lambda_2)\mathbf{x}.$$

But this implies $\mathbf{x} = \mathbf{0}$, which is impossible. Consequently, no such $\mathbf{x}$ can exist.

7.1.15. No—consider $\mathbf{A} = \begin{pmatrix} 1 & 0 & 0 \\ 0 & 1 & 0 \\ 0 & 0 & 2 \end{pmatrix}$ and $\mathbf{B} = \begin{pmatrix} 1 & 0 & 0 \\ 0 & 2 & 0 \\ 0 & 0 & 2 \end{pmatrix}$.

7.1.16. Almost any example with rather random entries will do the job, but avoid diagonal or triangular matrices—they are too special.

7.1.17. (a) $\mathbf{c} = (\mathbf{A} - \lambda\mathbf{I})^{-1}(\mathbf{A} - \lambda\mathbf{I})\mathbf{c} = (\mathbf{A} - \lambda\mathbf{I})^{-1}(\mathbf{A}\mathbf{c} - \lambda\mathbf{c}) = (\mathbf{A} - \lambda\mathbf{I})^{-1}(\lambda_k - \lambda)\mathbf{c}$.

(b) Use (6.2.3) to compute the characteristic polynomial for $\mathbf{A} + \mathbf{c}\mathbf{d}^T$ to be

$$\begin{aligned}
\det\left(\mathbf{A} + \mathbf{c}\mathbf{d}^T - \lambda\mathbf{I}\right) &= \det\left(\mathbf{A} - \lambda\mathbf{I} + \mathbf{c}\mathbf{d}^T\right) \\
&= \det\left(\mathbf{A} - \lambda\mathbf{I}\right)\left(1 + \mathbf{d}^T(\mathbf{A} - \lambda\mathbf{I})^{-1}\mathbf{c}\right) \\
&= \left(\pm\prod_{i=1}^{n}(\lambda_j - \lambda)\right)\left(1 + \frac{\mathbf{d}^T\mathbf{c}}{\lambda_k - \lambda}\right) \\
&= \left(\pm\prod_{j\neq k}(\lambda_j - \lambda)\right)\left(\lambda_k + \mathbf{d}^T\mathbf{c} - \lambda\right).
\end{aligned}$$

The roots of this polynomial are $\lambda_1, \ldots, \lambda_{k-1}, \lambda_k + \mathbf{d}^T\mathbf{c}, \lambda_{k+1}, \ldots, \lambda_n$.

(c) $\mathbf{d} = \dfrac{(\mu - \lambda_k)\mathbf{c}}{\mathbf{c}^T\mathbf{c}}$ will do the job.

7.1.18. (a) The transpose does not alter the determinant—recall (6.1.4)—so that

$$\det(\mathbf{A} - \lambda\mathbf{I}) = \det\left(\mathbf{A}^T - \lambda\mathbf{I}\right).$$

(b) We know from Exercise 6.1.8 that $\overline{\det(\mathbf{A})} = \det(\mathbf{A}^*)$, so

$$\lambda \in \sigma(\mathbf{A}) \iff 0 = \det(\mathbf{A} - \lambda\mathbf{I})$$
$$\iff 0 = \overline{\det(\mathbf{A} - \lambda\mathbf{I})} = \det((\mathbf{A} - \lambda\mathbf{I})^*) = \det\left(\mathbf{A}^* - \overline{\lambda}\mathbf{I}\right)$$
$$\iff \overline{\lambda} \in \sigma(\mathbf{A}^*).$$

(c) Yes.

(d) Apply the reverse order law for conjugate transposes to obtain

$$\mathbf{y}^*\mathbf{A} = \mu\mathbf{y}^* \implies \mathbf{A}^*\mathbf{y} = \overline{\mu}\mathbf{y} \implies \mathbf{A}^T\mathbf{y} = \overline{\mu}\mathbf{y} \implies \overline{\mu} \in \sigma\left(\mathbf{A}^T\right) = \sigma(\mathbf{A}),$$

and use the conclusion of part (c) insuring that the eigenvalues of real matrices must occur in conjugate pairs.

7.1.19. (a) When $m = n$, Exercise 6.2.16 insures that

$$\lambda^n \det(\mathbf{AB} - \lambda\mathbf{I}) = \lambda^n \det(\mathbf{BA} - \lambda\mathbf{I}) \quad \text{for all } \lambda,$$

so $\det(\mathbf{AB} - \lambda\mathbf{I}) = \det(\mathbf{BA} - \lambda\mathbf{I})$.

(b) If $m \neq n$, then the characteristic polynomials of $\mathbf{AB}$ and $\mathbf{BA}$ are of degrees m and n, respectively, so they must be different. When m and n are different—say $m > n$—Exercise 6.2.16 implies that

$$\det(\mathbf{AB} - \lambda\mathbf{I}) = (-\lambda)^{m-n}\det(\mathbf{BA} - \lambda\mathbf{I}).$$

Consequently, $\mathbf{AB}$ has $m - n$ more zero eigenvalues than $\mathbf{BA}$.

7.1.20. Suppose that $\mathbf{A}$ and $\mathbf{B}$ are $n \times n$, and suppose $\mathbf{X}$ is $n \times g$. The equation $(\mathbf{A} - \lambda\mathbf{I})\mathbf{BX} = \mathbf{0}$ says that the columns of $\mathbf{BX}$ are in $N(\mathbf{A} - \lambda\mathbf{I})$, and hence they are linear combinations of the basis vectors in $\mathbf{X}$. Thus

$$[\mathbf{BX}]_{*j} = \sum_i p_{ij}\mathbf{X}_{*j} \implies \mathbf{BX} = \mathbf{XP}, \quad \text{where } \mathbf{P}_{g \times g} = [p_{ij}].$$

If $(\mu, \mathbf{z})$ is any eigenpair for $\mathbf{P}$, then

$$\mathbf{B}(\mathbf{Xz}) = \mathbf{XPz} = \mu(\mathbf{Xz}) \quad \text{and} \quad \mathbf{AX} = \lambda\mathbf{X} \implies \mathbf{A}(\mathbf{Xz}) = \lambda(\mathbf{Xz}),$$

so $\mathbf{Xz}$ is a common eigenvector.

7.1.21. (a) If $\mathbf{Px} = \lambda\mathbf{x}$ and $\mathbf{y}^*\mathbf{Q} = \mu\mathbf{y}^*$, then $\mathbf{T}(xy^*) = \mathbf{P}xy^*\mathbf{Q} = \lambda\mu xy^*$.

(b) Since $\dim \mathcal{C}^{m \times n} = mn$, the operator $\mathbf{T}$ (as well as any coordinate matrix representation of $\mathbf{T}$) must have exactly mn eigenvalues (counting multiplicities), and since there are exactly mn products $\lambda\mu$, where $\lambda \in \sigma(\mathbf{P})$, $\mu \in \sigma(\mathbf{Q})$, it follows that $\sigma(\mathbf{T}) = \{\lambda\mu \mid \lambda \in \sigma(\mathbf{P}), \mu \in \sigma(\mathbf{Q})\}$. Use the fact that the trace is the sum of the eigenvalues (recall (7.1.7)) to conclude that $trace\,(\mathbf{T}) = \sum_{i,j} \lambda_i \mu_j = \sum_i \lambda_i \sum_j \mu_j = trace\,(\mathbf{P})\,trace\,(\mathbf{Q})$.

7.1.22. (a) Use (6.2.3) to compute the characteristic polynomial for $\mathbf{D} + \alpha\mathbf{v}\mathbf{v}^T$ to be

$$
\begin{aligned}
p(\lambda) &= \det\left(\mathbf{D} + \alpha\mathbf{v}\mathbf{v}^T - \lambda\mathbf{I}\right) \\
&= \det\left(\mathbf{D} - \lambda\mathbf{I} + \alpha\mathbf{v}\mathbf{v}^T\right) \\
&= \det\left(\mathbf{D} - \lambda\mathbf{I}\right)\left(1 + \alpha\mathbf{v}^T(\mathbf{D} - \lambda\mathbf{I})^{-1}\mathbf{v}\right) \qquad\qquad (\ddagger) \\
&= \left(\prod_{j=1}^n (\lambda - \lambda_j)\right)\left(1 + \alpha\sum_{i=1}^n \frac{v_i^2}{\lambda_i - \lambda}\right) \\
&= \prod_{j=1}^n (\lambda - \lambda_j) + \alpha\sum_{i=1}^n \left(v_i \prod_{j\neq i}(\lambda - \lambda_j)\right).
\end{aligned}
$$

For each λ_k, it is true that

$$
p(\lambda_k) = \alpha v_k \prod_{j\neq k}(\lambda_k - \lambda_j) \neq 0,
$$

and hence no λ_k can be an eigenvalue for $\mathbf{D} + \alpha\mathbf{v}\mathbf{v}^T$. Consequently, if ξ is an eigenvalue for $\mathbf{D} + \alpha\mathbf{v}\mathbf{v}^T$, then $\det\,(\mathbf{D} - \xi\mathbf{I}) \neq 0$, so $p(\xi) = 0$ and ($\ddagger$) imply that

$$
0 = 1 + \alpha\mathbf{v}^T(\mathbf{D} - \xi\mathbf{I})^{-1}\mathbf{v} = 1 + \alpha\sum_{i=1}^n \frac{v_i^2}{\lambda_i - \xi} = f(\xi).
$$

(b) Use the fact that $f(\xi_i) = 1 + \alpha\mathbf{v}^T(\mathbf{D} - \xi_i\mathbf{I})^{-1}\mathbf{v} = 0$ to write

$$
\begin{aligned}
\left(\mathbf{D} + \alpha\mathbf{v}\mathbf{v}^T\right)(\mathbf{D} - \xi_i\mathbf{I})^{-1}\mathbf{v} &= \mathbf{D}(\mathbf{D} - \xi_i\mathbf{I})^{-1}\mathbf{v} + \mathbf{v}\left(\alpha\mathbf{v}^T(\mathbf{D} - \xi_i\mathbf{I})^{-1}\mathbf{v}\right) \\
&= \mathbf{D}(\mathbf{D} - \xi_i\mathbf{I})^{-1}\mathbf{v} - \mathbf{v} \\
&= \left(\mathbf{D} - (\mathbf{D} - \xi_i\mathbf{I})\right)(\mathbf{D} - \xi_i\mathbf{I})^{-1}\mathbf{v} \\
&= \xi_i(\mathbf{D} - \xi_i\mathbf{I})^{-1}\mathbf{v}.
\end{aligned}
$$

7.1.23. (a) If $p(\lambda) = (\lambda - \lambda_1)(\lambda - \lambda_2)\cdots(\lambda - \lambda_n)$, then

$$
\ln p(\lambda) = \sum_{i=1}^n \ln(\lambda - \lambda_i) \implies \frac{p'(\lambda)}{p(\lambda)} = \sum_{i=1}^n \frac{1}{(\lambda - \lambda_i)}.
$$

(b) If $|\lambda_i/\lambda| < 1$, then we can write

$$
(\lambda - \lambda_i)^{-1} = \left(\lambda\left(1 - \frac{\lambda_i}{\lambda}\right)\right)^{-1} = \frac{1}{\lambda}\left(1 - \frac{\lambda_i}{\lambda}\right)^{-1} = \frac{1}{\lambda}\left(1 + \frac{\lambda_i}{\lambda} + \frac{\lambda_i^2}{\lambda^2} + \cdots\right).
$$

Consequently,

$$\sum_{i=1}^{n} \frac{1}{(\lambda - \lambda_i)} = \sum_{i=1}^{n} \left(\frac{1}{\lambda} + \frac{\lambda_i}{\lambda^2} + \frac{\lambda_i^2}{\lambda^3} + \cdots \right) = \frac{n}{\lambda} + \frac{\tau_1}{\lambda^2} + \frac{\tau_2}{\lambda^3} + \cdots.$$

(c) Combining these two results yields

$$n\lambda^{n-1} + (n-1)c_1\lambda^{n-2} + (n-2)c_2\lambda^{n-3} + \cdots + c_{n-1}$$
$$= \left(\lambda^n + c_1\lambda^{n-1} + c_2\lambda^{n-2} + \cdots + c_n \right) \left(\frac{n}{\lambda} + \frac{\tau_1}{\lambda^2} + \frac{\tau_2}{\lambda^3} + \cdots \right)$$
$$= n\lambda^{n-1} + (nc_1 + \tau_1)\lambda^{n-2} + (nc_2 + \tau_1 c_1 + \tau_2)\lambda^{n-3}$$
$$+ \cdots + (nc_{n-1} + \tau_1 c_{n-2} + \tau_2 c_{n-3} + \cdots + \tau_{n-1})$$
$$+ (nc_n + \tau_1 c_{n-1} + \tau_2 c_{n-2} \cdots + \tau_n) \frac{1}{\lambda} + \cdots,$$

and equating like powers of λ produces the desired conclusion.

7.1.24. We know from Exercise 7.1.10 that $\lambda \in \sigma(\mathbf{A}) \implies \lambda^k \in \sigma(\mathbf{A}^k)$, so (7.1.7) guarantees that $trace(\mathbf{A}^k) = \sum_i \lambda_i^k = \tau_k$. Proceed by induction. The result is true for $k = 1$ because (7.1.7) says that $c_1 = -trace(\mathbf{A})$. Assume that

$$c_i = -\frac{trace(\mathbf{AB}_{i-1})}{i} \quad \text{for} \quad i = 1, 2, \ldots, k-1,$$

and prove the result holds for $i = k$. Recursive application of the induction hypothesis produces

$$\mathbf{B}_1 = c_1\mathbf{I} + \mathbf{A}$$
$$\mathbf{B}_2 = c_2\mathbf{I} + c_1\mathbf{A} + \mathbf{A}^2$$
$$\vdots$$
$$\mathbf{B}_{k-1} = c_{k-1}\mathbf{I} + c_{k-2}\mathbf{A} + \cdots + c_1\mathbf{A}^{k-2} + \mathbf{A}^{k-1},$$

and therefore we can use Newton's identities given in Exercise 7.1.23 to obtain

$$trace(\mathbf{AB}_{k-1}) = trace\left(c_{k-1}\mathbf{A} + c_{k-2}\mathbf{A}^2 + \cdots + c_1\mathbf{A}^{k-1} + \mathbf{A}^k \right)$$
$$= c_{k-1}\tau_1 + c_{k-2}\tau_2 + \cdots + c_1\tau_{k-1} + \tau_k$$
$$= -kc_k.$$

Solutions for exercises in section 7. 2

7.2.1. The characteristic equation is $\lambda^2 - 2\lambda - 8 = (\lambda+2)(\lambda-4) = 0$, so the eigenvalues are $\lambda_1 = -2$ and $\lambda_2 = 4$. Since no eigenvalue is repeated, (7.2.6) insures $\mathbf{A}$ must be diagonalizable. A similarity transformation $\mathbf{P}$ that diagonalizes $\mathbf{A}$ is constructed from a complete set of independent eigenvectors. Compute a pair of eigenvectors associated with λ_1 and λ_2 to be

$$\mathbf{x}_1 = \begin{pmatrix} -1 \\ 1 \end{pmatrix}, \ \mathbf{x}_2 = \begin{pmatrix} -1 \\ 2 \end{pmatrix}, \quad \text{and set} \quad \mathbf{P} = \begin{pmatrix} -1 & -1 \\ 1 & 2 \end{pmatrix}.$$

Now verify that

$$\mathbf{P}^{-1}\mathbf{A}\mathbf{P} = \begin{pmatrix} -2 & -1 \\ 1 & 1 \end{pmatrix} \begin{pmatrix} -8 & -6 \\ 12 & 10 \end{pmatrix} \begin{pmatrix} -1 & -1 \\ 1 & 2 \end{pmatrix} = \begin{pmatrix} -2 & 0 \\ 0 & 4 \end{pmatrix} = \mathbf{D}.$$

7.2.2. (a) The characteristic equation is $\lambda^3 - 3\lambda - 2 = (\lambda - 2)(\lambda + 1)^2 = 0$, so the eigenvalues are $\lambda = 2$ and $\lambda = -1$. By reducing $\mathbf{A} - 2\mathbf{I}$ and $\mathbf{A} + \mathbf{I}$ to echelon form, compute bases for $N(\mathbf{A} - 2\mathbf{I})$ and $N(\mathbf{A} + \mathbf{I})$. One set of bases is

$$N(\mathbf{A} - 2\mathbf{I}) = span\left\{ \begin{pmatrix} -1 \\ 0 \\ 2 \end{pmatrix} \right\} \quad \text{and} \quad N(\mathbf{A} + \mathbf{I}) = span\left\{ \begin{pmatrix} -1 \\ 1 \\ 0 \end{pmatrix}, \begin{pmatrix} -1 \\ 0 \\ 1 \end{pmatrix} \right\}.$$

Therefore,

$$geo\ mult_{\mathbf{A}}(2) = \dim N(\mathbf{A} - 2\mathbf{I}) = 1 = alg\ mult_{\mathbf{A}}(2),$$
$$geo\ mult_{\mathbf{A}}(-1) = \dim N(\mathbf{A} + \mathbf{I}) = 2 = alg\ mult_{\mathbf{A}}(-1).$$

In other words, $\lambda = 2$ is a simple eigenvalue, and $\lambda = -1$ is a semisimple eigenvalue.

(b) A similarity transformation $\mathbf{P}$ that diagonalizes $\mathbf{A}$ is constructed from a complete set of independent eigenvectors, and these are obtained from the above bases. Set $\mathbf{P} = \begin{pmatrix} -1 & -1 & -1 \\ 0 & 1 & 0 \\ 2 & 0 & 1 \end{pmatrix}$, and compute $\mathbf{P}^{-1} = \begin{pmatrix} 1 & 1 & 1 \\ 0 & 1 & 0 \\ -2 & -2 & -1 \end{pmatrix}$ and verify that $\mathbf{P}^{-1}\mathbf{A}\mathbf{P} = \begin{pmatrix} 2 & 0 & 0 \\ 0 & -1 & 0 \\ 0 & 0 & -1 \end{pmatrix}$.

7.2.3. Consider the matrix $\mathbf{A}$ of Exercise 7.2.1. We know from its solution that $\mathbf{A}$ is similar to $\mathbf{D} = \begin{pmatrix} -2 & 0 \\ 0 & 4 \end{pmatrix}$, but the two eigenspaces for $\mathbf{A}$ are spanned by $\begin{pmatrix} -1 \\ 1 \end{pmatrix}$ and $\begin{pmatrix} -1 \\ 2 \end{pmatrix}$, whereas the eigenspaces for $\mathbf{D}$ are spanned by the unit vectors $\mathbf{e}_1$ and $\mathbf{e}_2$.

7.2.4. The characteristic equation of $\mathbf{A}$ is $p(\lambda) = (\lambda-1)(\lambda-2)^2$, so $alg\ mult_{\mathbf{A}}(2) = 2$. To find $geo\ mult_{\mathbf{A}}(2)$, reduce $\mathbf{A} - 2\mathbf{I}$ to echelon form to find that

$$N(\mathbf{A} - 2\mathbf{I}) = span\left\{ \begin{pmatrix} -1 \\ 0 \\ 1 \end{pmatrix} \right\},$$

so $geo\ mult_{\mathbf{A}}(2) = \dim N(\mathbf{A} - 2\mathbf{I}) = 1$. Since there exists at least one eigenvalue such that $geo\ mult_{\mathbf{A}}(\lambda) < alg\ mult_{\mathbf{A}}(\lambda)$, it follows (7.2.5) on p. 512 that $\mathbf{A}$ cannot be diagonalized by a similarity transformation.

7.2.5. A formal induction argument can be given, but it suffices to "do it with dots" by writing

$$\mathbf{B}^k = (\mathbf{P}^{-1}\mathbf{A}\mathbf{P})(\mathbf{P}^{-1}\mathbf{A}\mathbf{P})\cdots(\mathbf{P}^{-1}\mathbf{A}\mathbf{P})$$
$$= \mathbf{P}^{-1}\mathbf{A}(\mathbf{P}\mathbf{P}^{-1})\mathbf{A}(\mathbf{P}\mathbf{P}^{-1})\cdots(\mathbf{P}\mathbf{P}^{-1})\mathbf{A}\mathbf{P} = \mathbf{P}^{-1}\mathbf{A}\mathbf{A}\cdots\mathbf{A}\mathbf{P} = \mathbf{P}^{-1}\mathbf{A}^k\mathbf{P}.$$

7.2.6. $\lim_{n\to\infty}\mathbf{A}^n = \begin{pmatrix} 5 & 2 \\ -10 & -4 \end{pmatrix}$. Of course, you could compute $\mathbf{A}, \mathbf{A}^2, \mathbf{A}^3, \ldots$ in hopes of seeing a pattern, but this clumsy approach is not definitive. A better technique is to diagonalize $\mathbf{A}$ with a similarity transformation, and then use the result of Exercise 7.2.5. The characteristic equation is $0 = \lambda^2 - (19/10)\lambda + (9/10) = (\lambda - 1)(\lambda - (9/10))$, so the eigenvalues are $\lambda = 1$ and $\lambda = .9$. By reducing $\mathbf{A} - \mathbf{I}$ and $\mathbf{A} - .9\mathbf{I}$ to echelon form, we see that

$$N(\mathbf{A} - \mathbf{I}) = span\left\{ \begin{pmatrix} -1 \\ 2 \end{pmatrix} \right\} \quad \text{and} \quad N(\mathbf{A} - .9\mathbf{I}) = span\left\{ \begin{pmatrix} -2 \\ 5 \end{pmatrix} \right\},$$

so $\mathbf{A}$ is indeed diagonalizable, and $\mathbf{P} = \begin{pmatrix} -1 & -2 \\ 2 & 5 \end{pmatrix}$ is a matrix such that $\mathbf{P}^{-1}\mathbf{A}\mathbf{P} = \begin{pmatrix} 1 & 0 \\ 0 & .9 \end{pmatrix} = \mathbf{D}$ or, equivalently, $\mathbf{A} = \mathbf{P}\mathbf{D}\mathbf{P}^{-1}$. The result of Exercise 7.2.5 says that $\mathbf{A}^n = \mathbf{P}\mathbf{D}^n\mathbf{P}^{-1} = \mathbf{P}\begin{pmatrix} 1 & 0 \\ 0 & .9^n \end{pmatrix}\mathbf{P}^{-1}$, so

$$\lim_{n\to\infty}\mathbf{A}^n = \mathbf{P}\begin{pmatrix} 1 & 0 \\ 0 & 0 \end{pmatrix}\mathbf{P}^{-1} = \begin{pmatrix} -1 & -2 \\ 2 & 5 \end{pmatrix}\begin{pmatrix} 1 & 0 \\ 0 & 0 \end{pmatrix}\begin{pmatrix} -5 & -2 \\ 2 & 1 \end{pmatrix} = \begin{pmatrix} 5 & 2 \\ -10 & -4 \end{pmatrix}.$$

7.2.7. It follows from $\mathbf{P}^{-1}\mathbf{P} = \mathbf{I}$ that $\mathbf{y}_i^*\mathbf{x}_j = \begin{cases} 1 & \text{if } i = j, \\ 0 & \text{if } i \neq j, \end{cases}$ as well as $\mathbf{y}_i^*\mathbf{X} = \mathbf{0}$ and $\mathbf{Y}^*\mathbf{x}_i = \mathbf{0}$ for each $i = 1, \ldots, t$, so

$$\mathbf{P}^{-1}\mathbf{A}\mathbf{P} = \begin{pmatrix} \mathbf{y}_1^* \\ \vdots \\ \mathbf{y}_t^* \\ \mathbf{Y}^* \end{pmatrix} \mathbf{A}\left(\mathbf{x}_1 | \cdots | \mathbf{x}_t | \mathbf{X}\right) = \begin{pmatrix} \lambda_1 & \cdots & 0 & \mathbf{0} \\ \vdots & \ddots & \vdots & \vdots \\ 0 & \cdots & \lambda_t & \mathbf{0} \\ \mathbf{0} & \cdots & \mathbf{0} & \mathbf{Y}^*\mathbf{A}\mathbf{X} \end{pmatrix} = \mathbf{B}.$$

Therefore, examining the first t rows on both sides of $\mathbf{P}^{-1}\mathbf{A} = \mathbf{B}\mathbf{P}^{-1}$ yields $\mathbf{y}_i^*\mathbf{A} = \lambda_i\mathbf{y}_i^*$ for $i = 1, \ldots, t$.

7.2.8. If $\mathbf{P}^{-1}\mathbf{A}\mathbf{P} = \text{diag}\,(\lambda_1, \lambda_2, \ldots, \lambda_n)$, then $\mathbf{P}^{-1}\mathbf{A}^k\mathbf{P} = \text{diag}\,(\lambda_1^k, \lambda_2^k, \ldots, \lambda_n^k)$ for $k = 0, 1, 2, \ldots$ or, equivalently, $\mathbf{A}^k = \mathbf{P}\,\text{diag}\,(\lambda_1^k, \lambda_2^k, \ldots, \lambda_n^k)\,\mathbf{P}^{-1}$. Therefore, $\mathbf{A}^k \to \mathbf{0}$ if and only if each $\lambda_i^k \to 0$, which is equivalent to saying that $|\lambda_i| < 1$ for each i. Since $\rho(\mathbf{A}) = \max_{\lambda_i \in \sigma(\mathbf{A})} |\lambda_i|$ (recall Example 7.1.4 on p. 497), it follows that $\mathbf{A}^k \to \mathbf{0}$ if and only if $\rho(\mathbf{A}) < 1$.

7.2.9. The characteristic equation for $\mathbf{A}$ is $\lambda^2 - 2\lambda + 1$, so $\lambda = 1$ is the only distinct eigenvalue. By reducing $\mathbf{A} - \mathbf{I}$ to echelon form, we see that $\begin{pmatrix} 3 \\ 4 \end{pmatrix}$ is a basis for $N(\mathbf{A} - \mathbf{I})$, so $\mathbf{x} = (1/5)\begin{pmatrix} 3 \\ 4 \end{pmatrix}$ is an eigenvector of unit length. Following the procedure on p. 325, we find that $\mathbf{R} = \begin{pmatrix} 3/5 & 4/5 \\ 4/5 & -3/5 \end{pmatrix}$ is an elementary reflector having $\mathbf{x}$ as its first column, and $\mathbf{R}^T\mathbf{A}\mathbf{R} = \mathbf{R}\mathbf{A}\mathbf{R} = \begin{pmatrix} 1 & 25 \\ 0 & 1 \end{pmatrix}$.

7.2.10. From Example 7.2.1 on p. 507 we see that the characteristic equation for $\mathbf{A}$ is $p(\lambda) = \lambda^3 + 5\lambda^2 + 3\lambda - 9 = (\lambda - 1)(\lambda + 3)^2 = 0$. Straightforward computation shows that

$$p(\mathbf{A}) = (\mathbf{A} - \mathbf{I})(\mathbf{A} + 3\mathbf{I})^2 = \begin{pmatrix} 0 & -4 & -4 \\ 8 & -12 & -8 \\ -8 & 8 & 4 \end{pmatrix}\begin{pmatrix} 16 & -16 & -16 \\ 32 & -32 & -32 \\ -32 & 32 & 32 \end{pmatrix} = \mathbf{0}.$$

7.2.11. Rescale the observed eigenvector as $\mathbf{x} = (1/2)(1, 1, 1, 1)^T = \mathbf{y}$ so that $\mathbf{x}^T\mathbf{x} = 1$. Follow the procedure described in Example 5.6.3 (p. 325), and set $\mathbf{u} = \mathbf{x} - \mathbf{e}_1$ to construct

$$\mathbf{R} = \mathbf{I} - \frac{2\mathbf{u}\mathbf{u}^T}{\mathbf{u}^T\mathbf{u}} = \frac{1}{2}\begin{pmatrix} 1 & 1 & 1 & 1 \\ 1 & 1 & -1 & -1 \\ 1 & -1 & 1 & -1 \\ 1 & -1 & -1 & 1 \end{pmatrix} = \mathbf{P} = [\mathbf{x}\,|\,\mathbf{X}] \quad (\text{since } \mathbf{x} = \mathbf{y}).$$

Consequently, $\mathbf{B} = \mathbf{X}^T\mathbf{A}\mathbf{X} = \begin{pmatrix} -1 & 0 & -1 \\ 0 & 2 & 0 \\ -1 & 0 & 1 \end{pmatrix}$, and $\sigma(\mathbf{B}) = \{2, \sqrt{2}, -\sqrt{2}\}$.

7.2.12. Use the spectral theorem with properties $\mathbf{G}_i\mathbf{G}_j = \mathbf{0}$ for $i \neq j$ and $\mathbf{G}_i^2 = \mathbf{G}_i$ to write $\mathbf{A}\mathbf{G}_i = (\lambda_1\mathbf{G}_1 + \lambda_2\mathbf{G}_2 + \cdots + \lambda_k\mathbf{G}_k)\mathbf{G}_i = \lambda_i\mathbf{G}_i^2 = \lambda_i\mathbf{G}_i$. A similar argument shows $\mathbf{G}_i\mathbf{A} = \lambda_i\mathbf{G}_i$.

7.2.13. Use (6.2.3) to show that $\lambda^{n-1}(\lambda - \mathbf{d}^T\mathbf{c}) = 0$ is the characteristic equation for $\mathbf{A}$. Thus $\lambda = 0$ and $\lambda = \mathbf{d}^T\mathbf{c}$ are the eigenvalues of $\mathbf{A}$. We know from (7.2.5) that $\mathbf{A}$ is diagonalizable if and only if the algebraic and geometric multiplicities agree for each eigenvalue. Since $geo\ mult_\mathbf{A}(0) = \dim N(\mathbf{A}) = n - rank(\mathbf{A}) = n - 1$, and since

$$alg\ mult_\mathbf{A}(0) = \begin{cases} n - 1 & \text{if } \mathbf{d}^T\mathbf{c} \neq 0, \\ n & \text{if } \mathbf{d}^T\mathbf{c} = 0, \end{cases}$$

it follows that $\mathbf{A}$ is diagonalizable if and only if $\mathbf{d}^T\mathbf{c} \neq 0$.

7.2.14. If $\mathbf{W}$ and $\mathbf{Z}$ are diagonalizable—say $\mathbf{P}^{-1}\mathbf{W}\mathbf{P}$ and $\mathbf{Q}^{-1}\mathbf{Z}\mathbf{Q}$ are diagonal—then $\begin{pmatrix} \mathbf{P} & \mathbf{0} \\ \mathbf{0} & \mathbf{Q} \end{pmatrix}$ diagonalizes $\mathbf{A}$. Use an indirect argument for the converse. Suppose $\mathbf{A}$ is diagonalizable but $\mathbf{W}$ (or $\mathbf{Z}$) is not. Then there is an eigenvalue $\lambda \in \sigma(\mathbf{W})$ with *geo mult*$_{\mathbf{W}}(\lambda) < $ *alg mult*$_{\mathbf{W}}(\lambda)$. Since $\sigma(\mathbf{A}) = \sigma(\mathbf{W}) \cup \sigma(\mathbf{Z})$ (Exercise 7.1.4), this would mean that

$$
\begin{aligned}
\textit{geo mult}_{\mathbf{A}}(\lambda) &= \dim N(\mathbf{A} - \lambda\mathbf{I}) = (s+t) - \textit{rank}(\mathbf{A} - \lambda\mathbf{I}) \\
&= (s - \textit{rank}(\mathbf{W} - \lambda\mathbf{I})) + (t - \textit{rank}(\mathbf{Z} - \lambda\mathbf{I})) \\
&= \dim N(\mathbf{W} - \lambda\mathbf{I}) + \dim N(\mathbf{Z} - \lambda\mathbf{I}) \\
&= \textit{geo mult}_{\mathbf{W}}(\lambda) + \textit{geo mult}_{\mathbf{Z}}(\lambda) \\
&< \textit{alg mult}_{\mathbf{W}}(\lambda) + \textit{alg mult}_{\mathbf{Z}}(\lambda) \\
&< \textit{alg mult}_{\mathbf{A}}(\lambda),
\end{aligned}
$$

which contradicts the fact that $\mathbf{A}$ is diagonalizable.

7.2.15. If $\mathbf{AB} = \mathbf{BA}$, then, by Exercise 7.1.20 (p. 503), $\mathbf{A}$ and $\mathbf{B}$ have a common eigenvector—say $\mathbf{Ax} = \lambda\mathbf{x}$ and $\mathbf{Bx} = \mu\mathbf{x}$, where $\mathbf{x}$ has been scaled so that $\|\mathbf{x}\|_2 = 1$. If $\mathbf{R} = [\mathbf{x} \,|\, \mathbf{X}]$ is a unitary matrix having $\mathbf{x}$ as its first column (Example 5.6.3, p. 325), then

$$
\mathbf{R}^*\mathbf{AR} = \begin{pmatrix} \lambda & \mathbf{x}^*\mathbf{AX} \\ \mathbf{0} & \mathbf{X}^*\mathbf{AX} \end{pmatrix} \quad \text{and} \quad \mathbf{R}^*\mathbf{BR} = \begin{pmatrix} \mu & \mathbf{x}^*\mathbf{BX} \\ \mathbf{0} & \mathbf{X}^*\mathbf{BX} \end{pmatrix}.
$$

Since $\mathbf{A}$ and $\mathbf{B}$ commute, so do $\mathbf{R}^*\mathbf{AR}$ and $\mathbf{R}^*\mathbf{BR}$, which in turn implies $\mathbf{A}_2 = \mathbf{X}^*\mathbf{AX}$ and $\mathbf{B}_2 = \mathbf{X}^*\mathbf{BX}$ commute. Thus the problem is deflated, so the same argument can be applied inductively in a manner similar to the development of Schur's triangularization theorem (p. 508).

7.2.16. If $\mathbf{P}^{-1}\mathbf{AP} = \mathbf{D}_1$ and $\mathbf{P}^{-1}\mathbf{BP} = \mathbf{D}_2$ are both diagonal, then $\mathbf{D}_1\mathbf{D}_2 = \mathbf{D}_2\mathbf{D}_1$ implies that $\mathbf{AB} = \mathbf{BA}$. Conversely, suppose $\mathbf{AB} = \mathbf{BA}$. Let $\lambda \in \sigma(\mathbf{A})$ with *alg mult*$_{\mathbf{A}}(\lambda) = a$, and let $\mathbf{P}$ be such that $\mathbf{P}^{-1}\mathbf{AP} = \begin{pmatrix} \lambda\mathbf{I}_a & \mathbf{0} \\ \mathbf{0} & \mathbf{D} \end{pmatrix}$, where $\mathbf{D}$ is a diagonal matrix with $\lambda \notin \sigma(\mathbf{D})$. Since $\mathbf{A}$ and $\mathbf{B}$ commute, so do $\mathbf{P}^{-1}\mathbf{AP}$ and $\mathbf{P}^{-1}\mathbf{BP}$. Consequently, if $\mathbf{P}^{-1}\mathbf{BP} = \begin{pmatrix} \mathbf{W} & \mathbf{X} \\ \mathbf{Y} & \mathbf{Z} \end{pmatrix}$, then

$$
\begin{pmatrix} \lambda\mathbf{I}_a & \mathbf{0} \\ \mathbf{0} & \mathbf{D} \end{pmatrix}\begin{pmatrix} \mathbf{W} & \mathbf{X} \\ \mathbf{Y} & \mathbf{Z} \end{pmatrix} = \begin{pmatrix} \mathbf{W} & \mathbf{X} \\ \mathbf{Y} & \mathbf{Z} \end{pmatrix}\begin{pmatrix} \lambda\mathbf{I}_a & \mathbf{0} \\ \mathbf{0} & \mathbf{D} \end{pmatrix} \implies \begin{cases} \lambda\mathbf{X} = \mathbf{XD}, \\ \mathbf{DY} = \lambda\mathbf{Y}, \end{cases}
$$

so $(\mathbf{D} - \lambda\mathbf{I})\mathbf{X} = \mathbf{0}$ and $(\mathbf{D} - \lambda\mathbf{I})\mathbf{Y} = \mathbf{0}$. But $(\mathbf{D} - \lambda\mathbf{I})$ is nonsingular, so $\mathbf{X} = \mathbf{0}$ and $\mathbf{Y} = \mathbf{0}$, and thus $\mathbf{P}^{-1}\mathbf{BP} = \begin{pmatrix} \mathbf{W} & \mathbf{0} \\ \mathbf{0} & \mathbf{Z} \end{pmatrix}$. Since $\mathbf{B}$ is diagonalizable, so is

$\mathbf{P}^{-1}\mathbf{BP}$, and hence so are $\mathbf{W}$ and $\mathbf{Z}$ (Exercise 7.2.14). If $\mathbf{Q} = \begin{pmatrix} \mathbf{Q}_w & \mathbf{0} \\ \mathbf{0} & \mathbf{Q}_z \end{pmatrix}$,
where $\mathbf{Q}_w$ and $\mathbf{Q}_z$ are such that $\mathbf{Q}_w^{-1}\mathbf{W}\mathbf{Q}_w = \mathbf{D}_w$ and $\mathbf{Q}_z^{-1}\mathbf{Z}\mathbf{Q}_z = \mathbf{D}_z$ are
each diagonal, then

$$(\mathbf{PQ})^{-1}\mathbf{A}(\mathbf{PQ}) = \begin{pmatrix} \lambda\mathbf{I}_a & \mathbf{0} \\ \mathbf{0} & \mathbf{Q}_z^{-1}\mathbf{D}\mathbf{Q}_z \end{pmatrix} \quad \text{and} \quad (\mathbf{PQ})^{-1}\mathbf{B}(\mathbf{PQ}) = \begin{pmatrix} \mathbf{D}_w & \mathbf{0} \\ \mathbf{0} & \mathbf{D}_z \end{pmatrix}.$$

Thus the problem is deflated because $\mathbf{A}_2 = \mathbf{Q}_z^{-1}\mathbf{D}\mathbf{Q}_z$ and $\mathbf{B}_2 = \mathbf{D}_z$ commute
and are diagonalizable, so the same argument can be applied to them. If $\mathbf{A}$ has k
distinct eigenvalues, then the desired conclusion is attained after k repetitions.

7.2.17. It's not legitimate to equate $p(\mathbf{A})$ with $\det(\mathbf{A} - \mathbf{A}\mathbf{I})$ because the former is a
matrix while the latter is a scalar.

7.2.18. This follows from the eigenvalue formula developed in Example 7.2.5 (p. 514) by
using the identity $1 - \cos\theta = 2\sin^2(\theta/2)$.

7.2.19. (a) The result in Example 7.2.5 (p. 514) shows that the eigenvalues of $\mathbf{N} + \mathbf{N}^T$
and $\mathbf{N} - \mathbf{N}^T$ are $\lambda_j = 2\cos(j\pi/n + 1)$ and $\lambda_j = 2\mathrm{i}\cos(j\pi/n + 1)$, respectively.

(b) Since $\mathbf{N} - \mathbf{N}^T$ is skew symmetric, it follows from Exercise 6.1.12 (p. 473)
that $\mathbf{N} - \mathbf{N}^T$ is nonsingular if and only if n is even, which is equivalent to saying
$\mathbf{N} - \mathbf{N}^T$ has no zero eigenvalues (recall Exercise 7.1.6, p. 501), and hence, by
part (a), the same is true for $\mathbf{N} + \mathbf{N}^T$.

(b: Alternate) Since the eigenvalues of $\mathbf{N} + \mathbf{N}^T$ are $\lambda_j = 2\cos(j\pi/n + 1)$ you
can argue that $\mathbf{N} + \mathbf{N}^T$ has a zero eigenvalue (and hence is singular) if and only
if n is odd by showing that there exists an integer α such that $j\pi/n + 1 = \alpha\pi/2$
for some $1 \le j \le n$ if and only if n is odd.

(c) Since a determinant is the product of eigenvalues (recall (7.1.8), p. 494),
$\det(\mathbf{N} - \mathbf{N}^T)/\det(\mathbf{N} + \mathbf{N}^T) = (\mathrm{i}\lambda_1 \cdots \mathrm{i}\lambda_n)/(\lambda_1 \cdots \lambda_n) = \mathrm{i}^n = (-1)^{n/2}$.

7.2.20. The eigenvalues are $\{2, 0, 2, 0\}$. The columns of $\mathbf{F}_4 = \begin{pmatrix} 1 & 1 & 1 & 1 \\ 1 & -\mathrm{i} & -1 & \mathrm{i} \\ 1 & -1 & 1 & -1 \\ 1 & \mathrm{i} & -1 & -\mathrm{i} \end{pmatrix}$ are
corresponding eigenvectors.

7.2.21. $\mathbf{Ax} = \lambda\mathbf{x} \implies \mathbf{y}^*\mathbf{Ax} = \lambda\mathbf{y}^*\mathbf{x}$ and $\mathbf{y}^*\mathbf{A} = \mu\mathbf{y}^* \implies \mathbf{y}^*\mathbf{Ax} = \mu\mathbf{y}^*\mathbf{x}$.
Therefore, $\lambda\mathbf{y}^*\mathbf{x} = \mu\mathbf{y}^*\mathbf{x} \implies (\lambda - \mu)\mathbf{y}^*\mathbf{x} = 0 \implies \mathbf{y}^*\mathbf{x} = 0$ when $\lambda \ne \mu$.

7.2.22. (a) Suppose $\mathbf{P}$ is a nonsingular matrix such that $\mathbf{P}^{-1}\mathbf{AP} = \mathbf{D}$ is diagonal,
and suppose that λ is the kth diagonal entry in $\mathbf{D}$. If $\mathbf{x}$ and $\mathbf{y}^*$ are the kth
column and k^{th} row in $\mathbf{P}$ and $\mathbf{P}^{-1}$, respectively, then $\mathbf{x}$ and $\mathbf{y}^*$ must be
right-hand and left-hand eigenvectors associated with λ such that $\mathbf{y}^*\mathbf{x} = 1$.

(b) Consider $\mathbf{A} = \mathbf{I}$ with $\mathbf{x} = \mathbf{e}_i$ and $\mathbf{y} = \mathbf{e}_j$ for $i \ne j$.

(c) Consider $\mathbf{A} = \begin{pmatrix} 0 & 1 \\ 0 & 0 \end{pmatrix}$.

7.2.23. (a) Suppose not—i.e., suppose $\mathbf{y}^*\mathbf{x} = 0$. Then

$$\mathbf{x} \perp span(\mathbf{y}) = N(\mathbf{A} - \lambda\mathbf{I})^* \implies \mathbf{x} \in N(\mathbf{A} - \lambda\mathbf{I})^{*\perp} = R(\mathbf{A} - \lambda\mathbf{I}).$$

Also, $\mathbf{x} \in N(\mathbf{A} - \lambda\mathbf{I})$, so $\mathbf{x} \in R(\mathbf{A} - \lambda\mathbf{I}) \cap N(\mathbf{A} - \lambda\mathbf{I})$. However, because λ is a simple eigenvalue, the the core-nilpotent decomposition on p. 397 insures that $\mathbf{A} - \lambda\mathbf{I}$ is similar to a matrix of the form $\begin{pmatrix} \mathbf{C} & \mathbf{0} \\ \mathbf{0} & 0_{1\times 1} \end{pmatrix}$, and this implies that $R(\mathbf{A} - \lambda\mathbf{I}) \cap N(\mathbf{A} - \lambda\mathbf{I}) = \mathbf{0}$ (Exercise 5.10.12, p. 402), which is a contradiction. Thus $\mathbf{y}^*\mathbf{x} \neq 0$.

(b) Consider $\mathbf{A} = \mathbf{I}$ with $\mathbf{x} = \mathbf{e}_i$ and $\mathbf{y} = \mathbf{e}_j$ for $i \neq j$.

7.2.24. Let $\mathcal{B}_i$ be a basis for $N(\mathbf{A} - \lambda_i\mathbf{I})$, and suppose $\mathbf{A}$ is diagonalizable. Since $geo\ mult_{\mathbf{A}}(\lambda_i) = alg\ mult_{\mathbf{A}}(\lambda_i)$ for each i, (7.2.4) implies $\mathcal{B} = \mathcal{B}_1 \cup \mathcal{B}_2 \cup \cdots \cup \mathcal{B}_k$ is a set of n independent vectors—i.e., $\mathcal{B}$ is a basis for $\Re^n$. Exercise 5.9.14 now guarantees that $\Re^n = N(\mathbf{A} - \lambda_1\mathbf{I}) \oplus N(\mathbf{A} - \lambda_2\mathbf{I}) \oplus \cdots \oplus N(\mathbf{A} - \lambda_k\mathbf{I})$. Conversely, if this equation holds, then Exercise 5.9.14 says $\mathcal{B} = \mathcal{B}_1 \cup \mathcal{B}_2 \cup \cdots \cup \mathcal{B}_k$ is a basis for $\Re^n$, and hence $\mathbf{A}$ is diagonalizable because $\mathcal{B}$ is a complete independent set of eigenvectors.

7.2.25. Proceed inductively just as in the development of Schur's triangularization theorem. If the first eigenvalue λ is real, the reduction is exactly the same as described on p. 508 (with everything being real). If λ is complex, then $(\lambda, \mathbf{x})$ and $(\overline{\lambda}, \overline{\mathbf{x}})$ are both eigenpairs for $\mathbf{A}$, and, by (7.2.3), $\{\mathbf{x}, \overline{\mathbf{x}}\}$ is linearly independent. Consequently, if $\mathbf{x} = \mathbf{u} + i\mathbf{v}$, with $\mathbf{u}, \mathbf{v} \in \Re^{n\times 1}$, then $\{\mathbf{u}, \mathbf{v}\}$ is linearly independent—otherwise, $\mathbf{u} = \xi\mathbf{v}$ implies $\mathbf{x} = (1 + i\xi)\mathbf{u}$ and $\overline{\mathbf{x}} = (1 - i\xi)\mathbf{u}$, which is impossible. Let $\lambda = \alpha + i\beta$, $\alpha, \beta \in \Re$, and observe that $\mathbf{A}\mathbf{x} = \lambda\mathbf{x}$ implies $\mathbf{A}\mathbf{u} = \alpha\mathbf{u} - \beta\mathbf{v}$ and $\mathbf{A}\mathbf{v} = \beta\mathbf{u} + \alpha\mathbf{v}$, so $\mathbf{A}\mathbf{W} = \mathbf{W}\begin{pmatrix} \alpha & \beta \\ -\beta & \alpha \end{pmatrix}$, where $\mathbf{W} = [\mathbf{u}\,|\,\mathbf{v}]$. Let $\mathbf{W} = \mathbf{Q}_{n\times 2}\mathbf{R}_{2\times 2}$ be a rectangular QR factorization (p. 311), and let $\mathbf{B} = \mathbf{R}\begin{pmatrix} \alpha & \beta \\ -\beta & \alpha \end{pmatrix}\mathbf{R}^{-1}$ so that $\sigma(\mathbf{B}) = \sigma\begin{pmatrix} \alpha & \beta \\ -\beta & \alpha \end{pmatrix} = \{\lambda, \overline{\lambda}\}$, and

$$\mathbf{A}\mathbf{W} = \mathbf{A}\mathbf{Q}\mathbf{R} = \mathbf{Q}\mathbf{R}\begin{pmatrix} \alpha & \beta \\ -\beta & \alpha \end{pmatrix} \implies \mathbf{Q}^T\mathbf{A}\mathbf{Q} = \mathbf{R}\begin{pmatrix} \alpha & \beta \\ -\beta & \alpha \end{pmatrix}\mathbf{R}^{-1} = \mathbf{B}.$$

If $\mathbf{X}_{n\times n-2}$ is chosen so that $\mathbf{P} = [\mathbf{Q}\,|\,\mathbf{X}]$ is an orthogonal matrix (i.e., the columns of $\mathbf{X}$ complete the two columns of $\mathbf{Q}$ to an orthonormal basis for $\Re^n$), then $\mathbf{X}^T\mathbf{A}\mathbf{Q} = \mathbf{X}^T\mathbf{Q}\mathbf{B} = \mathbf{0}$, and

$$\mathbf{P}^T\mathbf{A}\mathbf{P} = \begin{pmatrix} \mathbf{Q}^T\mathbf{A}\mathbf{Q} & \mathbf{Q}^T\mathbf{A}\mathbf{X} \\ \mathbf{X}^T\mathbf{A}\mathbf{Q} & \mathbf{X}^T\mathbf{A}\mathbf{X} \end{pmatrix} = \begin{pmatrix} \mathbf{B} & \mathbf{Q}^T\mathbf{A}\mathbf{X} \\ \mathbf{0} & \mathbf{X}^T\mathbf{A}\mathbf{X} \end{pmatrix}.$$

Now repeat the argument on the $n - 2 \times n - 2$ matrix $\mathbf{X}^T\mathbf{A}\mathbf{X}$. Continuing in this manner produces the desired conclusion.

7.2.26. Let the columns $\mathbf{R}_{n\times r}$ be linearly independent eigenvectors corresponding to the real eigenvalues ρ_j, and let $\{\mathbf{x}_1, \overline{\mathbf{x}}_1, \mathbf{x}_2, \overline{\mathbf{x}}_2, \ldots, \mathbf{x}_t, \overline{\mathbf{x}}_t\}$ be a set of linearly independent eigenvectors associated with $\{\lambda_1, \overline{\lambda}_1, \lambda_2, \overline{\lambda}_2, \ldots, \lambda_t, \overline{\lambda}_t\}$ so that the matrix $\mathbf{Q} = [\mathbf{R}\,|\,\mathbf{x}_1\,|\,\overline{\mathbf{x}}_1\,|\cdots|\,\mathbf{x}_t\,|\,\overline{\mathbf{x}}_t]$ is nonsingular. Write $\mathbf{x}_j = \mathbf{u}_j + i\mathbf{v}_j$ for

$\mathbf{u}_j, \mathbf{v}_j \in \Re^{n \times 1}$ and $\lambda_j = \alpha_j + i\beta_j$ for $\alpha, \beta \in \Re$, and let $\mathbf{P}$ be the real matrix $\mathbf{P} = \begin{bmatrix} \mathbf{R} \,|\, \mathbf{u}_1 \,|\, \mathbf{v}_1 \,|\, \mathbf{u}_2 \,|\, \mathbf{v}_2 \,|\, \cdots \,|\, \mathbf{u}_t \,|\, \mathbf{v}_t \end{bmatrix}$. This matrix is nonsingular because Exercise 6.1.14 can be used to show that $\det(\mathbf{P}) = 2t(-i)^t \det(\mathbf{Q})$. For example, if $t = 1$, then $\mathbf{P} = \begin{bmatrix} \mathbf{R} \,|\, \mathbf{u}_1 \,|\, \mathbf{v}_1 \end{bmatrix}$ and

$$
\begin{aligned}
\det(\mathbf{Q}) &= \det\begin{bmatrix} \mathbf{R} \,|\, \mathbf{x}_1 \,|\, \overline{\mathbf{x}}_1 \end{bmatrix} = \det\begin{bmatrix} \mathbf{R} \,|\, \mathbf{u}_1 + i\mathbf{v}_1 \,|\, \mathbf{u}_1 - i\mathbf{v}_1 \end{bmatrix} \\
&= \det\begin{bmatrix} \mathbf{R} \,|\, \mathbf{u}_1 \,|\, \mathbf{u}_1 \end{bmatrix} + \det\begin{bmatrix} \mathbf{R} \,|\, \mathbf{u}_1 \,|\, -i\mathbf{v}_1 \end{bmatrix} \\
&\quad + \det\begin{bmatrix} \mathbf{R} \,|\, i\mathbf{v}_1 \,|\, \mathbf{u}_1 \end{bmatrix} + \det\begin{bmatrix} \mathbf{R} \,|\, i\mathbf{v}_1 \,|\, i\mathbf{v}_1 \end{bmatrix} \\
&= -i\det\begin{bmatrix} \mathbf{R} \,|\, \mathbf{u}_1 \,|\, \mathbf{v}_1 \end{bmatrix} + i\det\begin{bmatrix} \mathbf{R} \,|\, \mathbf{v}_1 \,|\, \mathbf{u}_1 \end{bmatrix} \\
&= -i\det\begin{bmatrix} \mathbf{R} \,|\, \mathbf{u}_1 \,|\, \mathbf{v}_1 \end{bmatrix} - i\det\begin{bmatrix} \mathbf{R} \,|\, \mathbf{u}_1 \,|\, \mathbf{v}_1 \end{bmatrix} = 2(-i)\det(\mathbf{P}).
\end{aligned}
$$

Induction can now be used. The equations $\mathbf{A}(\mathbf{u}_j + i\mathbf{v}_j) = (\alpha_j + i\beta_j)(\mathbf{u}_j + i\mathbf{v}_j)$ yield $\mathbf{A}\mathbf{u}_j = \alpha_j\mathbf{u}_j - \beta_j\mathbf{v}_j$ and $\mathbf{A}\mathbf{v}_j = \beta_j\mathbf{u}_j + \alpha_j\mathbf{v}_j$. Couple these with the fact that $\mathbf{AR} = \mathbf{RD}$ to conclude that

$$
\mathbf{AP} = \begin{bmatrix} \mathbf{RD} \,|\, \cdots \,|\, \alpha_j\mathbf{u}_j - \beta_j\mathbf{v}_j \,|\, \beta_j\mathbf{u}_j + \alpha_j\mathbf{v}_j \,|\, \cdots \end{bmatrix} = \mathbf{P}\begin{pmatrix} \mathbf{D} & 0 & \cdots & 0 \\ 0 & \mathbf{B}_1 & \cdots & 0 \\ \vdots & \vdots & \ddots & \vdots \\ 0 & 0 & \cdots & \mathbf{B}_t \end{pmatrix},
$$

where

$$
\mathbf{D} = \begin{pmatrix} \rho_1 & 0 & \cdots & 0 \\ 0 & \rho_2 & \cdots & 0 \\ \vdots & \vdots & \ddots & \vdots \\ 0 & 0 & \cdots & \rho_r \end{pmatrix} \quad \text{and} \quad \mathbf{B}_j = \begin{pmatrix} \alpha_j & \beta_j \\ -\beta_j & \alpha_j \end{pmatrix}.
$$

7.2.27. Schur's triangularization theorem says $\mathbf{U}^*\mathbf{A}\mathbf{U} = \mathbf{T}$ where $\mathbf{U}$ is unitary and $\mathbf{T}$ is upper triangular. Setting $\mathbf{x} = \mathbf{U}\mathbf{e}_i$ in $\mathbf{x}^*\mathbf{A}\mathbf{x} = 0$ yields that $t_{ii} = 0$ for each i, so $t_{ij} = 0$ for all $i \geq j$. Now set $\mathbf{x} = \mathbf{U}(\mathbf{e}_i + \mathbf{e}_j)$ with $i < j$ in $\mathbf{x}^*\mathbf{A}\mathbf{x} = 0$ to conclude that $t_{ij} = 0$ whenever $i < j$. Consequently, $\mathbf{T} = \mathbf{0}$, and thus $\mathbf{A} = \mathbf{0}$. To see that $\mathbf{x}^T\mathbf{A}\mathbf{x} = 0 \;\forall\; \mathbf{x} \in \Re^{n \times 1} \not\Rightarrow \mathbf{A} = \mathbf{0}$, consider $\mathbf{A} = \begin{pmatrix} 0 & -1 \\ 1 & 0 \end{pmatrix}$.

Solutions for exercises in section 7. 3

7.3.1. $\cos \mathbf{A} = \begin{pmatrix} 0 & 1 \\ 1 & 0 \end{pmatrix}$. The characteristic equation for $\mathbf{A}$ is $\lambda^2 + \pi\lambda = 0$, so the eigenvalues of $\mathbf{A}$ are $\lambda_1 = 0$ and $\lambda_2 = -\pi$. Note that $\mathbf{A}$ is diagonalizable because no eigenvalue is repeated. Associated eigenvectors are computed in the usual way to be

$$
\mathbf{x}_1 = \begin{pmatrix} 1 \\ 1 \end{pmatrix} \quad \text{and} \quad \mathbf{x}_2 = \begin{pmatrix} -1 \\ 1 \end{pmatrix},
$$

so

$$
\mathbf{P} = \begin{pmatrix} 1 & -1 \\ 1 & 1 \end{pmatrix} \quad \text{and} \quad \mathbf{P}^{-1} = \frac{1}{2}\begin{pmatrix} 1 & 1 \\ -1 & 1 \end{pmatrix}.
$$

Thus

$$\cos \mathbf{A} = \mathbf{P} \begin{pmatrix} \cos(0) & 0 \\ 0 & \cos(-\pi) \end{pmatrix} \mathbf{P}^{-1} = \frac{1}{2} \begin{pmatrix} 1 & -1 \\ 1 & 1 \end{pmatrix} \begin{pmatrix} 1 & 0 \\ 0 & -1 \end{pmatrix} \begin{pmatrix} 1 & 1 \\ -1 & 1 \end{pmatrix}$$

$$= \begin{pmatrix} 0 & 1 \\ 1 & 0 \end{pmatrix}.$$

7.3.2. From Example 7.3.3, the eigenvalues are $\lambda_1 = 0$ and $\lambda_2 = -(\alpha + \beta)$, and associated eigenvectors are computed in the usual way to be

$$\mathbf{x}_1 = \begin{pmatrix} \beta/\alpha \\ 1 \end{pmatrix} \quad \text{and} \quad \mathbf{x}_2 = \begin{pmatrix} -1 \\ 1 \end{pmatrix},$$

so

$$\mathbf{P} = \begin{pmatrix} \beta/\alpha & -1 \\ 1 & 1 \end{pmatrix} \quad \text{and} \quad \mathbf{P}^{-1} = \frac{1}{1 + \beta/\alpha} \begin{pmatrix} 1 & 1 \\ -1 & \beta/\alpha \end{pmatrix}.$$

Thus

$$\mathbf{P} \begin{pmatrix} e^{\lambda_1 t} & 0 \\ 0 & e^{\lambda_2 t} \end{pmatrix} \mathbf{P}^{-1} = \frac{\alpha}{\alpha + \beta} \begin{pmatrix} \beta/\alpha & -1 \\ 1 & 1 \end{pmatrix} \begin{pmatrix} 1 & 0 \\ 0 & e^{-(\alpha+\beta)t} \end{pmatrix} \begin{pmatrix} 1 & 1 \\ -1 & \beta/\alpha \end{pmatrix}$$

$$= \frac{1}{\alpha + \beta} \left[\begin{pmatrix} \beta & \beta \\ \alpha & \alpha \end{pmatrix} + e^{-(\alpha+\beta)t} \begin{pmatrix} \alpha & -\beta \\ -\alpha & \beta \end{pmatrix} \right]$$

$$= e^{\lambda_1 t} \mathbf{G}_1 + e^{\lambda_2 t} \mathbf{G}_2.$$

7.3.3. *Solution* 1: If $\mathbf{A} = \mathbf{P}\mathbf{D}\mathbf{P}^{-1}$, where $\mathbf{D} = \text{diag}(\lambda_1, \lambda_2, \ldots, \lambda_n)$, then

$$\sin^2 \mathbf{A} = \mathbf{P}\left(\sin^2 \mathbf{D}\right)\mathbf{P}^{-1} = \mathbf{P} \begin{pmatrix} \sin^2 \lambda_1 & 0 & \cdots & 0 \\ 0 & \sin^2 \lambda_2 & \cdots & 0 \\ \vdots & \vdots & \ddots & \vdots \\ 0 & 0 & \cdots & \sin^2 \lambda_n \end{pmatrix} \mathbf{P}^{-1}.$$

Similarly for $\cos^2 \mathbf{A}$, so $\sin^2 \mathbf{A} + \cos^2 \mathbf{A} = \mathbf{P}\left(\sin^2 \mathbf{D} + \cos^2 \mathbf{D}\right)\mathbf{P}^{-1} = \mathbf{P}\mathbf{I}\mathbf{P}^{-1} = \mathbf{I}$.

Solution 2: If $\sigma(\mathbf{A}) = \{\lambda_1, \lambda_2, \ldots, \lambda_k\}$, use the spectral representation (7.3.6) to write $\sin^2 \mathbf{A} = \sum_{i=1}^{k}(\sin^2 \lambda_i)\mathbf{G}_i$ and $\cos^2 \mathbf{A} = \sum_{i=1}^{k}(\cos^2 \lambda_i)\mathbf{G}_i$, so that $\sin^2 \mathbf{A} + \cos^2 \mathbf{A} = \sum_{i=1}^{k}(\sin^2 \lambda_i + \cos^2 \lambda_i)\mathbf{G}_i = \sum_{i=1}^{k} \mathbf{G}_i = \mathbf{I}$.

7.3.4. The infinite series representation of $e^{\mathbf{A}}$ readily yields this.

7.3.5. (a) Eigenvalues are invariant under a similarity transformation, so the eigenvalues of $f(\mathbf{A}) = \mathbf{P}f(\mathbf{D})\mathbf{P}^{-1}$ are the eigenvalues of $f(\mathbf{D})$, which are given by $\{f(\lambda_1), f(\lambda_2), \ldots, f(\lambda_n)\}$.

(b) If $(\lambda, \mathbf{x})$ is an eigenpair for $\mathbf{A}$, then $(\mathbf{A} - z_0\mathbf{I})^n \mathbf{x} = (\lambda - z_0)^n \mathbf{x}$ implies that $(f(\lambda), \mathbf{x})$ is an eigenpair for $f(\mathbf{A})$.

7.3.6. If $\{\lambda_1, \lambda_2, \ldots, \lambda_n\}$ are the eigenvalues of $\mathbf{A}_{n \times n}$, then $\{e^{\lambda_1}, e^{\lambda_2}, \ldots, e^{\lambda_n}\}$ are the eigenvalues of $e^{\mathbf{A}}$ by the spectral mapping property from Exercise 7.3.5. The trace is the sum of the eigenvalues, and the determinant is the product of the eigenvalues (p. 494), so $\det\left(e^{\mathbf{A}}\right) = e^{\lambda_1} e^{\lambda_2} \cdots e^{\lambda_n} = e^{\lambda_1 + \lambda_2 + \cdots + \lambda_n} = e^{trace(\mathbf{A})}$.

7.3.7. The Cayley–Hamilton theorem says that each $\mathbf{A}_{m \times m}$ satisfies its own characteristic equation, $0 = \det(\mathbf{A} - \lambda \mathbf{I}) = \lambda^m + c_1 \lambda^{m-1} + c_2 \lambda^{m-2} + \cdots + c_{m-1} \lambda + c_m$, so $\mathbf{A}^m = -c_1 \mathbf{A}^{m-1} - \cdots - c_{m-1} \mathbf{A} - c_m \mathbf{I}$. Consequently, $\mathbf{A}^m$ and every higher power of $\mathbf{A}$ is a polynomial in $\mathbf{A}$ of degree at most $m - 1$, and thus any expression involving powers of $\mathbf{A}$ can always be reduced to an expression involving at most $\mathbf{I}, \mathbf{A}, \ldots, \mathbf{A}^{m-1}$.

7.3.8. When $\mathbf{A}$ is diagonalizable, (7.3.11) insures $f(\mathbf{A}) = p(\mathbf{A})$ is a polynomial in $\mathbf{A}$, and $\mathbf{A}p(\mathbf{A}) = p(\mathbf{A})\mathbf{A}$. If $f(\mathbf{A})$ is defined by the series (7.3.7) in the nondiagonalizable case, then, by Exercise 7.3.7, it's still true that $f(\mathbf{A}) = p(\mathbf{A})$ is a polynomial in $\mathbf{A}$, and thus $\mathbf{A}f(\mathbf{A}) = f(\mathbf{A})\mathbf{A}$ holds in the nondiagonalizable case also.

7.3.9. If $\mathbf{A}$ and $\mathbf{B}$ are diagonalizable with $\mathbf{AB} = \mathbf{BA}$, Exercise 7.2.16 insures $\mathbf{A}$ and $\mathbf{B}$ can be simultaneously diagonalized. If $\mathbf{P}^{-1}\mathbf{AP} = \mathbf{D}_A = \mathrm{diag}\,(\lambda_1, \lambda_2, \ldots, \lambda_n)$ and $\mathbf{P}^{-1}\mathbf{BP} = \mathbf{D}_B = \mathrm{diag}\,(\mu_1, \mu_2, \ldots, \mu_n)$, then $\mathbf{A} + \mathbf{B} = \mathbf{P}(\mathbf{D}_A + \mathbf{D}_B)\mathbf{P}^{-1}$, so

$$
e^{\mathbf{A}+\mathbf{B}} = \mathbf{P}\left(e^{\mathbf{D}_A + \mathbf{D}_B}\right)\mathbf{P}^{-1} = \mathbf{P}\begin{pmatrix} e^{\lambda_1 + \mu_1} & 0 & \cdots & 0 \\ 0 & e^{\lambda_2 + \mu_2} & \cdots & 0 \\ \vdots & \vdots & \ddots & \vdots \\ 0 & 0 & \cdots & e^{\lambda_n + \mu_n} \end{pmatrix}\mathbf{P}^{-1}
$$

$$
= \mathbf{P}\begin{pmatrix} e^{\lambda_1} & 0 & \cdots & 0 \\ 0 & e^{\lambda_2} & \cdots & 0 \\ \vdots & \vdots & \ddots & \vdots \\ 0 & 0 & \cdots & e^{\lambda_n} \end{pmatrix}\mathbf{P}^{-1}\mathbf{P}\begin{pmatrix} e^{\mu_1} & 0 & \cdots & 0 \\ 0 & e^{\mu_2} & \cdots & 0 \\ \vdots & \vdots & \ddots & \vdots \\ 0 & 0 & \cdots & e^{\mu_n} \end{pmatrix}\mathbf{P}^{-1}
$$

$$
= e^{\mathbf{A}}e^{\mathbf{B}}.
$$

In general, the same brute force multiplication of scalar series that yields

$$
e^{x+y} = \sum_{n=0}^{\infty} \frac{(x+y)^n}{n!} = \left(\sum_{n=0}^{\infty} \frac{x^n}{n!}\right)\left(\sum_{n=0}^{\infty} \frac{y^n}{n!}\right) = e^x e^y
$$

holds for matrix series when $\mathbf{AB} = \mathbf{BA}$, but this is quite messy. A more elegant approach is to set $\mathbf{F}(t) = e^{\mathbf{A}t + \mathbf{B}t} - e^{\mathbf{A}t}e^{\mathbf{B}t}$ and note that $\mathbf{F}'(t) = \mathbf{0}$ for all t when $\mathbf{AB} = \mathbf{BA}$, so $\mathbf{F}(t)$ must be a constant matrix for all t. Since $\mathbf{F}(0) = \mathbf{0}$, it follows that $e^{(\mathbf{A}+\mathbf{B})t} = e^{\mathbf{A}t}e^{\mathbf{B}t}$ for all t. To see that $e^{\mathbf{A}+\mathbf{B}}$, $e^{\mathbf{A}}e^{\mathbf{B}}$, and $e^{\mathbf{B}}e^{\mathbf{A}}$ can be different when $\mathbf{AB} \neq \mathbf{BA}$, consider $\mathbf{A} = \begin{pmatrix} 1 & 0 \\ 0 & 0 \end{pmatrix}$ and $\mathbf{B} = \begin{pmatrix} 0 & 1 \\ 1 & 0 \end{pmatrix}$.

7.3.10. The infinite series representation of $e^{\mathbf{A}}$ shows that if $\mathbf{A}$ is skew symmetric, then $\left(e^{\mathbf{A}}\right)^T = e^{\mathbf{A}^T} = e^{-\mathbf{A}}$, and hence $e^{\mathbf{A}}\left(e^{\mathbf{A}}\right)^T = e^{\mathbf{A}-\mathbf{A}} = e^{\mathbf{0}} = \mathbf{I}$.

7.3.11. (a)   Draw a transition diagram similar to that in Figure 7.3.1 with North and South replaced by ON and OFF, respectively. Let x_k be the fraction of switches in the ON state and let y_k be the fraction of switches in the OFF state after k clock cycles have elapsed. According to the given information,

$$x_k = x_{k-1}(.1) + y_{k-1}(.3)$$
$$y_k = x_{k-1}(.9) + y_{k-1}(.7)$$

so that $\mathbf{p}_{k+1}^T = \mathbf{p}_k^T \mathbf{T}$, where $\mathbf{p}_k^T = (x_k \quad y_k)$ and $\mathbf{T} = \begin{pmatrix} .1 & .9 \\ .3 & .7 \end{pmatrix}$. Compute $\sigma(\mathbf{T}) = \{1, -1/5\}$, and use the methods of Example 7.3.4 to determine the steady-state (or limiting) distribution as

$$\mathbf{p}_\infty^T = \lim_{k\to\infty} \mathbf{p}_k^T = \lim_{k\to\infty} \mathbf{p}_0^T \mathbf{T}^k = \mathbf{p}_0^T \lim_{k\to\infty} \mathbf{T}^k = (x_0 \quad y_0) \begin{pmatrix} 1/4 & 3/4 \\ 1/4 & 3/4 \end{pmatrix}$$

$$= \left(\frac{x_0 + y_0}{4} \quad \frac{3(x_0 + y_0)}{4} \right) = (1/4 \quad 3/4).$$

Alternately, (7.3.15) can be used with $\mathbf{x}_1 = \begin{pmatrix} 1 \\ 1 \end{pmatrix}$ and $\mathbf{y}_1 = (1 \quad 3)$ to obtain

$$\mathbf{p}_\infty^T = \mathbf{p}_0^T \lim_{k\to\infty} \mathbf{T}^k = \mathbf{p}_0^T \lim_{k\to\infty} \mathbf{G}_1 = \frac{(\mathbf{p}_0^T \mathbf{x}_1)\mathbf{y}_1^T}{\mathbf{y}_1^T \mathbf{x}_1} = \frac{\mathbf{y}_1^T}{\mathbf{y}_1^T \mathbf{x}_1} = (1/4 \quad 3/4).$$

(b)   Computing a few powers of $\mathbf{T}$ reveals that

$$\mathbf{T}^2 = \begin{pmatrix} .280 & .720 \\ .240 & .760 \end{pmatrix}, \quad \mathbf{T}^3 = \begin{pmatrix} .244 & .756 \\ .252 & .748 \end{pmatrix},$$

$$\mathbf{T}^4 = \begin{pmatrix} .251 & .749 \\ .250 & .750 \end{pmatrix}, \quad \mathbf{T}^5 = \begin{pmatrix} .250 & .750 \\ .250 & .750 \end{pmatrix},$$

so, for practical purposes, the device can be considered to be in equilibrium after about 5 clock cycles, regardless of the initial configuration.

7.3.12. Let $\sigma(\mathbf{A}) = \{\lambda_1, \lambda_2, \ldots, \lambda_k\}$ with $|\lambda_1| \geq |\lambda_2| \geq \cdots \geq |\lambda_k|$, and assume $\lambda_1 \neq 0$; otherwise $\mathbf{A} = \mathbf{0}$ and there is nothing to prove. Set

$$\nu_n = \frac{\|\mathbf{A}^n\|}{|\lambda_1^n|} = \frac{\|\lambda_1^n \mathbf{G}_1 + \lambda_2^n \mathbf{G}_2 + \cdots + \lambda_k^n \mathbf{G}_k\|}{|\lambda_1^n|} = \left\| \frac{\lambda_1^n \mathbf{G}_1 + \lambda_2^n \mathbf{G}_2 + \cdots + \lambda_k^n \mathbf{G}_k}{\lambda_1^n} \right\|$$

$$= \left\| \mathbf{G}_1 + \left(\frac{\lambda_2}{\lambda_1}\right)^n \mathbf{G}_2 + \cdots + \left(\frac{\lambda_k}{\lambda_1}\right)^n \mathbf{G}_k \right\| \quad \text{and let} \quad \nu = \sum_{i=1}^{k} \|\mathbf{G}_i\|.$$

Observe that $1 \leq \nu_n \leq \nu$ for every positive integer n—the first inequality follows because $\lambda_1^n \in \sigma(\mathbf{A}^n)$ implies $|\lambda_1^n| \leq \|\mathbf{A}^n\|$ by (7.1.12) on p. 497, and the second is the result of the triangle inequality. Consequently,

$$1^{1/n} \leq \nu_n^{1/n} \leq \nu^{1/n} \implies 1 \leq \lim_{n \to \infty} \nu_n^{1/n} \leq 1 \implies 1 = \lim_{n \to \infty} \nu_n^{1/n} = \lim_{n \to \infty} \frac{\|\mathbf{A}^n\|^{1/n}}{|\lambda_1|}.$$

7.3.13. The dominant eigenvalue is $\lambda_1 = 4$, and all corresponding eigenvectors are multiples of $(-1, 0, 1)^T$.

7.3.15. Consider

$$\mathbf{x}_n = \begin{pmatrix} 1 - 1/n \\ -1 \end{pmatrix} \to \mathbf{x} = \begin{pmatrix} 1 \\ -1 \end{pmatrix},$$

but $m(\mathbf{x}_n) = -1$ for all $n = 1, 2, \ldots$, and $m(\mathbf{x}) = 1$, so $m(\mathbf{x}_n) \not\to m(\mathbf{x})$. Nevertheless, if $\lim_{n \to \infty} \mathbf{x}_n \neq \mathbf{0}$, then $\lim_{n \to \infty} m(\mathbf{x}_n) \neq 0$ because the function $\tilde{m}(\mathbf{v}) = |m(\mathbf{v})| = \|\mathbf{v}\|_\infty$ is continuous.

7.3.16. (a) The "vanilla" QR iteration fails to converge.

(b) $\mathbf{H} - \mathbf{I} = \mathbf{QR} = \begin{pmatrix} 0 & 0 & 1 \\ -1 & 0 & 0 \\ 0 & 1 & 0 \end{pmatrix} \begin{pmatrix} 1 & 3 & 1 \\ 0 & 2 & 0 \\ 0 & 0 & 0 \end{pmatrix}$ and $\mathbf{RQ} + \mathbf{I} = \begin{pmatrix} -2 & 1 & 1 \\ -2 & 1 & 0 \\ 0 & 0 & 1 \end{pmatrix}$.

Solutions for exercises in section 7. 4

7.4.1. The unique solution to $\mathbf{u}' = \mathbf{Au}$, $\mathbf{u}(0) = \mathbf{c}$, is

$$\mathbf{u} = e^{\mathbf{A}t}\mathbf{c} = \mathbf{P} \begin{pmatrix} e^{\lambda_1 t} & 0 & \cdots & 0 \\ 0 & e^{\lambda_2 t} & \cdots & 0 \\ \vdots & \vdots & \ddots & \vdots \\ 0 & 0 & \cdots & e^{\lambda_n t} \end{pmatrix} \mathbf{P}^{-1}\mathbf{c}$$

$$= [\mathbf{x}_1 \,|\, \mathbf{x}_2 \,|\, \cdots \,|\, \mathbf{x}_n] \begin{pmatrix} e^{\lambda_1 t} & 0 & \cdots & 0 \\ 0 & e^{\lambda_2 t} & \cdots & 0 \\ \vdots & \vdots & \ddots & \vdots \\ 0 & 0 & \cdots & e^{\lambda_n t} \end{pmatrix} \begin{pmatrix} \xi_1 \\ \xi_2 \\ \vdots \\ \xi_n \end{pmatrix}$$

$$= \xi_1 e^{\lambda_1 t}\mathbf{x}_1 + \xi_2 e^{\lambda_2 t}\mathbf{x}_2 + \cdots + \xi_n e^{\lambda_n t}\mathbf{x}_n.$$

7.4.2. (a) All eigenvalues in $\sigma(\mathbf{A}) = \{-1, -3\}$ are negative, so the system is stable.

(b) All eigenvalues in $\sigma(\mathbf{A}) = \{1, 3\}$ are positive, so the system is unstable.

(c) $\sigma(\mathbf{A}) = \{\pm i\}$, so the system is semistable. If $\mathbf{c} \neq \mathbf{0}$, then the components in $\mathbf{u}(t)$ will oscillate indefinitely.

7.4.3. (a) If $u_k(t)$ denotes the number in population k at time t, then

$$\begin{aligned} u_1' &= 2u_1 - u_2, & u_1(0) &= 100, \\ u_2' &= -u_1 + 2u_2, & u_2(0) &= 200, \end{aligned}$$

or $\mathbf{u}' = \mathbf{A}\mathbf{u}$, $\mathbf{u}(0) = \mathbf{c}$, where $\mathbf{A} = \begin{pmatrix} 2 & -1 \\ -1 & 2 \end{pmatrix}$ and $\mathbf{c} = \begin{pmatrix} 100 \\ 200 \end{pmatrix}$. The characteristic equation for $\mathbf{A}$ is $p(\lambda) = \lambda^2 - 4\lambda + 3 = (\lambda - 1)(\lambda - 3) = 0$, so the eigenvalues for $\mathbf{A}$ are $\lambda_1 = 1$ and $\lambda_2 = 3$. We know from (7.4.7) that

$$\mathbf{u}(t) = e^{\lambda_1 t}\mathbf{v}_1 + e^{\lambda_2 t}\mathbf{v}_2 \quad (\text{where } \mathbf{v}_i = \mathbf{G}_i\mathbf{c})$$

is the solution to $\mathbf{u}' = \mathbf{A}\mathbf{u}$, $\mathbf{u}(0) = \mathbf{c}$. The spectral theorem on p. 517 implies $\mathbf{A} - \lambda_2\mathbf{I} = (\lambda_1 - \lambda_2)\mathbf{G}_1$ and $\mathbf{I} = \mathbf{G}_1 + \mathbf{G}_2$, so $(\mathbf{A} - \lambda_2\mathbf{I})\mathbf{c} = (\lambda_1 - \lambda_2)\mathbf{v}_1$ and $\mathbf{c} = \mathbf{v}_1 + \mathbf{v}_2$, and consequently

$$\mathbf{v}_1 = \frac{(\mathbf{A} - \lambda_2\mathbf{I})\mathbf{c}}{(\lambda_1 - \lambda_2)} = \begin{pmatrix} 150 \\ 150 \end{pmatrix} \quad \text{and} \quad \mathbf{v}_2 = \mathbf{c} - \mathbf{v}_1 = \begin{pmatrix} -50 \\ 50 \end{pmatrix},$$

so

$$u_1(t) = 150e^t - 50e^{3t} \quad \text{and} \quad u_2(t) = 150e^t + 50e^{3t}.$$

(b)   As $t \to \infty$, $u_1(t) \to -\infty$ and $u_2(t) \to +\infty$. But a population can't become negative, so species I is destined to become extinct, and this occurs at the value of t for which $u_1(t) = 0$—i.e., when

$$e^t\left(e^{2t} - 3\right) = 0 \implies e^{2t} = 3 \implies t = \frac{\ln 3}{2}.$$

7.4.4. If $u_k(t)$ denotes the number in population k at time t, then the hypothesis says

$$\begin{aligned} u_1' &= -u_1 + u_2, & u_1(0) &= 200, \\ u_2' &= u_1 - 2u_2, & u_2(0) &= 400, \end{aligned}$$

or $\mathbf{u}' = \mathbf{A}\mathbf{u}$, $\mathbf{u}(0) = \mathbf{c}$, where $\mathbf{A} = \begin{pmatrix} -1 & 1 \\ 1 & -1 \end{pmatrix}$ and $\mathbf{c} = \begin{pmatrix} 200 \\ 400 \end{pmatrix}$. The characteristic equation for $\mathbf{A}$ is $p(\lambda) = \lambda^2 + 2\lambda = \lambda(\lambda + 2) = 0$, so the eigenvalues for $\mathbf{A}$ are $\lambda_1 = 0$ and $\lambda_2 = -2$. We know from (7.4.7) that

$$\mathbf{u}(t) = e^{\lambda_1 t}\mathbf{v}_1 + e^{\lambda_2 t}\mathbf{v}_2 \quad (\text{where } \mathbf{v}_i = \mathbf{G}_i\mathbf{c})$$

is the solution to $\mathbf{u}' = \mathbf{A}\mathbf{u}$, $\mathbf{u}(0) = \mathbf{c}$. The spectral theorem on p. 517 implies $\mathbf{A} - \lambda_2\mathbf{I} = (\lambda_1 - \lambda_2)\mathbf{G}_1$ and $\mathbf{I} = \mathbf{G}_1 + \mathbf{G}_2$, so $(\mathbf{A} - \lambda_2\mathbf{I})\mathbf{c} = (\lambda_1 - \lambda_2)\mathbf{v}_1$ and $\mathbf{c} = \mathbf{v}_1 + \mathbf{v}_2$, and consequently

$$\mathbf{v}_1 = \frac{(\mathbf{A} - \lambda_2\mathbf{I})\mathbf{c}}{(\lambda_1 - \lambda_2)} = \begin{pmatrix} 300 \\ 300 \end{pmatrix} \quad \text{and} \quad \mathbf{v}_2 = \mathbf{c} - \mathbf{v}_1 = \begin{pmatrix} -100 \\ 100 \end{pmatrix},$$

so

$$u_1(t) = 300 - 100e^{-2t} \quad \text{and} \quad u_2(t) = 300 + 100e^{-2t}.$$

As $t \to \infty$, $u_1(t) \to 300$ and $u_2(t) \to 300$, so both populations will stabilize at 300.

Solutions for exercises in section 7. 5

7.5.1. Yes, because $\mathbf{A}^*\mathbf{A} = \mathbf{A}\mathbf{A}^* = \begin{pmatrix} 30 & 6-6\,\mathrm{i} \\ 6+6\,\mathrm{i} & 24 \end{pmatrix}$.

7.5.2. Real skew-symmetric and orthogonal matrices are examples.

7.5.3. We already know from (7.5.3) that real-symmetric matrices are normal and have real eigenvalues, so only the converse needs to be proven. If $\mathbf{A}$ is real and normal with real eigenvalues, then there is a complete orthonormal set of real eigenvectors, so using them as columns in $\mathbf{P} \in \Re^{n \times n}$ results in an orthogonal matrix such that $\mathbf{P}^T\mathbf{A}\mathbf{P} = \mathbf{D}$ is diagonal or, equivalently, $\mathbf{A} = \mathbf{P}\mathbf{D}\mathbf{P}^T$, and thus $\mathbf{A} = \mathbf{A}^T$.

7.5.4. If $(\lambda, \mathbf{x})$ is an eigenpair for $\mathbf{A} = -\mathbf{A}^*$ then $\mathbf{x}^*\mathbf{x} \neq 0$, and $\lambda\mathbf{x} = \mathbf{A}\mathbf{x}$ implies $\overline{\lambda}\mathbf{x}^* = \mathbf{x}^*\mathbf{A}^*$, so

$$\mathbf{x}^*\mathbf{x}(\lambda + \overline{\lambda}) = \mathbf{x}^*(\lambda + \overline{\lambda})\mathbf{x} = \mathbf{x}^*\mathbf{A}\mathbf{x} + \mathbf{x}^*\mathbf{A}^*\mathbf{x} = 0 \implies \lambda = -\overline{\lambda} \implies \Re e(\lambda) = 0.$$

7.5.5. If $\mathbf{A}$ is skew hermitian (real skew symmetric), then $\mathbf{A}$ is normal, and hence $\mathbf{A}$ is unitarily (orthogonally) similar to a diagonal matrix—say $\mathbf{A} = \mathbf{U}\mathbf{D}\mathbf{U}^*$. Moreover, the eigenvalues λ_j in $\mathbf{D} = \mathrm{diag}\,(\lambda_1, \lambda_2, \ldots, \lambda_n)$ are pure imaginary numbers (Exercise 7.5.4). Since $f(z) = (1-z)(1+z)^{-1}$ maps the imaginary axis in the complex plane to points on the unit circle, each $f(\lambda_j)$ is on the unit circle, so there is some θ_j such that $f(\lambda_j) = \mathrm{e}^{\mathrm{i}\theta_j} = \cos\theta_j + \mathrm{i}\sin\theta_j$. Consequently,

$$f(\mathbf{A}) = \mathbf{U}\begin{pmatrix} f(\lambda_1) & 0 & \cdots & 0 \\ 0 & f(\lambda_2) & \cdots & 0 \\ \vdots & \vdots & \ddots & \vdots \\ 0 & 0 & \cdots & f(\lambda_n) \end{pmatrix}\mathbf{U}^* = \mathbf{U}\begin{pmatrix} \mathrm{e}^{\mathrm{i}\theta_1} & 0 & \cdots & 0 \\ 0 & \mathrm{e}^{\mathrm{i}\theta_2} & \cdots & 0 \\ \vdots & \vdots & \ddots & \vdots \\ 0 & 0 & \cdots & \mathrm{e}^{\mathrm{i}\theta_n} \end{pmatrix}\mathbf{U}^*$$

together with $\mathrm{e}^{\mathrm{i}\theta_j}\overline{\mathrm{e}^{\mathrm{i}\theta_j}} = \mathrm{e}^{\mathrm{i}\theta_j}\mathrm{e}^{-\mathrm{i}\theta_j} = 1$ yields $f(\mathbf{A})^*f(\mathbf{A}) = \mathbf{I}$. **Note:** The fact that $(\mathbf{I} - \mathbf{A})(\mathbf{I} + \mathbf{A})^{-1} = (\mathbf{I} + \mathbf{A})^{-1}(\mathbf{I} - \mathbf{A})$ follows from Exercise 7.3.8. See the solution to Exercise 5.6.6 for an alternate approach.

7.5.6. Consider the identity matrix—every nonzero vector is an eigenvector, so not every complete independent set of eigenvectors needs to be orthonormal. Given a complete independent set of eigenvectors for a normal $\mathbf{A}$ with $\sigma(\mathbf{A}) = \{\lambda_1, \lambda_2, \ldots, \lambda_k\}$, use the Gram–Schmidt procedure to form an orthonormal basis for $N(\mathbf{A} - \lambda_i\mathbf{I})$ for each i. Since $N(\mathbf{A} - \lambda_i\mathbf{I}) \perp N(\mathbf{A} - \lambda_j\mathbf{I})$ for $\lambda_i \neq \lambda_j$ (by (7.5.2)), the union of these orthonormal bases will be a complete orthonormal set of eigenvectors for $\mathbf{A}$.

7.5.7. Consider $\mathbf{A} = \begin{pmatrix} 0 & 1 & 0 \\ 0 & 0 & 0 \\ 0 & 0 & 1 \end{pmatrix}$.

7.5.8. Suppose $\mathbf{T}_{n \times n}$ is an upper-triangular matrix such that $\mathbf{T}^*\mathbf{T} = \mathbf{T}\mathbf{T}^*$. The (1,1)-entry of $\mathbf{T}^*\mathbf{T}$ is $|t_{11}|^2$, and the (1,1)-entry of $\mathbf{T}\mathbf{T}^*$ is $\sum_{k=1}^{n} |t_{1k}|^2$. Equating these implies $t_{12} = t_{13} = \cdots = t_{1n} = 0$. Now use this and compare the (2,2)-entries to get $t_{23} = t_{24} = \cdots = t_{2n} = 0$. Repeating this argument for each row produces the conclusion that $\mathbf{T}$ must be diagonal. Conversely, if $\mathbf{T}$ is diagonal, then $\mathbf{T}$ is normal because $\mathbf{T}^*\mathbf{T} = \text{diag}\,(|t_{11}|^2 \cdots |t_{nn}|^2) = \mathbf{TT}^*$.

7.5.9. Schur's triangularization theorem on p. 508 says every square matrix is unitarily similar to an upper-triangular matrix—say $\mathbf{U}^*\mathbf{AU} = \mathbf{T}$. If $\mathbf{A}$ is normal, then so is $\mathbf{T}$. Exercise 7.5.8 therefore insures that $\mathbf{T}$ must be diagonal. Conversely, if $\mathbf{T}$ is diagonal, then it is normal, and thus so is $\mathbf{A}$.

7.5.10. If $\mathbf{A}$ is normal, so is $\mathbf{A} - \lambda\mathbf{I}$. Consequently, $\mathbf{A} - \lambda\mathbf{I}$ is RPN, and hence $N\,(\mathbf{A} - \lambda\mathbf{I}) = N\,(\mathbf{A} - \lambda\mathbf{I})^*$ (p. 408), so $(\mathbf{A} - \lambda\mathbf{I})\mathbf{x} = \mathbf{0} \Longleftrightarrow (\mathbf{A}^* - \bar{\lambda}\mathbf{I})\mathbf{x} = \mathbf{0}$.

7.5.11. Just as in the proof of the min-max part, it suffices to prove

$$\lambda_i = \max_{\dim \mathcal{V}=i} \; \min_{\substack{\mathbf{y} \in \mathcal{V} \\ \|\mathbf{y}\|_2 = 1}} \mathbf{y}^*\mathbf{Dy}.$$

For each subspace $\mathcal{V}$ of dimension i, let $\mathcal{S}_\mathcal{V} = \{\mathbf{y} \in \mathcal{V}, \|\mathbf{y}\|_2 = 1\}$, and let

$$\mathcal{S}'_\mathcal{V} = \{\mathbf{y} \in \mathcal{V} \cap \mathcal{F}^\perp, \|\mathbf{y}\|_2 = 1\}, \quad \text{where} \quad \mathcal{F} = \{\mathbf{e}_1, \mathbf{e}_2, \ldots, \mathbf{e}_{i-1}\}.$$

($\mathcal{V} \cap \mathcal{F}^\perp \neq \mathbf{0}$ —otherwise $\dim(\mathcal{V} + \mathcal{F}^\perp) = \dim\mathcal{V} + \dim\mathcal{F}^\perp = n + 1$, which is impossible.) So $\mathcal{S}'_\mathcal{V}$ contains vectors of $\mathcal{S}_\mathcal{V}$ of the form $\mathbf{y} = (0, \ldots, 0, y_i, \ldots, y_n)^T$ with $\sum_{j=i}^n |y_j|^2 = 1$, and for each subspace $\mathcal{V}$ with $\dim\mathcal{V} = i$,

$$\mathbf{y}^*\mathbf{Dy} = \sum_{j=i}^n \lambda_j |y_j|^2 \leq \lambda_i \sum_{j=i}^n |y_j|^2 = \lambda_i \quad \text{for all } \mathbf{y} \in \mathcal{S}'_\mathcal{V}.$$

Since $\mathcal{S}'_\mathcal{V} \subseteq \mathcal{S}_\mathcal{V}$, it follows that $\min_{\mathcal{S}_\mathcal{V}} \mathbf{y}^*\mathbf{Dy} \leq \min_{\mathcal{S}'_\mathcal{V}} \mathbf{y}^*\mathbf{Dy} \leq \lambda_i$, and hence

$$\max_\mathcal{V} \min_{\mathcal{S}_\mathcal{V}} \mathbf{y}^*\mathbf{Dy} \leq \lambda_i.$$

To reverse this inequality, let $\tilde{\mathcal{V}} = span\,\{\mathbf{e}_1, \mathbf{e}_2, \ldots, \mathbf{e}_i\}$, and observe that

$$\mathbf{y}^*\mathbf{Dy} = \sum_{j=1}^i \lambda_j |y_j|^2 \geq \lambda_i \sum_{j=1}^i |y_j|^2 = \lambda_i \quad \text{for all } \mathbf{y} \in \mathcal{S}_{\tilde{\mathcal{V}}},$$

so $\max_\mathcal{V} \min_{\mathcal{S}_\mathcal{V}} \mathbf{y}^*\mathbf{Dy} \geq \max_{\mathcal{S}_{\tilde{\mathcal{V}}}} \mathbf{y}^*\mathbf{Dy} \geq \lambda_i$.

7.5.12. Just as before, it suffices to prove $\lambda_i = \min_{\mathbf{v}_1,\ldots,\mathbf{v}_{i-1}\in\mathcal{C}^n} \; \max_{\substack{\mathbf{y}\perp\mathbf{v}_1,\ldots,\mathbf{v}_{i-1} \\ \|\mathbf{y}\|_2=1}} \mathbf{y}^*\mathbf{Dy}$. For each set $\mathcal{V} = \{\mathbf{v}_1, \mathbf{v}_2, \ldots, \mathbf{v}_{i-1}\}$, let $\mathcal{S}_\mathcal{V} = \{\mathbf{y} \in \mathcal{V}^\perp, \|\mathbf{y}\|_2 = 1\}$, and let

$$\mathcal{S}'_\mathcal{V} = \{\mathbf{y} \in \mathcal{V}^\perp \cap \mathcal{T}^\perp, \|\mathbf{y}\|_2 = 1\}, \quad \text{where} \quad \mathcal{T} = \{\mathbf{e}_{i+1}, \ldots, \mathbf{e}_n\}$$

$(\mathcal{V}^\perp \cap \mathcal{T}^\perp \neq \mathbf{0}$ —otherwise $\dim(\mathcal{V}^\perp + \mathcal{T}^\perp) = \dim \mathcal{V}^\perp + \dim \mathcal{T}^\perp = n+1$, which is impossible.) So $\mathcal{S}'_\mathcal{V}$ contains vectors of $\mathcal{S}_\mathcal{V}$ of the form $\mathbf{y} = (y_1, \ldots, y_i, 0, \ldots, 0)^T$ with $\sum_{j=1}^{i} |y_j|^2 = 1$, and for each $\mathcal{V} = \{\mathbf{v}_1, \ldots, \mathbf{v}_{i-1}\}$,

$$\mathbf{y}^*\mathbf{D}\mathbf{y} = \sum_{j=1}^{i} \lambda_j |y_j|^2 \geq \lambda_i \sum_{j=1}^{i} |y_j|^2 = \lambda_i \quad \text{for all } \mathbf{y} \in \mathcal{S}'_\mathcal{V}.$$

Since $\mathcal{S}'_\mathcal{V} \subseteq \mathcal{S}_\mathcal{V}$, it follows that $\max_{\mathcal{S}_\mathcal{V}} \mathbf{y}^*\mathbf{D}\mathbf{y} \geq \max_{\mathcal{S}'_\mathcal{V}} \mathbf{y}^*\mathbf{D}\mathbf{y} \geq \lambda_i$, and hence

$$\min_\mathcal{V} \max_{\mathcal{S}_\mathcal{V}} \mathbf{y}^*\mathbf{D}\mathbf{y} \geq \lambda_i.$$

This inequality is reversible because if $\tilde{\mathcal{V}} = \{\mathbf{e}_1, \mathbf{e}_2, \ldots, \mathbf{e}_{i-1}\}$, then every $\mathbf{y} \in \tilde{\mathcal{V}}$ has the form $\mathbf{y} = (0, \ldots, 0, y_i, \ldots, y_n)^T$, so

$$\mathbf{y}^*\mathbf{D}\mathbf{y} = \sum_{j=i}^{n} \lambda_j |y_j|^2 \leq \lambda_i \sum_{j=i}^{n} |y_j|^2 = \lambda_i \quad \text{for all } \mathbf{y} \in \mathcal{S}_{\tilde{\mathcal{V}}},$$

and thus $\min_\mathcal{V} \max_{\mathcal{S}_\mathcal{V}} \mathbf{y}^*\mathbf{D}\mathbf{y} \leq \max_{\mathcal{S}_{\tilde{\mathcal{V}}}} \mathbf{y}^*\mathbf{D}\mathbf{y} \leq \lambda_i$. The solution for Exercise 7.5.11 can be adapted in a similar fashion to prove the alternate max-min expression.

7.5.13. (a) Unitary matrices are unitarily diagonalizable because they are normal. Furthermore, if $(\lambda, \mathbf{x})$ is an eigenpair for a unitary $\mathbf{U}$, then

$$\|\mathbf{x}\|_2^2 = \|\mathbf{U}\mathbf{x}\|_2^2 = \|\lambda\mathbf{x}\|_2^2 = |\lambda|^2 \|\mathbf{x}\|_2^2 \implies |\lambda| = 1 \implies \lambda = \cos\theta + i\sin\theta = e^{i\theta}.$$

(b) This is a special case of Exercise 7.2.26 whose solution is easily adapted to provide the solution for the case at hand.

Solutions for exercises in section 7. 6

7.6.1. Check the pivots in the $\mathbf{LDL}^T$ factorization to see that $\mathbf{A}$ and $\mathbf{C}$ are positive definite. $\mathbf{B}$ is positive semidefinite.

7.6.2. (a) Examining Figure 7.6.7 shows that the force on m_1 to the left, by Hooke's law, is $F_1^{(l)} = kx_1$, and the force to the right is $F_1^{(r)} = k(x_2 - x_1)$, so the total force on m_1 is $F_1 = F_1^{(l)} - F_1^{(r)} = k(2x_1 - x_2)$. Similarly, the total force on m_2 is $F_2 = k(-x_1 + 2x_2)$. Using Newton's laws $F_1 = m_1 a_1 = m_1 x_1''$ and $F_2 = m_2 a_2 = m_2 x_2''$ yields the two second-order differential equations

$$\begin{matrix} m_1 x_1''(t) = k(2x_1 - x_2) \\ m_2 x_2''(t) = k(-x_1 + 2x_2) \end{matrix} \implies \mathbf{M}\mathbf{x}'' = \mathbf{K}\mathbf{x},$$

where $\mathbf{M} = \begin{pmatrix} m_1 & 0 \\ 0 & m_2 \end{pmatrix}$, and $\mathbf{K} = k\begin{pmatrix} 2 & -1 \\ -1 & 2 \end{pmatrix}$.

(b) $\lambda = (3 \pm \sqrt{3})/2$, and the normal modes are determined by the corresponding eigenvectors, which are found in the usual way by solving

$$(\mathbf{K} - \lambda\mathbf{M})\mathbf{v} = \mathbf{0}.$$

They are

$$\mathbf{v}_1 = \begin{pmatrix} -1 - \sqrt{3} \\ 1 \end{pmatrix} \quad \text{and} \quad \mathbf{v}_2 = \begin{pmatrix} -1 + \sqrt{3} \\ 1 \end{pmatrix}$$

(c)   This part is identical to that in Example 7.6.1 (p. 559) except a 2×2 matrix is used in place of a 3×3 matrix.

7.6.3.   Each mass "feels" only the spring above and below it, so

$$m_1 y_1'' = \text{Force up} - \text{Force down} = ky_1 - k(y_2 - y_1) = k(2y_1 - y_2)$$
$$m_2 y_2'' = \text{Force up} - \text{Force down} = k(y_2 - y_1) - k(y_3 - y_2) = k(-y_1 + 2y_2 - y_3)$$
$$m_3 y_3'' = \text{Force up} - \text{Force down} = k(y_3 - y_2)$$

(b)   Gerschgorin's theorem (p. 498) shows that the eigenvalues are nonnegative, as since $\det(\mathbf{K}) \neq 0$, it follows that $\mathbf{K}$ is positive definite.

(c)   The same technique used in the vibrating beads problem in Example 7.6.1 (p. 559) shows that modes are determined by the eigenvectors. Some computation is required to produce $\lambda_1 \approx .198$, $\lambda_2 \approx 1.55$, and $\lambda_3 \approx 3.25$. The modes are defined by the associated eigenvectors

$$\mathbf{x}_1 = \begin{pmatrix} \gamma \\ \alpha \\ \beta \end{pmatrix} \approx \begin{pmatrix} .328 \\ .591 \\ .737 \end{pmatrix}, \quad \mathbf{x}_2 = \begin{pmatrix} -\beta \\ -\gamma \\ \alpha \end{pmatrix}, \quad \text{and} \quad \mathbf{x}_3 = \begin{pmatrix} -\alpha \\ \beta \\ -\gamma \end{pmatrix}.$$

7.6.4.   Write the quadratic form as $13x^2 + 10xy + 13y^2 = \begin{pmatrix} x & y \end{pmatrix} \begin{pmatrix} 13 & 5 \\ 5 & 13 \end{pmatrix} \begin{pmatrix} x \\ y \end{pmatrix} = \mathbf{z}^T\mathbf{A}\mathbf{z}$. We know from Example 7.6.3 on p. 567 that if $\mathbf{Q}$ is an orthogonal matrix such that $\mathbf{Q}^T\mathbf{A}\mathbf{Q} = \mathbf{D} = \begin{pmatrix} \lambda_1 & 0 \\ 0 & \lambda_2 \end{pmatrix}$, and if $\mathbf{w} = \mathbf{Q}^T\mathbf{z} = \begin{pmatrix} u \\ v \end{pmatrix}$, then

$$13x^2 + 10xy + 13y^2 = \mathbf{z}^T\mathbf{A}\mathbf{z} = \mathbf{w}^T\mathbf{D}\mathbf{w} = \lambda_1 u^2 + \lambda_2 v^2.$$

Computation reveals that $\lambda_1 = 8$, $\lambda_2 = 18$, and $\mathbf{Q} = \frac{1}{\sqrt{2}}\begin{pmatrix} 1 & 1 \\ -1 & 1 \end{pmatrix}$, so the graph of $13x^2 + 10xy + 13y^2 = 72$ is the same as that for $8u^2 + 18v^2 = 72$ or, equivalently, $u^2/9 + v^2/4 = 1$. It follows from (5.6.13) on p. 326 that the uv-coordinate system results from rotating the standard xy-coordinate system counterclockwise by $45°$.

7.6.5.   Since $\mathbf{A}$ is symmetric, the LDU factorization is really $\mathbf{A} = \mathbf{L}\mathbf{D}\mathbf{L}^T$ (see Exercise 3.10.9 on p. 157). In other words, $\mathbf{A} \cong \mathbf{D}$, so Sylvester's law of inertia guarantees that the inertia of $\mathbf{A}$ is the same as the inertia of $\mathbf{D}$.

7.6.6. (a) Notice that, in general, when $\mathbf{x}^T\mathbf{A}\mathbf{x}$ is expanded, the coefficient of x_ix_j is given by $(a_{ij} + a_{ji})/2$. Therefore, for the problem at hand, we can take

$$\mathbf{A} = \frac{1}{9}\begin{pmatrix} -2 & 2 & 8 \\ 2 & 7 & 10 \\ 8 & 10 & 4 \end{pmatrix}.$$

(b) Gaussian elimination provides $\mathbf{A} = \mathbf{L}\mathbf{D}\mathbf{L}^T$, where

$$\mathbf{L} = \begin{pmatrix} 1 & 0 & 0 \\ -1 & 1 & 0 \\ -4 & 2 & 1 \end{pmatrix} \quad \text{and} \quad \mathbf{D} = \begin{pmatrix} -2/9 & 0 & 0 \\ 0 & 1 & 0 \\ 0 & 0 & 0 \end{pmatrix},$$

so the inertia of $\mathbf{A}$ is $(1,1,1)$. Setting $\mathbf{y} = \mathbf{L}^T\mathbf{x}$ (or, $\mathbf{x} = (\mathbf{L}^T)^{-1}\mathbf{y}$) yields

$$\mathbf{x}^T\mathbf{A}\mathbf{x} = \mathbf{y}^T\mathbf{D}\mathbf{y} = -\frac{2}{9}y_1^2 + y_2^2.$$

(c) No, the form is indefinite.

(d) The eigenvalues of $\mathbf{A}$ are $\{2, -1, 0\}$, and hence the inertia is $(1,1,1)$.

7.6.7. $\mathbf{A}\mathbf{A}^*$ is positive definite (because $\mathbf{A}$ is nonsingular), so its eigenvalues λ_i are real and positive. Consequently, the spectral decomposition (p. 517) allows us to write $\mathbf{A}\mathbf{A}^* = \sum_{i=1}^{k}\lambda_i\mathbf{G}_i$. Use the results on (p. 526), and set

$$\mathbf{R} = (\mathbf{A}\mathbf{A}^*)^{1/2} = \sum_{i=1}^{k}\lambda_i^{1/2}\mathbf{G}_i, \quad \text{and} \quad \mathbf{R}^{-1} = (\mathbf{A}\mathbf{A}^*)^{-1/2} = \sum_{i=1}^{k}\lambda_i^{-1/2}\mathbf{G}_i.$$

It now follows that $\mathbf{R}$ is positive definite, and $\mathbf{A} = \mathbf{R}(\mathbf{R}^{-1}\mathbf{A}) = \mathbf{R}\mathbf{U}$, where $\mathbf{U} = \mathbf{R}^{-1}\mathbf{A}$. Finally, $\mathbf{U}$ is unitary because

$$\mathbf{U}\mathbf{U}^* = (\mathbf{A}\mathbf{A}^*)^{-1/2}\mathbf{A}\mathbf{A}^*(\mathbf{A}\mathbf{A}^*)^{-1/2} = \mathbf{I}.$$

If $\mathbf{R}_1\mathbf{U}_1 = \mathbf{A} = \mathbf{R}_2\mathbf{U}_2$, then $\mathbf{R}_2^{-1}\mathbf{R}_1 = \mathbf{U}_2\mathbf{U}_1^*$, which is unitary, so

$$\mathbf{R}_2^{-1}\mathbf{R}_1\mathbf{R}_1\mathbf{R}_2^{-1} = \mathbf{I} \implies \mathbf{R}_1^2 = \mathbf{R}_2^2 \implies \mathbf{R}_1 = \mathbf{R}_2 \text{ (because the } \mathbf{R}_i\text{'s are pd)}.$$

7.6.8. The 2-norm condition number is the ratio of the largest to smallest singular values. Since $\mathbf{L}$ is symmetric and positive definite, the singular values are the eigenvalues, and, by (7.6.8), $\max\lambda_{ij} \to 8$ and $\min\lambda_{ij} \to 0$ as $n \to \infty$.

7.6.9. The procedure is essentially identical to that in Example 7.6.2. The only difference is that when (7.6.6) is applied, the result is

$$\frac{-4u_{ij} + (u_{i-1,j} + u_{i+1,j} + u_{i,j-1} + u_{i,j+1})}{h^2} + O(h^2) = f_{ij}$$

or, equivalently,

$$4u_{ij} - (u_{i-1,j} + u_{i+1,j} + u_{i,j-1} + u_{i,j+1}) + O(h^4) = -h^2 f_{ij} \quad \text{for} \quad i,j = 1,2,\ldots,n.$$

If the $O(h^4)$ terms are neglected, and if the boundary values g_{ij} are taken to the right-hand side, then, with the same ordering as indicated in Example 7.6.2, the system $\mathbf{Lu} = \mathbf{g} - h^2 \mathbf{f}$ is produced.

7.6.10. $\mathbf{I}_n \otimes \mathbf{A}_n = \begin{pmatrix} \mathbf{A}_n & 0 & \cdots & 0 \\ 0 & \mathbf{A}_n & \cdots & 0 \\ \vdots & \vdots & \ddots & \vdots \\ 0 & 0 & \cdots & \mathbf{A}_n \end{pmatrix}$, $\mathbf{A}_n \otimes \mathbf{I}_n = \begin{pmatrix} 2\mathbf{I}_n & -\mathbf{I}_n & & & \\ -\mathbf{I}_n & 2\mathbf{I}_n & -\mathbf{I}_n & & \\ & \ddots & \ddots & \ddots & \\ & & -\mathbf{I}_n & 2\mathbf{I}_n & -\mathbf{I}_n \\ & & & -\mathbf{I}_n & 2\mathbf{I}_n \end{pmatrix}$,

$$\mathbf{A}_n + 2\mathbf{I}_n = \mathbf{T}_n = \begin{pmatrix} 4 & -1 & & & \\ -1 & 4 & -1 & & \\ & \ddots & \ddots & \ddots & \\ & & -1 & 4 & -1 \\ & & & -1 & 4 \end{pmatrix}_{n \times n}, \text{ so}$$

$$(\mathbf{I}_n \otimes \mathbf{A}_n) + (\mathbf{A}_n \otimes \mathbf{I}_n) = \begin{pmatrix} \mathbf{T}_n & -\mathbf{I}_n & & & \\ -\mathbf{I}_n & \mathbf{T}_n & -\mathbf{I}_n & & \\ & \ddots & \ddots & \ddots & \\ & & -\mathbf{I}_n & \mathbf{T}_n & -\mathbf{I}_n \\ & & & -\mathbf{I}_n & \mathbf{T}_n \end{pmatrix} = \mathbf{L}_{n^2 \times n^2}.$$

Solutions for exercises in section 7. 7

7.7.1. No. This can be deduced directly from the definition of index given on p. 395, or it can be seen by looking at the Jordan form (7.7.6) on p. 579.

7.7.2. Since the index k of a 4×4 nilpotent matrix cannot exceed 4, consider the different possibilities for $k = 1,2,3,4$. For $k = 1$, $\mathbf{N} = \mathbf{0}_{4\times 4}$ is the only possibility. If $k = 2$, the largest Jordan block in $\mathbf{N}$ is 2×2, so

$$\mathbf{N} = \left(\begin{array}{cc|cc} 0 & 1 & 0 & 0 \\ 0 & 0 & 0 & 0 \\ \hline 0 & 0 & 0 & 1 \\ 0 & 0 & 0 & 0 \end{array}\right) \quad \text{and} \quad \mathbf{N} = \left(\begin{array}{cc|c|c} 0 & 1 & 0 & 0 \\ 0 & 0 & 0 & 0 \\ \hline 0 & 0 & 0 & 0 \\ \hline 0 & 0 & 0 & 0 \end{array}\right)$$

are the only possibilities. If $k = 3$ or $k = 4$, then the largest Jordan block is 3×3 or 4×4, so

$$\mathbf{N} = \left(\begin{array}{ccc|c} 0 & 1 & 0 & 0 \\ 0 & 0 & 1 & 0 \\ 0 & 0 & 0 & 0 \\ \hline 0 & 0 & 0 & 0 \end{array}\right) \quad \text{and} \quad \mathbf{N} = \begin{pmatrix} 0 & 1 & 0 & 0 \\ 0 & 0 & 1 & 0 \\ 0 & 0 & 0 & 1 \\ 0 & 0 & 0 & 0 \end{pmatrix}$$

are the only respective possibilities.

7.7.3. Let $k = index(\mathbf{L})$, and let ζ_i denote the number of blocks of size $i \times i$ or larger. This number is determined by the number of chains of length i or larger, and such chains emanate from the vectors in $\mathcal{S}_{k-1} \cup \mathcal{S}_{k-2} \cup \cdots \cup \mathcal{S}_{i-1} = \mathcal{B}_{i-1}$. Since $\mathcal{B}_{i-1}$ is a basis for $\mathcal{M}_{i-1}$, it follows that $\zeta_i = \dim \mathcal{M}_{i-1} = r_{i-1} - r_i$, where $r_i = rank(\mathbf{L}^i)$ —recall (7.7.3).

7.7.4. $\mathbf{x} \in \mathcal{M}_i = R(\mathbf{L}^i) \cap N(\mathbf{L}) \implies \mathbf{x} = \mathbf{L}^i\mathbf{y}$ for some $\mathbf{y}$ and $\mathbf{Lx} = \mathbf{0} \implies \mathbf{x} = \mathbf{L}^{i-1}(\mathbf{Ly})\mathbf{y}$ and $\mathbf{Lx} = \mathbf{0} \implies \mathbf{x} \in R(\mathbf{L}^{i-1}) \cap N(\mathbf{L}) = \mathcal{M}_{i-1}$.

7.7.5. It suffices to prove that $R(\mathbf{L}^{k-1}) \subseteq N(\mathbf{L})$, and this is accomplished by writing

$$\mathbf{x} \in R(\mathbf{L}^{k-1}) \implies \mathbf{x} = \mathbf{L}^{k-1}\mathbf{y} \text{ for some } \mathbf{y} \implies \mathbf{Lx} = \mathbf{L}^k\mathbf{y} = \mathbf{0} \implies \mathbf{x} \in N(\mathbf{L}).$$

7.7.6. This follows from the result on p. 211.

7.7.7. $\mathbf{L}^2 = \mathbf{0}$ means that $\mathbf{L}$ is nilpotent of index $k = 2$. Consequently, the size of the largest Jordan block in $\mathbf{N}$ is 2×2. Since $r_1 = 2$ and $r_i = 0$ for $i \geq 2$, the number of 2×2 blocks is $r_1 - 2r_2 + r_3 = 2$, so the Jordan form is

$$\mathbf{N} = \left(\begin{array}{cc|cc} 0 & 1 & 0 & 0 \\ 0 & 0 & 0 & 0 \\ \hline 0 & 0 & 0 & 1 \\ 0 & 0 & 0 & 0 \end{array}\right).$$

In this case, $\mathcal{M}_0 = N(\mathbf{L}) = R(\mathbf{L}) = \mathcal{M}_1$ because $\mathbf{L}^2 = \mathbf{0} \implies R(\mathbf{L}) \subseteq N(\mathbf{L})$ and $\dim R(\mathbf{L}) = 2 = \dim N(\mathbf{L})$. Now, $\mathcal{S}_1 = \{\mathbf{L}_{*1}, \mathbf{L}_{*2}\}$ (the basic columns in $\mathbf{L}$) is a basis for $\mathcal{M}_1 = R(\mathbf{L})$, and $\mathcal{S}_0 = \phi$. Since $\mathbf{x}_1 = \mathbf{e}_1$ and $\mathbf{x}_2 = \mathbf{e}_2$ are solutions for $\mathbf{Lx}_1 = \mathbf{L}_{*1}$ and $\mathbf{Lx}_2 = \mathbf{L}_{*1}$, respectively, there are two Jordan chains, namely $\{\mathbf{Lx}_1, \mathbf{x}_1\} = \{\mathbf{L}_{*1}, \mathbf{e}_1\}$ and $\{\mathbf{Lx}_2, \mathbf{x}_2\} = \{\mathbf{L}_{*2}, \mathbf{e}_2\}$, so

$$\mathbf{P} = [\mathbf{L}_{*1} \,|\, \mathbf{e}_1 \,|\, \mathbf{L}_{*2} \,|\, \mathbf{e}_2] = \begin{pmatrix} 3 & 1 & 3 & 0 \\ -2 & 0 & -1 & 1 \\ 1 & 0 & -1 & 0 \\ -5 & 0 & -4 & 0 \end{pmatrix}.$$

Use direct computation to verify that $\mathbf{P}^{-1}\mathbf{LP} = \mathbf{N}$.

7.7.8. Computing $r_i = rank(\mathbf{L}^i)$ reveals that $r_1 = 4$, $r_2 = 1$, and $r_3 = 0$, so the index of $\mathbf{L}$ is $k = 3$, and

$$\begin{aligned} \text{the number of } 3 \times 3 \text{ blocks} &= r_2 - 2r_3 + r_4 = 1, \\ \text{the number of } 2 \times 2 \text{ blocks} &= r_1 - 2r_2 + r_3 = 2, \\ \text{the number of } 1 \times 1 \text{ blocks} &= r_0 - 2r_1 + r_2 = 1. \end{aligned}$$

Consequently, the Jordan form of $\mathbf{L}$ is

$$
\mathbf{N} = \left(\begin{array}{ccc|cc|cc|c}
0 & 1 & 0 & 0 & 0 & 0 & 0 & 0 \\
0 & 0 & 1 & 0 & 0 & 0 & 0 & 0 \\
0 & 0 & 0 & 0 & 0 & 0 & 0 & 0 \\
\hline
0 & 0 & 0 & 0 & 1 & 0 & 0 & 0 \\
0 & 0 & 0 & 0 & 0 & 0 & 0 & 0 \\
\hline
0 & 0 & 0 & 0 & 0 & 0 & 1 & 0 \\
0 & 0 & 0 & 0 & 0 & 0 & 0 & 0 \\
\hline
0 & 0 & 0 & 0 & 0 & 0 & 0 & 0
\end{array}\right).
$$

Notice that four Jordan blocks were found, and this agrees with the fact that $\dim N(\mathbf{L}) = 8 - rank(\mathbf{L}) = 4$.

7.7.9. If $\mathbf{N}_i$ is an $n_i \times n_i$, nilpotent block as described in (7.7.5), and if $\mathbf{D}_i$ is the diagonal matrix $\mathbf{D}_i = \text{diag}\left(1, \epsilon_i, \dots, \epsilon^{n_i-1}\right)$, then $\mathbf{D}_i^{-1}\mathbf{N}_i\mathbf{D}_i = \epsilon_i\mathbf{N}_i$. Therefore, if $\mathbf{P}^{-1}\mathbf{L}\mathbf{P} = \mathbf{N}$ is in Jordan form, and if $\mathbf{Q} = \mathbf{P}\mathbf{D}$, where $\mathbf{D}$ is the block-diagonal matrix $\mathbf{D} = \text{diag}\left(\mathbf{D}_1, \mathbf{D}_2, \dots, \mathbf{D}_t\right)$, then $\mathbf{Q}^{-1}\mathbf{L}\mathbf{Q} = \tilde{\mathbf{N}}$.

Solutions for exercises in section 7. 8

7.8.1. Since $rank(\mathbf{A}) = 7$, $rank\left(\mathbf{A}^2\right) = 6$, and $rank\left(\mathbf{A}^3\right) = 5 = rank\left(\mathbf{A}^{3+i}\right)$, there is one 3×3 Jordan block associates with $\lambda = 0$. Since $rank(\mathbf{A} + \mathbf{I}) = 6$ and $rank\left((\mathbf{A} + \mathbf{I})^2\right) = 5 = rank\left((\mathbf{A} + \mathbf{I})^{2+i}\right)$, there is one 1×1 and one 2×2 Jordan block associated with $\lambda = -1$. Finally, $rank(\mathbf{A} - \mathbf{I}) = rank\left((\mathbf{A} - \mathbf{I})^{1+i}\right)$ implies there are two 1×1 blocks associated with $\lambda = 1$—i.e., $\lambda = 1$ is a semisimple eigenvalue. Therefore, the Jordan form for $\mathbf{A}$ is

$$
\mathbf{J} = \left(\begin{array}{ccc|cc|c|c|c}
0 & 1 & 0 & & & & & \\
 & 0 & 1 & & & & & \\
 & & 0 & & & & & \\
\hline
 & & & -1 & 1 & & & \\
 & & & 0 & -1 & & & \\
\hline
 & & & & & -1 & & \\
\hline
 & & & & & & 1 & \\
\hline
 & & & & & & & 1
\end{array}\right).
$$

7.8.2. As noted in Example 7.8.3, $\sigma(\mathbf{A}) = \{1\}$ and $k = index(1) = 2$. Use the procedure on p. 211 to determine a basis for $\mathcal{M}_{k-1} = \mathcal{M}_1 = R(\mathbf{A} - \mathbf{I}) \cap N(\mathbf{A} - \mathbf{I})$ to be $\mathcal{S}_1 = \left\{\begin{pmatrix} 1 \\ -2 \\ -2 \end{pmatrix} = \mathbf{b}_1\right\}$. (You might also determine this just by inspection.) A basis for $N(\mathbf{A} - \mathbf{I})$ is easily found to be $\left\{\begin{pmatrix} 0 \\ 1 \\ 0 \end{pmatrix}, \begin{pmatrix} 1 \\ 0 \\ -2 \end{pmatrix}\right\}$, so examining the basic columns of $\begin{pmatrix} 1 & 0 & 1 \\ -2 & 1 & 0 \\ -2 & 0 & -2 \end{pmatrix}$ yields the extension set $\mathcal{S}_0 = \left\{ \begin{pmatrix} 0 \\ 1 \\ 0 \end{pmatrix} = \mathbf{b}_2 \right\}$.

Solving $(\mathbf{A} - \mathbf{I})\mathbf{x} = \mathbf{b}_1$ produces $\mathbf{x} = \mathbf{e}_3$, so the associated Jordan chain is $\{\mathbf{b}_1, \mathbf{e}_3\}$, and thus $\mathbf{P} = (\mathbf{b}_1 \mid \mathbf{e}_3 \mid \mathbf{b}_2) = \begin{pmatrix} 1 & 0 & 0 \\ -2 & 0 & 1 \\ -2 & 1 & 0 \end{pmatrix}$. It's easy to check that

$\mathbf{P}^{-1}\mathbf{A}\mathbf{P} = \begin{pmatrix} 1 & 1 & 0 \\ 0 & 1 & 0 \\ 0 & 0 & 1 \end{pmatrix}$ is indeed the Jordan form for $\mathbf{A}$.

7.8.3. If $k = index(\lambda)$, then the size of the largest Jordan block associated with λ is $k \times k$. This insures that λ must be repeated at least k times, and thus $index(\lambda) \leq alg\ mult(\lambda)$.

7.8.4. $index(\lambda) = 1$ if and only if every Jordan block is 1×1, which happens if and only if the number of eigenvectors associated with λ in $\mathbf{P}$ such that $\mathbf{P}^{-1}\mathbf{A}\mathbf{P} = \mathbf{J}$ is the same as the number of Jordan blocks, and this is just another way to say that $alg\ mult_{\mathbf{A}}(\lambda) = geo\ mult_{\mathbf{A}}(\lambda)$, which is the definition of λ being a semisimple eigenvalue (p. 510).

7.8.5. Notice that $\mathbf{R}^2 = \mathbf{I}$, so $\mathbf{R}^{-1} = \mathbf{R} = \mathbf{R}^T$, and if $\mathbf{J}_\star = \begin{pmatrix} \lambda & 1 & \\ & \ddots & \ddots \\ & & \lambda \end{pmatrix}$ is a generic Jordan block, then $\mathbf{R}^{-1}\mathbf{J}_\star\mathbf{R} = \mathbf{R}\mathbf{J}_\star\mathbf{R} = \mathbf{J}_\star^T$. Thus every Jordan block is similar to its transpose. Given any Jordan form, reversal matrices of appropriate sizes can be incorporated into a block-diagonal matrix $\widetilde{\mathbf{R}}$ such that $\widetilde{\mathbf{R}}^{-1}\mathbf{J}\widetilde{\mathbf{R}} = \mathbf{J}^T$ showing that $\mathbf{J}$ is similar to $\mathbf{J}^T$. Consequently, if $\mathbf{A} = \mathbf{P}\mathbf{J}\mathbf{P}^{-1}$, then

$$\mathbf{A}^T = \mathbf{P}^{-1^T}\mathbf{J}^T\mathbf{P}^T = \mathbf{P}^{-1^T}\widetilde{\mathbf{R}}^{-1}\mathbf{J}\widetilde{\mathbf{R}}\mathbf{P}^T = \mathbf{P}^{-1^T}\widetilde{\mathbf{R}}^{-1}\mathbf{P}^{-1}\mathbf{A}\mathbf{P}\widetilde{\mathbf{R}}\mathbf{P}^T = \mathbf{Q}^{-1}\mathbf{A}\mathbf{Q},$$

where $\mathbf{Q} = \mathbf{P}\widetilde{\mathbf{R}}\mathbf{P}^T$.

7.8.6. If $\sigma(\mathbf{A}) = \{\lambda_1, \lambda_2, \ldots, \lambda_s\}$, where $alg\ mult(\lambda_i) = a_i$, then the characteristic equation for $\mathbf{A}$ is $0 = (x - \lambda_1)^{a_1}(x - \lambda_2)^{a_2} \cdots (x - \lambda_s)^{a_s} = c(x)$. If

$$\mathbf{J} = \begin{pmatrix} \ddots & & \\ & \mathbf{J}_\star & \\ & & \ddots \end{pmatrix} = \mathbf{P}^{-1}\mathbf{A}\mathbf{P} \text{ is in Jordan form with } \mathbf{J}_\star = \begin{pmatrix} \lambda_i & 1 & \\ & \ddots & \ddots \\ & & \lambda_i \end{pmatrix}$$

representing a generic Jordan block, then

$$c(\mathbf{A}) = c(\mathbf{P}\mathbf{J}\mathbf{P}^{-1}) = \mathbf{P}c(\mathbf{J})\mathbf{P}^{-1} = \mathbf{P}\begin{pmatrix} \ddots & & \\ & c(\mathbf{J}_\star) & \\ & & \ddots \end{pmatrix}\mathbf{P}^{-1}.$$

Notice that if $\mathbf{J}_\star$ is $r \times r$, then $(\mathbf{J}_\star - \lambda_i\mathbf{I})^r = \begin{pmatrix} 0 & 1 & \\ & \ddots & \ddots \\ & & 0 \end{pmatrix}^r = \mathbf{0}$. Since the size of the largest Jordan block associated with λ_i is $k_i \times k_i$, where $k_i = index(\lambda_i) \leq alg\ mult(\lambda_i) = a_i$, it follows that $(\mathbf{J}_\star - \lambda_i\mathbf{I})^{a_i} = \mathbf{0}$. Consequently $c(\mathbf{J}_\star) = \mathbf{0}$ for every Jordan block, and thus $c(\mathbf{A}) = \mathbf{0}$.

7.8.7. By using the Jordan form for $\mathbf{A}$, one can find a similarity transformation $\mathbf{P}$ such that $\mathbf{P}^{-1}(\mathbf{A} - \lambda\mathbf{I})\mathbf{P} = \begin{pmatrix} \mathbf{L}_{m \times m} & \mathbf{0} \\ \mathbf{0} & \mathbf{C} \end{pmatrix}$ with $\mathbf{L}^k = \mathbf{0}$ and $\mathbf{C}$ nonsingular. Therefore, $\mathbf{P}^{-1}(\mathbf{A} - \lambda\mathbf{I})^k\mathbf{P} = \begin{pmatrix} \mathbf{0}_{m \times m} & \mathbf{0} \\ \mathbf{0} & \mathbf{C}^k \end{pmatrix}$, and thus

$$\dim N\big((\mathbf{A} - \lambda\mathbf{I})^k\big) = n - rank\big((\mathbf{A} - \lambda\mathbf{I})^k\big) = n - rank\big(\mathbf{C}^k\big) = m.$$

It is also true that $\dim N\big((\mathbf{A} - \lambda\mathbf{I})^m\big) = m$ because the nullspace remains the same for all powers beyond the index (p. 395).

7.8.8. To prove $\mathcal{M}_{k_j}(\lambda_j) = \mathbf{0}$, suppose $\mathbf{x} \in \mathcal{M}_{k_j}(\lambda_j)$ so that $(\mathbf{A} - \lambda_j\mathbf{I})\mathbf{x} = \mathbf{0}$ and $\mathbf{x} = (\mathbf{A} - \lambda_j\mathbf{I})^{k_j}\mathbf{z}$ for some $\mathbf{z}$. Combine these with the properties of $index(\lambda_j)$ (p. 587) to obtain

$$(\mathbf{A} - \lambda_j\mathbf{I})^{k_j+1}\mathbf{z} = \mathbf{0} \implies (\mathbf{A} - \lambda_j\mathbf{I})^{k_j}\mathbf{z} = \mathbf{0} \implies \mathbf{x} = \mathbf{0}.$$

The fact that the subspaces are nested follows from the observation that if $\mathbf{x} \in \mathcal{M}_{i+1}(\lambda_j)$, then $\mathbf{x} = (\mathbf{A} - \lambda_j\mathbf{I})^{i+1}\mathbf{z}$ and $(\mathbf{A} - \lambda_j\mathbf{I})\mathbf{x} = \mathbf{0}$ implies $\mathbf{x} = (\mathbf{A} - \lambda_j\mathbf{I})^i\big((\mathbf{A} - \lambda_j\mathbf{I})\mathbf{z}\big)$ and $(\mathbf{A} - \lambda_j\mathbf{I})\mathbf{x} = \mathbf{0}$, so $\mathcal{M}_{i+1}(\lambda_j) \subseteq \mathcal{M}_i(\lambda_j)$.

7.8.9. $\mathbf{b}(\lambda_j) \in \mathcal{S}_i(\lambda_j) \subseteq \mathcal{M}_i(\lambda_j) = R\big((\mathbf{A} - \lambda_j\mathbf{I})^i\big) \cap N(\mathbf{A} - \lambda_j\mathbf{I}) \subseteq R\big((\mathbf{A} - \lambda_j\mathbf{I})^i\big)$.

7.8.10. No—consider $\mathbf{A} = \begin{pmatrix} 1 & 1 & 0 \\ 0 & 1 & 0 \\ 0 & 0 & 2 \end{pmatrix}$ and $\lambda = 1$.

7.8.11. (a) All of these facts are established by straightforward arguments using elementary properties of matrix algebra, so the details are omitted here.

(b) To show that the eigenvalues of $\mathbf{A} \otimes \mathbf{B}$ are $\{\lambda_i\mu_j\}_{i=1}^{m}{}_{j=1}^{n}$, let $\mathbf{J}_A = \mathbf{P}^{-1}\mathbf{A}\mathbf{P}$ and $\mathbf{J}_B = \mathbf{Q}^{-1}\mathbf{B}\mathbf{Q}$ be the respective Jordan forms for $\mathbf{A}$ and $\mathbf{B}$, and use properties from (a) to establish that $\mathbf{A} \otimes \mathbf{B}$ is similar to $\mathbf{J}_A \otimes \mathbf{J}_B$ by writing

$$\mathbf{J}_A \otimes \mathbf{J}_B = (\mathbf{P}^{-1}\mathbf{A}\mathbf{P}) \otimes (\mathbf{Q}^{-1}\mathbf{B}\mathbf{Q}) = (\mathbf{P}^{-1} \otimes \mathbf{Q}^{-1})(\mathbf{A} \otimes \mathbf{B})(\mathbf{P} \otimes \mathbf{Q})$$
$$= (\mathbf{P} \otimes \mathbf{Q})^{-1}(\mathbf{A} \otimes \mathbf{B})(\mathbf{P} \otimes \mathbf{Q})$$

Thus the eigenvalues of $\mathbf{A} \otimes \mathbf{B}$ are the same as those of $\mathbf{J}_A \otimes \mathbf{J}_B$, and because $\mathbf{J}_A$ and $\mathbf{J}_B$ are upper triangular with the λ_i's and μ_i's on the diagonal, it's clear that $\mathbf{J}_A \otimes \mathbf{J}_B$ is also upper triangular with diagonal entries being $\lambda_i\mu_j$. To show that the eigenvalues of $(\mathbf{A} \otimes \mathbf{I}_n) + (\mathbf{I}_m \otimes \mathbf{B})$ are $\{\lambda_i + \mu_j\}_{i=1}^{m}{}_{j=1}^{n}$, show that $(\mathbf{A} \otimes \mathbf{I}_n) + (\mathbf{I}_m \otimes \mathbf{B})$ is similar to $(\mathbf{J}_A \otimes \mathbf{I}_n) + (\mathbf{I}_m \otimes \mathbf{J}_B)$ by writing

$$(\mathbf{J}_A \otimes \mathbf{I}_n) + (\mathbf{I}_m \otimes \mathbf{J}_B) = (\mathbf{P}^{-1}\mathbf{A}\mathbf{P}) \otimes (\mathbf{Q}^{-1}\mathbf{I}\mathbf{Q}) + (\mathbf{P}^{-1}\mathbf{I}\mathbf{P}) \otimes (\mathbf{Q}^{-1}\mathbf{B}\mathbf{Q})$$
$$= (\mathbf{P}^{-1} \otimes \mathbf{Q}^{-1})(\mathbf{A} \otimes \mathbf{I})(\mathbf{P} \otimes \mathbf{Q})$$
$$+ (\mathbf{P}^{-1} \otimes \mathbf{Q}^{-1})(\mathbf{I} \otimes \mathbf{B})(\mathbf{P} \otimes \mathbf{Q})$$
$$= (\mathbf{P}^{-1} \otimes \mathbf{Q}^{-1})\big[(\mathbf{A} \otimes \mathbf{I}) + (\mathbf{I} \otimes \mathbf{B})\big](\mathbf{P} \otimes \mathbf{Q})$$
$$= (\mathbf{P} \otimes \mathbf{Q})^{-1}\big[(\mathbf{A} \otimes \mathbf{I}) + (\mathbf{I} \otimes \mathbf{B})\big](\mathbf{P} \otimes \mathbf{Q}).$$

Thus $(\mathbf{A}\otimes\mathbf{I}_n)+(\mathbf{I}_m\otimes\mathbf{B})$ and $(\mathbf{J}_A\otimes\mathbf{I}_n)+(\mathbf{I}_m\otimes\mathbf{J}_B)$ have the same eigenvalues, and the latter matrix is easily seen to be an upper-triangular matrix whose diagonal entries are $\{\lambda_i+\mu_j\}_{i=1}^{m}\,_{j=1}^{n}$.

7.8.12. It was established in Exercise 7.6.10 (p. 573) that $\mathbf{L}_{n^2\times n^2}=(\mathbf{I}_n\otimes\mathbf{A}_n)+(\mathbf{A}_n\otimes\mathbf{I}_n)$, where

$$\mathbf{A}_n=\begin{pmatrix} 2 & -1 & & & \\ -1 & 2 & -1 & & \\ & \ddots & \ddots & \ddots & \\ & & -1 & 2 & -1 \\ & & & -1 & 2 \end{pmatrix}_{n\times n}$$

is the finite difference matrix of Example 1.4.1 (p. 19). The eigenvalues of $\mathbf{A}_n$ were determined in Exercise 7.2.18 (p. 522) to be $\mu_j=4\sin^2[j\pi/2(n+1)]$ for $j=1,2,\ldots,n$, so it follows from the last property in Exercise 7.8.11 that the n^2 eigenvalues of $\mathbf{L}_{n^2\times n^2}=(\mathbf{I}_n\otimes\mathbf{A}_n)+(\mathbf{A}_n\otimes\mathbf{I}_n)$ are

$$\lambda_{ij}=\mu_i+\mu_j=4\left[\sin^2\left(\frac{i\pi}{2(n+1)}\right)+\sin^2\left(\frac{j\pi}{2(n+1)}\right)\right],\quad i,j=1,2,\ldots,n.$$

7.8.13. The same argument given in the solution of the last part of Exercise 7.8.11 applies to show that if $\mathbf{J}$ is the Jordan form for $\mathbf{A}$, then $\mathbf{L}$ is similar to $(\mathbf{I}\otimes\mathbf{I}\otimes\mathbf{J})+(\mathbf{I}\otimes\mathbf{J}\otimes\mathbf{I})+(\mathbf{J}\otimes\mathbf{I}\otimes\mathbf{I})$, and since $\mathbf{J}$ is upper triangular with the eigenvalues $\mu_j=4\sin^2[j\pi/2(n+1)]$ of $\mathbf{A}$ (recall Exercise 7.2.18 (p. 522)) on the diagonal of $\mathbf{J}$, it follows that the eigenvalues of $\mathbf{L}_{n^3\times n^3}$ are the n^3 numbers

$$\lambda_{ijk}=\mu_i+\mu_j+\mu_k=4\left[\sin^2\left(\frac{i\pi}{2(n+1)}\right)+\sin^2\left(\frac{j\pi}{2(n+1)}\right)+\sin^2\left(\frac{k\pi}{2(n+1)}\right)\right]$$

for $i,j,k=1,2,\ldots,n$.

Solutions for exercises in section 7.9

7.9.1. If $u_i(t)$ denotes the number of pounds of pollutant in lake i at time $t>0$, then the concentration of pollutant in lake i at time t is $u_i(t)/V$ lbs/gal, so the model $u_i'(t)=$ (lbs/sec) coming in$-$(lbs/sec) going out produces the system

$$\begin{aligned} u_1' &= \frac{4r}{V}u_2-\frac{4r}{V}u_1 \\ u_2' &= \frac{2r}{V}u_1+\frac{3r}{V}u_3-\frac{5r}{V}u_2 \\ u_3' &= \frac{2r}{V}u_1+\frac{r}{V}u_2-\frac{3r}{V}u_3 \end{aligned} \quad\text{or}\quad \begin{pmatrix} u_1' \\ u_2' \\ u_3' \end{pmatrix}=\frac{r}{V}\begin{pmatrix} -4 & 4 & 0 \\ 2 & -5 & 3 \\ 2 & 1 & -3 \end{pmatrix}\begin{pmatrix} u_1(t) \\ u_2(t) \\ u_3(t) \end{pmatrix}.$$

The solution of $\mathbf{u}'=\mathbf{A}\mathbf{u}$ with $\mathbf{u}(0)=\mathbf{c}$ is $\mathbf{u}=e^{\mathbf{A}t}\mathbf{c}$. The eigenvalues of $\mathbf{A}$ are $\lambda_1=0$ with $alg\;mult(\lambda_1)=1$ and $\lambda_2=-6r/V$ with $index(\lambda_2)=2$, so

$$\mathbf{u}=e^{\mathbf{A}t}\mathbf{c}=e^{\lambda_1 t}\mathbf{G}_1\mathbf{c}+e^{\lambda_2 t}\mathbf{G}_2\mathbf{c}+te^{\lambda_2 t}(\mathbf{A}-\lambda_2\mathbf{I})\mathbf{G}_2\mathbf{c}.$$

Since $\lambda_1 = 0$ is a simple eigenvalue, it follows from (7.2.12) on p. 518 that $\mathbf{G}_1 = \mathbf{x}\mathbf{y}^T/\mathbf{y}^T\mathbf{x}$, where $\mathbf{x}$ and $\mathbf{y}^T$ are any pair of respective right-hand and left-hand eigenvectors associated with $\lambda_1 = 0$. By observing that $\mathbf{A}\mathbf{e} = \mathbf{0}$ and $\mathbf{e}^T\mathbf{A} = \mathbf{0}^T$ for $\mathbf{e} = (1,\, 1,\, 1)^T$ (this is a result of being a closed system), and by using $\mathbf{G}_1 + \mathbf{G}_2 = \mathbf{I}$, we have (by using $\mathbf{x} = \mathbf{y} = \mathbf{e}$)

$$\mathbf{G}_1 = \frac{\mathbf{e}\mathbf{e}^T}{3}, \quad \mathbf{G}_2 = \mathbf{I} - \frac{\mathbf{e}\mathbf{e}^T}{3}, \quad \text{and} \quad (\mathbf{A} - \lambda_2\mathbf{I})\mathbf{G}_2 = \mathbf{A} - \lambda_2\mathbf{I} + \frac{\lambda_2}{3}\mathbf{e}\mathbf{e}^T.$$

If $\alpha = (c_1 + c_2 + c_3)/3 = \mathbf{e}^T\mathbf{c}/3$ denotes the average of the initial values, then $\mathbf{G}_1\mathbf{c} = \alpha\mathbf{e}$ and $\mathbf{G}_2\mathbf{c} = \mathbf{c} - \alpha\mathbf{e}$, so

$$\mathbf{u}(t) = \alpha\mathbf{e} + e^{\lambda_2 t}(\mathbf{c} - \alpha\mathbf{e}) + te^{\lambda_2 t}\big[\mathbf{A}\mathbf{c} - \lambda_2(\mathbf{c} - \alpha\mathbf{e})\big] \quad \text{for} \quad \lambda_2 = -6r/V.$$

Since $\lambda_2 < 0$, it follows that $\mathbf{u}(t) \to \alpha\mathbf{e}$ as $t \to \infty$. In other words, the long-run amount of pollution in each lake is the same—namely α lbs—and this is what common sense would dictate.

7.9.2. It follows from (7.9.9) that $f_i(\mathbf{A}) = \mathbf{G}_i$.

7.9.3. We know from Exercise 7.9.2 that $\mathbf{G}_i = f_i(\mathbf{A})$ for $f_i(z) = \begin{cases} 1 & \text{when } z = \lambda_i, \\ 0 & \text{otherwise}, \end{cases}$ and from Example 7.9.4 (p. 606) we know that every function of $\mathbf{A}$ is a polynomial in $\mathbf{A}$.

7.9.4. Using $f(z) = z^k$ in (7.9.9) on p. 603 produces the desired result.

7.9.5. Using $f(z) = z^k$ in (7.9.2) on p. 600 produces the desired result.

7.9.6. $\mathbf{A}$ is the matrix in Example 7.9.2, so the results derived there imply that

$$e^{\mathbf{A}} = e^2\mathbf{G}_1 + e^4\mathbf{G}_2 + e^4(\mathbf{A} - 4\mathbf{I})\mathbf{G}_2 = \begin{pmatrix} 3e^4 & 2e^4 & 7e^4 - e^2 \\ -2e^4 & -e^4 & -4e^4 - 2e^2 \\ 0 & 0 & e^2 \end{pmatrix}.$$

7.9.7. The eigenvalues of $\mathbf{A}$ are $\lambda_1 = 1$ and $\lambda_2 = 4$ with $alg\ mult(1) = 1$ and $index(4) = 2$, so

$$f(\mathbf{A}) = f(1)\mathbf{G}_1 + f(4)\mathbf{G}_2 + f'(4)(\mathbf{A} - 4\mathbf{I})\mathbf{G}_2$$

Since $\lambda_1 = 1$ is a simple eigenvalue, it follows from formula (7.2.12) on p. 518 that $\mathbf{G}_1 = \mathbf{x}\mathbf{y}^T/\mathbf{y}^T\mathbf{x}$, where $\mathbf{x}$ and $\mathbf{y}^T$ are any pair of respective right-hand and left-hand eigenvectors associated with $\lambda_1 = 1$. Using $\mathbf{x} = (-2,\, 1,\, 0)^T$ and $\mathbf{y} = (1,\, 1,\, 1)^T$ produces

$$\mathbf{G}_1 = \begin{pmatrix} 2 & 2 & 2 \\ -1 & -1 & -1 \\ 0 & 0 & 0 \end{pmatrix} \quad \text{and} \quad \mathbf{G}_2 = \mathbf{I} - \mathbf{G}_1 = \begin{pmatrix} -1 & -2 & -2 \\ 1 & 2 & 1 \\ 0 & 0 & 1 \end{pmatrix}$$

Therefore,

$$f(\mathbf{A}) = 4\sqrt{\mathbf{A}} - \mathbf{I} = 3\mathbf{G}_1 + 7\mathbf{G}_2 + (\mathbf{A} - 4\mathbf{I})\mathbf{G}_2 = \begin{pmatrix} -2 & -10 & -11 \\ 6 & 15 & 10 \\ -1 & -2 & 4 \end{pmatrix}.$$

7.9.8. (a) The only point at which derivatives of $f(z) = z^{1/2}$ fail to exist are at $\lambda = 0$, so as long as $\mathbf{A}$ is nonsingular, $f(\mathbf{A}) = \sqrt{\mathbf{A}}$ is defined.

(b) If $\mathbf{A}$ is singular so that $0 \in \sigma(\mathbf{A})$ it's clear from (7.9.9) that $\mathbf{A}^{1/2}$ exists if and only if derivatives of $f(z) = z^{1/2}$ need not be evaluated at $\lambda = 0$, and this is the case if and only if $index(0) = 1$.

7.9.9. If $\mathbf{0} \neq \mathbf{x}_h \in N(\mathbf{A} - \lambda_h \mathbf{I})$, then (7.9.11) guarantees that

$$\mathbf{G}_i \mathbf{x}_h = \begin{cases} \mathbf{0} & \text{if } h \neq i, \\ \mathbf{x}_h & \text{if } h = i, \end{cases}$$

so (7.9.9) can be used to conclude that $f(\mathbf{A})\mathbf{x}_h = f(\lambda_h)\mathbf{x}_h$. It's an immediate consequence of (7.9.3) that $alg\ mult_{\mathbf{A}}(\lambda) = alg\ mult_{f(\mathbf{A})}(f(\lambda))$.

7.9.10. (a) If $\mathbf{A}_{k \times k}$ (with $k > 1$) is a Jordan block associated with $\lambda = 0$, and if $f(z) = z^k$, then $f(\mathbf{A}) = \mathbf{0}$ is not similar to $\mathbf{A} \neq \mathbf{0}$.

(b) Also, $geo\ mult_{\mathbf{A}}(0) = 1$ but $geo\ mult_{f(\mathbf{A})}(f(0)) = geo\ mult_{\mathbf{0}}(0) = k$.

(c) And $index_{\mathbf{A}}(0) = k$ while $index_{f(\mathbf{A})}(f(\lambda)) = index_{\mathbf{0}}(0) = 1$.

7.9.11. This follows because, as explained in Example 7.9.4 (p. 606), there is always a polynomial $p(z)$ such that $f(\mathbf{A}) = p(\mathbf{A})$, and $\mathbf{A}$ commutes with $p(\mathbf{A})$.

7.9.12. Because every square matrix is similar to its transpose (recall Exercise 7.8.5 on p. 596), and because similar matrices have the same Jordan structure, transposition doesn't change the eigenvalues or their indicies. So $f(\mathbf{A})$ exists if and only if $f(\mathbf{A}^T)$ exists. As proven in Example 7.9.4 (p. 606), there is a polynomial $p(z)$ such that $f(\mathbf{A}) = p(\mathbf{A})$, so $[f(\mathbf{A})]^T = [p(\mathbf{A})]^T = p(\mathbf{A}^T) = f(\mathbf{A}^T)$. While transposition doesn't change eigenvalues, conjugate transposition does—it conjugates them—so it's possible that f can exist at $\mathbf{A}$ but not at $\mathbf{A}^*$. Furthermore, you can't replace $(\star)^T$ by $(\star)^*$ in the above argument because if $p(z)$ has some complex coefficients, then $[p(\mathbf{A})]^* \neq p(\mathbf{A}^*)$.

7.9.13. (a) If $f_1(z) = e^z$, $f_2(z) = e^{-z}$, and $p(x,y) = xy - 1$, then

$$h(z) = p(f_1(z), f_2(z)) = e^z e^{-z} - 1 = 0 \quad \text{for all} \quad z \in \mathcal{C},$$

so $h(\mathbf{A}) = p(f_1(\mathbf{A}), f_2(\mathbf{A})) = \mathbf{0}$ for all $\mathbf{A} \in \mathcal{C}^{n \times n}$, and thus $e^{\mathbf{A}} e^{-\mathbf{A}} - \mathbf{I} = \mathbf{0}$.

(b) Use $f_1(z) = e^{\alpha z}$, $f_2(z) = (e^z)^{\alpha}$, and $p(x,y) = x - y$. Since

$$h(z) = p(f_1(z), f_2(z)) = e^{\alpha z} - (e^z)^{\alpha} = 0 \quad \text{for all } z \in \mathcal{C},$$

$h(\mathbf{A}) = p(f_1(\mathbf{A}), f_2(\mathbf{A})) = \mathbf{0}$ for all $\mathbf{A} \in \mathcal{C}^{n \times n}$, and thus $e^{\alpha \mathbf{A}} = (e^{\mathbf{A}})^{\alpha}$.

(c) Using $f_1(z) = e^{iz}$, $f_2(z) = \cos z + i \sin z$, and $p(x,y) = x - y$ produces $h(z) = p(f_1(z), f_2(z))$, which is zero for all z, so $h(\mathbf{A}) = \mathbf{0}$ for all $\mathbf{A} \in \mathcal{C}^{n \times n}$.

7.9.14. (a) The representation $e^z = \sum_{n=0}^{\infty} z^n / n!$ together with $\mathbf{AB} = \mathbf{BA}$ yields

$$e^{\mathbf{A}+\mathbf{B}} = \sum_{n=0}^{\infty} \frac{(\mathbf{A}+\mathbf{B})^n}{n!} = \sum_{n=0}^{\infty} \frac{1}{n!} \sum_{j=0}^{n} \binom{n}{j} \mathbf{A}^j \mathbf{B}^{n-j} = \sum_{n=0}^{\infty} \sum_{j=0}^{n} \frac{\mathbf{A}^j \mathbf{B}^{n-j}}{j!(n-j)!}$$

$$= \sum_{r=0}^{\infty} \sum_{s=0}^{\infty} \frac{1}{r!} \frac{1}{s!} \mathbf{A}^r \mathbf{B}^s = \left(\sum_{r=0}^{\infty} \frac{\mathbf{A}^r}{r!} \right) \left(\sum_{s=0}^{\infty} \frac{\mathbf{B}^s}{s!} \right) = e^{\mathbf{A}} e^{\mathbf{B}}.$$

(b) If $\mathbf{A} = \begin{pmatrix} 0 & 1 \\ 0 & 0 \end{pmatrix}$ and $\mathbf{B} = \begin{pmatrix} 0 & 0 \\ 1 & 0 \end{pmatrix}$, then

$$e^{\mathbf{A}} e^{\mathbf{B}} = \begin{pmatrix} 1 & 1 \\ 0 & 1 \end{pmatrix} \begin{pmatrix} 1 & 0 \\ 1 & 1 \end{pmatrix} = \begin{pmatrix} 2 & 1 \\ 1 & 1 \end{pmatrix}, \quad \text{but} \quad e^{\mathbf{A}+\mathbf{B}} = \frac{1}{2} \begin{pmatrix} e + e^{-1} & e - e^{-1} \\ e - e^{-1} & e + e^{-1} \end{pmatrix}.$$

7.9.15. The characteristic equation for $\mathbf{A}$ is $\lambda^3 = 0$, and the number of 2×2 Jordan blocks associated with $\lambda = 0$ is $\nu_2 = rank(\mathbf{A}) - 2\,rank(\mathbf{A}^2) + rank(\mathbf{A}^3) = 1$ (from the formula on p. 590), so $index(\lambda = 0) = 2$. Therefore, for $f(z) = e^z$ we are looking for a polynomial $p(z) = \alpha_0 + \alpha_1 z$ such that $p(0) = f(0) = 1$ and $p'(0) = f'(0) = 1$. This yields the Hermite interpolation polynomial as

$$p(z) = 1 + z, \quad \text{so} \quad e^{\mathbf{A}} = p(\mathbf{A}) = \mathbf{I} + \mathbf{A}.$$

Note: Since $\mathbf{A}^2 = 0$, this agrees with the infinite series representation for $e^{\mathbf{A}}$.

7.9.16. (a) The advantage is that the only the algebraic multiplicity and not the index of each eigenvalue is required—determining index generally requires more effort. The disadvantage is that a higher-degree polynomial might be required, so a larger system might have to be solved. Another disadvantage is the fact that f may not have enough derivatives defined at some eigenvalue for this method to work in spite of the fact that $f(\mathbf{A})$ exists.

(b) The characteristic equation for $\mathbf{A}$ is $\lambda^3 = 0$, so, for $f(z) = e^z$, we are looking for a polynomial $p(z) = \alpha_0 + \alpha_1 z + \alpha_2 z^2$ such that $p(0) = f(0) = 1$, $p'(0) = f'(0) = 1$, and $p''(0) = f''(0) = 1$. This yields

$$p(z) = 1 + z + \frac{z^2}{2}, \quad \text{so} \quad e^{\mathbf{A}} = p(\mathbf{A}) = \mathbf{I} + \mathbf{A} + \frac{\mathbf{A}^2}{2}.$$

Note: Since $\mathbf{A}^2 = 0$, this agrees with the result in Exercise 7.9.15.

7.9.17. Since $\sigma(\mathbf{A}) = \{\alpha\}$ with $index(\alpha) = 3$, it follows from (7.9.9) that

$$f(\mathbf{A}) = f(\alpha)\mathbf{G}_1 + f'(\alpha)(\mathbf{A} - \alpha\mathbf{I})\mathbf{G}_1 + \frac{f''(\alpha)}{2!}(\mathbf{A} - \alpha\mathbf{I})^2 \mathbf{G}_1.$$

The desired result is produced by using $\mathbf{G}_1 = \mathbf{I}$ (because of (7.9.10)), and $\mathbf{A} - \alpha\mathbf{I} = \beta\mathbf{N} + \gamma\mathbf{N}^2$, and $\mathbf{N}^3 = \mathbf{0}$.

7.9.18. Since

$$g(\mathbf{J}_\star) = \begin{pmatrix} g(\lambda) & g'(\lambda) & g''(\lambda)/2! \\ 0 & g(\lambda) & g'(\lambda) \\ 0 & 0 & g(\lambda) \end{pmatrix},$$

using Exercise 7.9.17 with $\alpha = g(\lambda)$, $\beta = g'(\lambda)$, and $\gamma = g''(\lambda)/2!$ yields

$$f\big(g(\mathbf{J}_\star)\big) = f(g(\lambda))\mathbf{I} + g'(\lambda)f'(g(\lambda))\mathbf{N} + \left[\frac{g''(\lambda)f'(g(\lambda))}{2!} + \frac{[g'(\lambda)]^2 f''(g(\lambda))}{2!}\right]\mathbf{N}^2.$$

Observing that

$$h(\mathbf{J}_\star) = \begin{pmatrix} h(\lambda) & h'(\lambda) & h''(\lambda)/2! \\ 0 & h(\lambda) & h'(\lambda) \\ 0 & 0 & h(\lambda) \end{pmatrix} = h(\lambda)\mathbf{I} + h'(\lambda)\mathbf{N} + \frac{h''(\lambda)}{2!}\mathbf{N}^2,$$

where $h(\lambda) = f(g(\lambda))$, $h'(\lambda) = f'(g(\lambda))g'(\lambda)$, and

$$h''(\lambda) = g''(\lambda)f'(g(\lambda)) + f''(g(\lambda))[g'(\lambda)]^2$$

proves that $h(\mathbf{J}_\star) = f\big(g(\mathbf{J}_\star)\big)$.

7.9.19. For the function

$$f_i(z) = \begin{cases} 1 & \text{in a small circle about } \lambda_i \text{ that is interior to } \Gamma_i, \\ 0 & \text{elsewhere,} \end{cases}$$

it follows, just as in Exercise 7.9.2, that $f_i(\mathbf{A}) = \mathbf{G}_i$. But using f_i in (7.9.22) produces $f_i(\mathbf{A}) = \frac{1}{2\pi i}\int_{\Gamma_i}(\xi\mathbf{I} - \mathbf{A})^{-1}d\xi$, and thus the result is proven.

7.9.20. For a $k \times k$ Jordan block $\mathbf{J}_\star = \begin{pmatrix} \lambda & 1 & \\ & \ddots & \ddots \\ & & \lambda \end{pmatrix}$, it's straightforward to verify that

$$\mathbf{J}_\star^{-1} = \begin{pmatrix} \lambda^{-1} & -\lambda^{-2} & \lambda^{-3} & \cdots & -1^{(k-1)}\lambda^{-k} \\ & \lambda^{-1} & -\lambda^{-2} & \ddots & \vdots \\ & & \ddots & \ddots & \lambda^{-3} \\ & & & \lambda^{-1} & -\lambda^{-2} \\ & & & & \lambda^{-1} \end{pmatrix} = \begin{pmatrix} f(\lambda) & f'(\lambda) & \frac{f''(\lambda)}{2!} & \cdots & \frac{f^{(k-1)}(\lambda)}{(k-1)!} \\ & f(\lambda) & f'(\lambda) & \ddots & \vdots \\ & & \ddots & \ddots & \frac{f''(\lambda)}{2!} \\ & & & f(\lambda) & f'(\lambda) \\ & & & & f(\lambda) \end{pmatrix}$$

for $f(z) = z^{-1}$, and thus if $\mathbf{J} = \begin{pmatrix} \ddots & \\ & \mathbf{J}_\star \\ & & \ddots \end{pmatrix}$ is the Jordan form for $\mathbf{A} = \mathbf{PJP}^{-1}$,

then the representation of $\mathbf{A}^{-1}$ as $\mathbf{A}^{-1} = \mathbf{PJ}^{-1}\mathbf{P}^{-1}$ agrees with the expression for $f(\mathbf{A}) = \mathbf{P}f(\mathbf{J})\mathbf{P}^{-1}$ given in (7.9.3) when $f(z) = z^{-1}$.

7.9.21. $\dfrac{1}{2\pi i}\displaystyle\int_{\Gamma}\xi^{-1}(\xi\mathbf{I}-\mathbf{A})^{-1}d\xi=\mathbf{A}^{-1}.$

7.9.22. Partition the Jordan form for $\mathbf{A}$ as $\mathbf{J}=\begin{pmatrix}\mathbf{C}&\mathbf{0}\\\mathbf{0}&\mathbf{N}\end{pmatrix}$ in which $\mathbf{C}$ contains all Jordan segments associated with nonzero eigenvalues and $\mathbf{N}$ contains all Jordan segments associated with the zero eigenvalue (if one exists). Observe that $\mathbf{N}$ is nilpotent, so $g(\mathbf{N})=\mathbf{0}$, and consequently

$$\mathbf{A}=\mathbf{P}\begin{pmatrix}\mathbf{C}&\mathbf{0}\\\mathbf{0}&\mathbf{N}\end{pmatrix}\mathbf{P}^{-1}\Rightarrow g(\mathbf{A})=\mathbf{P}\begin{pmatrix}g(\mathbf{C})&\mathbf{0}\\\mathbf{0}&g(\mathbf{N})\end{pmatrix}\mathbf{P}^{-1}=\mathbf{P}\begin{pmatrix}\mathbf{C}^{-1}&\mathbf{0}\\\mathbf{0}&\mathbf{0}\end{pmatrix}\mathbf{P}^{-1}=\mathbf{A}^{D}.$$

It follows from Exercise 5.12.17 (p. 428) that $g(\mathbf{A})$ is the Moore–Penrose pseudoinverse $\mathbf{A}^{\dagger}$ if and only if $\mathbf{A}$ is an RPN matrix.

7.9.23. Use the Cauchy–Goursat theorem to observe that $\int_{\Gamma}\xi^{-j}d\xi=0$ for $j=2,3,\ldots,$ and follow the argument given in Example 7.9.8 (p. 611) with $\lambda_1=0$ along with the result of Exercise 7.9.22 to write

$$\frac{1}{2\pi i}\int_{\Gamma}\xi^{-1}(\xi\mathbf{I}-\mathbf{A})^{-1}d\xi=\frac{1}{2\pi i}\int_{\Gamma}\xi^{-1}\mathbf{R}(\xi)d\xi$$

$$=\frac{1}{2\pi i}\int_{\Gamma}\sum_{i=1}^{s}\sum_{j=0}^{k_i-1}\frac{\xi^{-1}}{(\xi-\lambda_i)^{j+1}}(\mathbf{A}-\lambda_i\mathbf{I})^{j}\mathbf{G}_i\,d\xi$$

$$=\sum_{i=1}^{s}\sum_{j=0}^{k_i-1}\left[\frac{1}{2\pi i}\int_{\Gamma}\frac{\xi^{-1}}{(\xi-\lambda_i)^{j+1}}d\xi\right](\mathbf{A}-\lambda_i\mathbf{I})^{j}\mathbf{G}_i$$

$$=\sum_{i=1}^{s}\sum_{j=0}^{k_i-1}\frac{g^{(j)}(\lambda_i)}{j!}(\mathbf{A}-\lambda_i\mathbf{I})^{j}\mathbf{G}_i=g(\mathbf{A})=\mathbf{A}^{D}.$$

Solutions for exercises in section 7. 10

7.10.1. The characteristic equation for $\mathbf{A}$ is $0=x^3-(3/4)x-(1/4)=(x-1)(x-1/2)^2$, so (7.10.33) guarantees that $\mathbf{A}$ is convergent (and hence also summable).

The characteristic equation for $\mathbf{B}$ is $x^3-1=0$, so the eigenvalues are the three cube roots of unity, and thus (7.10.33) insures $\mathbf{B}$ is not convergent, but $\mathbf{B}$ is summable because $\rho(\mathbf{B})=1$ and each eigenvalue on the unit circle is semisimple (in fact, each eigenvalue is simple).

The characteristic equation for $\mathbf{C}$ is

$$0=x^3-(5/2)x^2+2x-(1/2)=(x-1)^2(x-1/2),$$

but $index\,(\lambda=1)=2$ because $rank\,(\mathbf{C}-\mathbf{I})=2$ while $1=rank\,(\mathbf{C}-\mathbf{I})^2=rank\,(\mathbf{C}-\mathbf{I})^3=\cdots,$ so $\mathbf{C}$ is neither convergent nor summable.

7.10.2. Since $\mathbf{A}$ is convergent, (7.10.41) says that a full-rank factorization $\mathbf{I} - \mathbf{A} = \mathbf{BC}$ can be used to compute $\lim_{k\to\infty} \mathbf{A}^k = \mathbf{G} = \mathbf{I} - \mathbf{B}(\mathbf{CB})^{-1}\mathbf{C}$. One full-rank factorization is obtained by placing the basic columns of $\mathbf{I} - \mathbf{A}$ in $\mathbf{B}$ and the nonzero rows of $\mathbf{E}_{(\mathbf{I}-\mathbf{A})}$ in $\mathbf{C}$. This yields

$$\mathbf{B} = \begin{pmatrix} -3/2 & -3/2 \\ 1 & -1/2 \\ 1 & -1/2 \end{pmatrix}, \quad \mathbf{C} = \begin{pmatrix} 1 & -1 & 0 \\ 0 & 0 & 1 \end{pmatrix}, \quad \text{and} \quad \mathbf{G} = \begin{pmatrix} 0 & 1 & -1 \\ 0 & 1 & -1 \\ 0 & 0 & 0 \end{pmatrix}.$$

Alternately, since $\lambda = 1$ is a simple eigenvalue, the limit $\mathbf{G}$ can also be determined by computing right- and left-hand eigenvectors, $\mathbf{x} = (1,1,0)^T$ and $\mathbf{y}^T = (0,-1,1)$, associated with $\lambda = 1$ and setting $\mathbf{G} = \mathbf{xy}^T/(\mathbf{y}^T\mathbf{x})$ as described in (7.2.12) on p. 518. The matrix $\mathbf{B}$ is not convergent but it is summable, and since the unit eigenvalue is simple, the Cesàro limit $\mathbf{G}$ can be determined as described in (7.2.12) on p. 518 by computing right- and left-hand eigenvectors, $\mathbf{x} = (1,1,1)^T$ and $\mathbf{y}^T = (1,1,1)$, associated with $\lambda = 1$ and setting

$$\mathbf{G} = \mathbf{xy}^T/(\mathbf{y}^T\mathbf{x}) = \frac{1}{3}\begin{pmatrix} 1 & 1 & 1 \\ 1 & 1 & 1 \\ 1 & 1 & 1 \end{pmatrix}.$$

7.10.3. To see that $\mathbf{x}(k) = \mathbf{A}^k\mathbf{x}(0)$ solves $\mathbf{x}(k+1) = \mathbf{Ax}(k)$, use successive substitution to write $\mathbf{x}(1) = \mathbf{Ax}(0)$, $\mathbf{x}(2) = \mathbf{Ax}(1) = \mathbf{A}^2\mathbf{x}(0)$, $\mathbf{x}(3) = \mathbf{Ax}(2) = \mathbf{A}^3\mathbf{x}(0)$, etc. Of course you could build a formal induction argument, but it's not necessary. To see that $\mathbf{x}(k) = \mathbf{A}^k\mathbf{x}(0) + \sum_{j=0}^{k-1}\mathbf{A}^{k-j-1}\mathbf{b}(j)$ solves the nonhomogeneous equation $\mathbf{x}(k+1) = \mathbf{Ax}(k) + \mathbf{b}(k)$, use successive substitution to write

$$\mathbf{x}(1) = \mathbf{Ax}(0) + \mathbf{b}(0),$$
$$\mathbf{x}(2) = \mathbf{Ax}(1) + \mathbf{b}(1) = \mathbf{A}^2\mathbf{x}(0) + \mathbf{Ab}(0) + \mathbf{b}(0),$$
$$\mathbf{x}(3) = \mathbf{Ax}(2) + \mathbf{b}(2) = \mathbf{A}^3\mathbf{x}(0) + \mathbf{A}^2\mathbf{b}(0) + \mathbf{Ab}(0) + \mathbf{b}(0),$$

etc.

7.10.4. Put the basic columns of $\mathbf{I} - \mathbf{A}$ in $\mathbf{B}$ and the nonzero rows of the reduced row echelon form $\mathbf{E}_{(\mathbf{I}-\mathbf{A})}$ in $\mathbf{C}$ to build a full-rank factorization of $\mathbf{I} - \mathbf{A} = \mathbf{BC}$ (Exercise 3.9.8, p. 140), and use (7.10.41).

$$\mathbf{p} = \mathbf{Gp}(0) = (\mathbf{I} - \mathbf{B}(\mathbf{CB})^{-1}\mathbf{C})\mathbf{p}(0) = \begin{pmatrix} 1/6 & 1/6 & 1/6 & 1/6 \\ 1/3 & 1/3 & 1/3 & 1/3 \\ 1/3 & 1/3 & 1/3 & 1/3 \\ 1/6 & 1/6 & 1/6 & 1/6 \end{pmatrix} \begin{pmatrix} p_1(0) \\ p_2(0) \\ p_3(0) \\ p_4(0) \end{pmatrix} = \begin{pmatrix} 1/6 \\ 1/3 \\ 1/3 \\ 1/6 \end{pmatrix}.$$

7.10.5. To see that $\mathbf{x}(k) = \mathbf{A}^k\mathbf{x}(0)$ solves $\mathbf{x}(k+1) = \mathbf{Ax}(k)$, use successive substitution to write $\mathbf{x}(1) = \mathbf{Ax}(0)$, $\mathbf{x}(2) = \mathbf{Ax}(1) = \mathbf{A}^2\mathbf{x}(0)$, $\mathbf{x}(3) = \mathbf{Ax}(2) = \mathbf{A}^3\mathbf{x}(0)$, etc. Of course you could build a formal induction argument, but it's not necessary.

To see that $\mathbf{x}(k) = \mathbf{A}^k\mathbf{x}(0) + \sum_{j=0}^{k-1} \mathbf{A}^{k-j-1}\mathbf{b}(j)$ solves the nonhomogeneous equation $\mathbf{x}(k+1) = \mathbf{A}\mathbf{x}(k) + \mathbf{b}(k)$, use successive substitution to write

$$\mathbf{x}(1) = \mathbf{A}\mathbf{x}(0) + \mathbf{b}(0),$$
$$\mathbf{x}(2) = \mathbf{A}\mathbf{x}(1) + \mathbf{b}(1) = \mathbf{A}^2\mathbf{x}(0) + \mathbf{A}\mathbf{b}(0) + \mathbf{b}(0),$$
$$\mathbf{x}(3) = \mathbf{A}\mathbf{x}(2) + \mathbf{b}(2) = \mathbf{A}^3\mathbf{x}(0) + \mathbf{A}^2\mathbf{b}(0) + \mathbf{A}\mathbf{b}(0) + \mathbf{b}(0),$$

etc.

7.10.6. Use (7.1.12) on p. 497 along with (7.10.5) on p. 617.

7.10.7. For $\mathbf{A}_1$, the respective iteration matrices for Jacobi and Gauss–Seidel are

$$\mathbf{H}_J = \begin{pmatrix} 0 & -2 & 2 \\ -1 & 0 & -1 \\ -2 & -2 & 0 \end{pmatrix} \quad \text{and} \quad \mathbf{H}_{GS} = \begin{pmatrix} 0 & -2 & 2 \\ 0 & 2 & -3 \\ 0 & 0 & 2 \end{pmatrix}.$$

$\mathbf{H}_J$ is nilpotent of index three, so $\sigma(\mathbf{H}_J) = \{0\}$, and hence $\rho(\mathbf{H}_J) = 0 < 1$. Clearly, $\mathbf{H}_{GS}$ is triangular, so $\rho(\mathbf{H}_{GS}) = 2. > 1$ Therefore, for arbitrary right-hand sides, Jacobi's method converges after two steps, whereas the Gauss–Seidel method diverges. On the other hand, for $\mathbf{A}_2$,

$$\mathbf{H}_J = \frac{1}{2}\begin{pmatrix} 0 & 1 & -1 \\ -2 & 0 & -2 \\ 1 & 1 & 0 \end{pmatrix} \quad \text{and} \quad \mathbf{H}_{GS} = \frac{1}{2}\begin{pmatrix} 0 & 1 & -1 \\ 0 & -1 & -1 \\ 0 & 0 & -1 \end{pmatrix},$$

and a little computation reveals that $\sigma(\mathbf{H}_J) = \{\pm i\sqrt{5}/2\}$, so $\rho(\mathbf{H}_J) > 1$, while $\rho(\mathbf{H}_{GS}) = 1/2 < 1$. These examples show that there is no universal superiority enjoyed by either method because there is no universal domination of $\rho(\mathbf{H}_J)$ by $\rho(\mathbf{H}_{GS})$, or vise versa.

7.10.8. (a) $\det(\alpha\mathbf{D} - \mathbf{L} - \mathbf{U}) = \det(\alpha\mathbf{D} - \beta\mathbf{L} - \beta^{-1}\mathbf{U}) = 8\alpha^3 - 4\alpha$ for all real α and $\beta \neq 0$. Furthermore, the Jacobi iteration matrix is

$$\mathbf{H}_J = \begin{pmatrix} 0 & 1/2 & 0 \\ 1/2 & 0 & 1/2 \\ 0 & 1/2 & 0 \end{pmatrix},$$

and Example 7.2.5, p. 514, shows $\sigma(\mathbf{H}_J) = \{\cos(\pi/4), \cos(2\pi/4), \cos(3\pi/4)\}$. Clearly, these eigenvalues are real and $\rho(\mathbf{H}_J) = (1/\sqrt{2}) \approx .707 < 1$.

(b)   According to (7.10.24),

$$\omega_{\text{opt}} = \frac{2}{1 + \sqrt{1 - \rho^2(\mathbf{H}_J)}} \approx 1.172, \quad \text{and} \quad \rho(\mathbf{H}_{\omega_{\text{opt}}}) = \omega_{\text{opt}} - 1 \approx .172.$$

(c) $R_J = -\log_{10}\rho(\mathbf{H}_J) = \log_{10}(\sqrt{2}) \approx .1505$, $R_{GS} = 2R_J \approx .301$, and $R_{\text{opt}} = -\log_{10}\rho(\mathbf{H}_{\text{opt}}) \approx .766$.

(d) I used standard IEEE 64-bit floating-point arithmetic (i.e., about 16 decimal digits of precision) for all computations, but I rounded the results to 3 places to report the answers given below. Depending on your own implementation, your answers may vary slightly.

Jacobi with 21 iterations:

1	1.5	2.5	3.25	3.75	4.12	4.37	4.56	4.69	4.78	4.84	4.89	4.92	4.95	4.96	4.97	4.98	4.99	4.99	4.99	5	5
1	3	4.5	5.5	6.25	6.75	7.12	7.37	7.56	7.69	7.78	7.84	7.89	7.92	7.95	7.96	7.97	7.98	7.99	7.99	7.99	8
1	3.5	4.5	5.25	5.75	6.12	6.37	6.56	6.69	6.78	6.84	6.89	6.92	6.95	6.96	6.97	6.98	6.99	6.99	6.99	7	7

Gauss–Seidel with 11 iterations:

1	1.5	2.62	3.81	4.41	4.7	4.85	4.93	4.96	4.98	4.99	5
1	3.25	5.62	6.81	7.41	7.7	7.85	7.93	7.96	7.98	7.99	8
1	4.62	5.81	6.41	6.7	6.85	6.93	6.96	6.98	6.99	7	7

SOR (optimum) with 6 iterations:

1	1.59	3.06	4.59	4.89	4.98	5
1	3.69	6.73	7.69	7.93	7.99	8
1	5.5	6.51	6.9	6.98	7	7

7.10.9. The product rule for determinants produces

$$\det(\mathbf{H}_\omega) = \det\left[(\mathbf{D} - \omega\mathbf{L})^{-1}\right]\det\left[(1-\omega)\mathbf{D} + \omega\mathbf{U}\right] = \prod_{i=1}^{n} a_{ii}^{-1} \prod_{i=1}^{n}(1-\omega)a_{ii} = (1-\omega)^n.$$

But it's also true that $\det(\mathbf{H}_\omega) = \prod_{i=1}^{n}\lambda_i$, where the λ_i's are the eigenvalues of $\mathbf{H}_\omega$. Consequently, $|\lambda_k| \geq |1-\omega|$ for some k because if $|\lambda_i| < |1-\omega|$ for all i, then $|1-\omega|^n = |\det(\mathbf{H}_\omega)| = \prod_i |\lambda_i| < |1-\omega|^n$, which is impossible. Therefore, $|1-\omega| \leq |\lambda_k| \leq \rho(\mathbf{H}_\omega) < 1$ implies $0 < \omega < 2$.

7.10.10. Observe that $\mathbf{H}_J$ is the block-triangular matrix

$$\mathbf{H}_J = \frac{1}{4}\begin{pmatrix} \mathbf{K} & \mathbf{I} & & & \\ \mathbf{I} & \mathbf{K} & \mathbf{I} & & \\ & \ddots & \ddots & \ddots & \\ & & \mathbf{I} & \mathbf{K} & \mathbf{I} \\ & & & \mathbf{I} & \mathbf{K} \end{pmatrix}_{n^2 \times n^2} \quad \text{with} \quad \mathbf{K} = \begin{pmatrix} 0 & 1 & & & \\ 1 & 0 & 1 & & \\ & \ddots & \ddots & \ddots & \\ & & 1 & 0 & 1 \\ & & & 1 & 0 \end{pmatrix}_{n \times n}.$$

Proceed along the same lines as in Example 7.6.2, to argue that $\mathbf{H}_J$ is similar to the block-diagonal matrix

$$\begin{pmatrix} \mathbf{T}_1 & 0 & \cdots & 0 \\ 0 & \mathbf{T}_2 & \cdots & 0 \\ \vdots & \vdots & \ddots & \vdots \\ 0 & 0 & \cdots & \mathbf{T}_n \end{pmatrix}, \quad \text{where} \quad \mathbf{T}_i = \begin{pmatrix} \kappa_i & 1 & & & \\ 1 & \kappa_i & 1 & & \\ & \ddots & \ddots & \ddots & \\ & & 1 & \kappa_i & 1 \\ & & & 1 & \kappa_i \end{pmatrix}_{n \times n}$$

in which $\kappa_i \in \sigma(\mathbf{K})$. Use the result of Example 7.2.5 (p. 514) to infer that the eigenvalues of $\mathbf{T}_i$ are $\kappa_i + 2\cos j\pi/(n+1)$ for $j = 1, 2, \ldots, n$ and, similarly, the eigenvalues of $\mathbf{K}$ are $\kappa_i = 2\cos i\pi/(n+1)$ for $i = 1, 2, \ldots, n$. Consequently the n^2 eigenvalues of $\mathbf{H}_J$ are $\lambda_{ij} = (1/4)\big[2\cos i\pi/(n+1) + 2\cos j\pi/(n+1)\big]$, so $\rho(\mathbf{H}_J) = \max_{i,j} \lambda_{ij} = \cos \pi/(n+1)$.

7.10.11. If $\lim_{n\to\infty} \alpha_n = \alpha$, then for each $\epsilon > 0$ there is a natural number $N = N(\epsilon)$ such that $|\alpha_n - \alpha| < \epsilon/2$ for all $n \geq N$. Furthermore, there exists a real number β such that $|\alpha_n - \alpha| < \beta$ for all n. Consequently, for all $n \geq N$,

$$|\mu_n - \alpha| = \left|\frac{\alpha_1 + \alpha_2 + \cdots + \alpha_n}{n} - \alpha\right| = \frac{1}{n}\left|\sum_{k=1}^{N}(\alpha_k - \alpha) + \sum_{k=N+1}^{n}(\alpha_k - \alpha)\right|$$

$$\leq \frac{1}{n}\sum_{k=1}^{N}|\alpha_k - \alpha| + \frac{1}{n}\sum_{k=N+1}^{n}|\alpha_k - \alpha| < \frac{N\beta}{n} + \frac{n-N}{n}\frac{\epsilon}{2} \leq \frac{N\beta}{n} + \frac{\epsilon}{2}.$$

When n is sufficiently large, $N\beta/n \leq \epsilon/2$ so that $|\mu_n - \alpha| < \epsilon$, and therefore, $\lim_{n\to\infty}\mu_n = \alpha$. **Note:** The same proof works for vectors and matrices by replacing $|\star|$ with a vector or matrix norm.

7.10.12. Prove that (a) $\Rightarrow$ (b) $\Rightarrow$ (c) $\Rightarrow$ (d) $\Rightarrow$ (e) $\Rightarrow$ (f) $\Rightarrow$ (a).

(a) $\Rightarrow$ (b):   This is a consequence of (7.10.28).

(b) $\Rightarrow$ (c):   Use induction on the size of $\mathbf{A}_{n\times n}$. For $n = 1$, the result is trivial. Suppose the result holds for $n = k$ —i.e., suppose positive leading minors insures the existence of LU factors which are M-matrices when $n = k$. For $n = k+1$, use the induction hypothesis to write

$$\mathbf{A}_{(k+1)\times(k+1)} = \begin{pmatrix} \widetilde{\mathbf{A}} & \mathbf{c} \\ \mathbf{d}^T & \alpha \end{pmatrix} = \begin{pmatrix} \widetilde{\mathbf{L}}\widetilde{\mathbf{U}} & \mathbf{c} \\ \mathbf{d}^T & \alpha \end{pmatrix} = \begin{pmatrix} \widetilde{\mathbf{L}} & \mathbf{0} \\ \mathbf{d}^T\widetilde{\mathbf{U}}^{-1} & 1 \end{pmatrix}\begin{pmatrix} \widetilde{\mathbf{U}} & \widetilde{\mathbf{L}}^{-1}\mathbf{c} \\ \mathbf{0} & \sigma \end{pmatrix} = \mathbf{LU},$$

where $\widetilde{\mathbf{L}}$ and $\widetilde{\mathbf{U}}$ are M-matrices. Notice that $\sigma > 0$ because $\det(\widetilde{\mathbf{U}}) > 0$ and $0 < \det(\mathbf{A}) = \sigma \det(\widetilde{\mathbf{L}})\det(\widetilde{\mathbf{U}})$. Consequently, $\mathbf{L}$ and $\mathbf{U}$ are M-matrices because

$$\mathbf{L}^{-1} = \begin{pmatrix} \widetilde{\mathbf{L}}^{-1} & \mathbf{0} \\ -\mathbf{d}^T\widetilde{\mathbf{U}}^{-1}\widetilde{\mathbf{L}}^{-1} & 1 \end{pmatrix} \geq \mathbf{0} \quad \text{and} \quad \mathbf{U}^{-1} = \begin{pmatrix} \widetilde{\mathbf{U}}^{-1} & -\sigma^{-1}\widetilde{\mathbf{U}}^{-1}\widetilde{\mathbf{L}}^{-1}\mathbf{c} \\ \mathbf{0} & \sigma^{-1} \end{pmatrix} \geq \mathbf{0}.$$

(c) $\Rightarrow$ (d): $\mathbf{A} = \mathbf{LU}$ with $\mathbf{L}$ and $\mathbf{U}$ M-matrices implies $\mathbf{A}^{-1} = \mathbf{U}^{-1}\mathbf{L}^{-1} \geq \mathbf{0}$, so if $\mathbf{x} = \mathbf{A}^{-1}\mathbf{e}$, where $\mathbf{e} = (1, 1, \ldots, 1)^T$, then $\mathbf{x} > \mathbf{0}$ (otherwise $\mathbf{A}^{-1}$ would have a zero row, which would force $\mathbf{A}$ to be singular), and $\mathbf{Ax} = \mathbf{e} > \mathbf{0}$.

(d) $\Rightarrow$ (e):   If $\mathbf{x} > \mathbf{0}$ is such that $\mathbf{Ax} > \mathbf{0}$, define $\mathbf{D} = \text{diag}(x_1, x_2, \ldots, x_n)$ and set $\mathbf{B} = \mathbf{AD}$, which is clearly another Z-matrix. For $\mathbf{e} = (1, 1, \ldots, 1)^T$, notice that $\mathbf{Be} = \mathbf{ADe} = \mathbf{Ax} > \mathbf{0}$ says each row sum of $\mathbf{B} = \mathbf{AD}$ is positive. In other words, for each $i = 1, 2, \ldots, n$,

$$0 < \sum_j b_{ij} = \sum_{j\neq i} b_{ij} + b_{ii} \Rightarrow b_{ii} > \sum_{j\neq i} -b_{ij} = \sum_{j\neq i} |b_{ij}| \quad \text{for each } i = 1, 2, \ldots, n.$$

(e) $\Rightarrow$ (f): Suppose that $\mathbf{AD}$ is diagonally dominant for a diagonal matrix $\mathbf{D}$ with positive entries, and suppose each $a_{ii} > 0$. If $\mathbf{E} = \text{diag}\,(a_{11}, a_{22}, \ldots, a_{nn})$ and $-\mathbf{N}$ is the matrix containing the off-diagonal entries of $\mathbf{A}$, then $\mathbf{A} = \mathbf{E} - \mathbf{N}$ is the Jacobi splitting for $\mathbf{A}$ as described in Example 7.10.4 on p. 622, and $\mathbf{AD} = \mathbf{ED} + \mathbf{ND}$ is the Jacobi splitting for $\mathbf{AD}$ with the iteration matrix $\mathbf{H} = \mathbf{D}^{-1}\mathbf{E}^{-1}\mathbf{ND}$. It was shown in Example 7.10.4 that diagonal dominance insures convergence of Jacobi's method (i.e., $\rho(\mathbf{H}) < 1$), so, by (7.10.14), p. 620,

$$\mathbf{A} = \mathbf{ED}(\mathbf{I} - \mathbf{H})\mathbf{D}^{-1} \implies \mathbf{A}^{-1} = \mathbf{D}(\mathbf{I} - \mathbf{H})^{-1}\mathbf{D}^{-1}\mathbf{E}^{-1} \geq \mathbf{0},$$

and this guarantees that if $\mathbf{Ax} \geq \mathbf{0}$, then $\mathbf{x} \geq \mathbf{0}$.

(f) $\Rightarrow$ (a): Let $r \geq \max |a_{ii}|$ so that $\mathbf{B} = r\mathbf{I} - \mathbf{A} \geq \mathbf{0}$, and first show that the condition $(\mathbf{Ax} \geq \mathbf{0} \Rightarrow \mathbf{x} \geq \mathbf{0})$ insures the existence of $\mathbf{A}^{-1}$. For any $\mathbf{x} \in N(\mathbf{A})$,

$$(r\mathbf{I} - \mathbf{B})\mathbf{x} = \mathbf{0} \Rightarrow r\mathbf{x} = \mathbf{Bx} \Rightarrow r|\mathbf{x}| \leq |\mathbf{B}||\mathbf{x}| \Rightarrow \mathbf{A}(-|\mathbf{x}|) \geq \mathbf{0} \Rightarrow -|\mathbf{x}| \geq \mathbf{0} \Rightarrow \mathbf{x} = \mathbf{0},$$

so $N(\mathbf{A}) = \mathbf{0}$. Now, $\mathbf{A}[\mathbf{A}^{-1}]_{*i} = \mathbf{e}_i \geq \mathbf{0} \Rightarrow [\mathbf{A}^{-1}]_{*i} \geq \mathbf{0}$, and thus $\mathbf{A}^{-1} \geq \mathbf{0}$.

7.10.13. (a) If $\mathbf{M}_i$ is $n_i \times n_i$ with $rank(\mathbf{M}_i) = r_i$, then $\mathbf{B}_i$ is $n_i \times r_i$ and $\mathbf{C}_i$ is $r_i \times n_i$ with $rank(\mathbf{B}_i) = rank(\mathbf{C}_i) = r_i$. This means that $\mathbf{M}_{i+1} = \mathbf{C}_i\mathbf{B}_i$ is $r_i \times r_i$, so if $r_i < n_i$, then $\mathbf{M}_{i+1}$ has smaller size than $\mathbf{M}_i$. Since this can't happen indefinitely, there must be a point in the process at which $r_k = n_k$ or $r_k = 0$ and thus some $\mathbf{M}_k$ is either nonsingular or zero.

(b) Let $\mathbf{M} = \mathbf{M}_1 = \mathbf{A} - \lambda\mathbf{I}$, and notice that

$$\mathbf{M}^2 = \mathbf{B}_1\mathbf{C}_1\mathbf{B}_1\mathbf{C}_1 = \mathbf{B}_1\mathbf{M}_2\mathbf{C}_1,$$
$$\mathbf{M}^3 = \mathbf{B}_1\mathbf{C}_1\mathbf{B}_1\mathbf{C}_1\mathbf{B}_1\mathbf{C}_1 = \mathbf{B}_1(\mathbf{B}_2\mathbf{C}_2)(\mathbf{B}_2\mathbf{C}_2)\mathbf{C}_1 = \mathbf{B}_1\mathbf{B}_2\mathbf{M}_3\mathbf{C}_2\mathbf{C}_1,$$
$$\vdots$$
$$\mathbf{M}^i = \mathbf{B}_1\mathbf{B}_2\cdots\mathbf{B}_{i-1}\mathbf{M}_i\mathbf{C}_{i-1}\cdots\mathbf{C}_2\mathbf{C}_1.$$

In general, it's true that $rank(\mathbf{XYZ}) = rank(\mathbf{Y})$ whenever $\mathbf{X}$ has full column rank and $\mathbf{Z}$ has full row rank (Exercise 4.5.12, p. 220), so applying this yields $rank(\mathbf{M}^i) = rank(\mathbf{M}_i)$ for each $i = 1, 2, \ldots$. Suppose that some $\mathbf{M}_i = \mathbf{C}_{i-1}\mathbf{B}_{i-1}$ is $n_i \times n_i$ and nonsingular. For this to happen, we must have $\mathbf{M}_{i-1} = \mathbf{B}_{i-1}\mathbf{C}_{i-1}$, where $\mathbf{B}_{i-1}$ is $n_{i-1} \times n_i$, $\mathbf{C}_{i-1}$ is $n_i \times n_{i-1}$, and

$$rank(\mathbf{M}_{i-1}) = rank(\mathbf{B}_{i-1}) = rank(\mathbf{C}_{i-1}) = n_i = rank(\mathbf{M}_i).$$

Therefore, if k is the smallest positive integer such that $\mathbf{M}_k^{-1}$ exists, then k is the smallest positive integer such that $rank(\mathbf{M}_{k-1}) = rank(\mathbf{M}_k)$, and thus k is the smallest positive integer such that $rank(\mathbf{M}^{k-1}) = rank(\mathbf{M}^k)$, which means that $index(\mathbf{M}) = k - 1$ or, equivalently, $index(\lambda) = k - 1$. On the other hand, if some $\mathbf{M}_i = \mathbf{0}$, then $rank(\mathbf{M}^i) = rank(\mathbf{M}_i)$ insures that $\mathbf{M}^i = \mathbf{0}$.

Consequently, if k is the smallest positive integer such that $\mathbf{M}_k = \mathbf{0}$, then k is the smallest positive integer such that $\mathbf{M}^k = \mathbf{0}$. Therefore, $\mathbf{M}$ is nilpotent of index k, and this implies that $index(\lambda) = k$.

7.10.14. $\mathbf{M} = \mathbf{A} - 4\mathbf{I} = \begin{pmatrix} -7 & -8 & -9 \\ 5 & 7 & 9 \\ -1 & -2 & -3 \end{pmatrix} \longrightarrow \begin{pmatrix} 1 & 0 & -1 \\ 0 & 1 & 2 \\ 0 & 0 & 0 \end{pmatrix} \Rightarrow \mathbf{B}_1 = \begin{pmatrix} -7 & -8 \\ 5 & 7 \\ -1 & -2 \end{pmatrix}$

and $\mathbf{C}_1 = \begin{pmatrix} 1 & 0 & -1 \\ 0 & 1 & 2 \end{pmatrix}$, so $\mathbf{M}_2 = \mathbf{C}_1 \mathbf{B}_1 = \begin{pmatrix} -6 & -6 \\ 3 & 3 \end{pmatrix} \longrightarrow \begin{pmatrix} 1 & 1 \\ 0 & 0 \end{pmatrix} \Rightarrow$

$\mathbf{B}_2 = \begin{pmatrix} -6 \\ 3 \end{pmatrix}$ and $\mathbf{C}_2 = \begin{pmatrix} 1 & 1 \end{pmatrix}$, so $\mathbf{M}_3 = \mathbf{C}_2 \mathbf{B}_2 = -3$. Since $\mathbf{M}_3$ is the first $\mathbf{M}_i$ to be nonsingular, $index(4) = 3 - 1 = 2$. Now, $index(1)$ if forced to be 1 because $1 = alg\ mult(1) \geq index(1) \geq 1$.

7.10.15. (a) Since $\sigma(\mathbf{A}) = \{1, 4\}$ with $index(1) = 1$ and $index(4) = 2$, the Jordan form for $\mathbf{A}$ is $\mathbf{J} = \begin{pmatrix} 1 & 0 & 0 \\ 0 & 4 & 1 \\ 0 & 0 & 4 \end{pmatrix}$.

(b) The Hermite interpolation polynomial $p(z) = \alpha_0 + \alpha_1 z + \alpha_2 z^2$ is determined by solving $p(1) = f(1)$, $p(4) = f(4)$, and $p'(4) = f'(4)$ for α_i's. So

$$\begin{pmatrix} 1 & 1 & 1 \\ 1 & 4 & 16 \\ 0 & 1 & 8 \end{pmatrix} \begin{pmatrix} \alpha_0 \\ \alpha_1 \\ \alpha_2 \end{pmatrix} = \begin{pmatrix} f(1) \\ f(4) \\ f'(4) \end{pmatrix} \implies \begin{pmatrix} \alpha_0 \\ \alpha_1 \\ \alpha_2 \end{pmatrix} = \begin{pmatrix} 1 & 1 & 1 \\ 1 & 4 & 16 \\ 0 & 1 & 8 \end{pmatrix}^{-1} \begin{pmatrix} f(1) \\ f(4) \\ f'(4) \end{pmatrix}$$

$$= -\frac{1}{9} \begin{pmatrix} -16 & 7 & -12 \\ 8 & -8 & 15 \\ -1 & 1 & -3 \end{pmatrix} \begin{pmatrix} f(1) \\ f(4) \\ f'(4) \end{pmatrix}$$

$$= -\frac{1}{9} \begin{pmatrix} -16f(1) + 7f(4) - 12f'(4) \\ 8f(1) - 8f(4) + 15f'(4) \\ -f(1) + f(4) - 3f'(4) \end{pmatrix}.$$

Writing $f(\mathbf{A}) = p(\mathbf{A})$ produces

$$f(\mathbf{A}) = \left[\frac{-16\mathbf{I} + 8\mathbf{A} - \mathbf{A}^2}{-9} \right] f(1) + \left[\frac{7\mathbf{I} - 8\mathbf{A} + \mathbf{A}^2}{-9} \right] f(4)$$

$$+ \left[\frac{-12\mathbf{I} + 15\mathbf{A} - 3\mathbf{A}^2}{-9} \right] f'(4).$$

7.10.16. Suppose that $\lim_{k \to \infty} \mathbf{A}^k$ exists and is nonzero. It follows from (7.10.33) that $\lambda = 1$ is a semisimple eigenvalue of $\mathbf{A}$, so the Jordan form for $\mathbf{B}$ looks like $\mathbf{B} = \mathbf{I} - \mathbf{A} = \mathbf{P} \begin{pmatrix} 0 & 0 \\ 0 & \mathbf{I} - \mathbf{K} \end{pmatrix} \mathbf{P}^{-1}$, where $\mathbf{I} - \mathbf{K}$ is nonsingular. Therefore, $\mathbf{B}$ belongs to a matrix group and

$$\mathbf{B}^\# = \mathbf{P} \begin{pmatrix} 0 & 0 \\ 0 & (\mathbf{I} - \mathbf{K})^{-1} \end{pmatrix} \mathbf{P}^{-1} \implies \mathbf{I} - \mathbf{B}\mathbf{B}^\# = \mathbf{P} \begin{pmatrix} \mathbf{I} & 0 \\ 0 & 0 \end{pmatrix} \mathbf{P}^{-1}.$$

Comparing $\mathbf{I} - \mathbf{BB}^{\#}$ with (7.10.32) shows that $\lim_{k\to\infty} \mathbf{A}^k = \mathbf{I} - \mathbf{BB}^{\#}$. If $\lim_{k\to\infty} \mathbf{A}^k = \mathbf{0}$, then $\rho(\mathbf{A}) < 1$, and hence $\mathbf{B}$ is nonsingular, so $\mathbf{B}^{\#} = \mathbf{B}^{-1}$ and $\mathbf{I} - \mathbf{BB}^{\#} = \mathbf{0}$. In other words, it's still true that $\lim_{k\to\infty} \mathbf{A}^k = \mathbf{I} - \mathbf{BB}^{\#}$.

7.10.17. We already know from the development of (7.10.41) that if $rank(\mathbf{M}) = r$, then $\mathbf{CB}$ and $\mathbf{V}_1^* \mathbf{U}_1$ are $r \times r$ nonsingular matrices. It's a matter of simple algebra to verify that $\mathbf{MM}^{\#}\mathbf{M} = \mathbf{M}$, $\mathbf{M}^{\#}\mathbf{MM}^{\#} = \mathbf{M}^{\#}$, and $\mathbf{MM}^{\#} = \mathbf{M}^{\#}\mathbf{M}$.

Solutions for exercises in section 7. 11

7.11.1. $m(x) = x^2 - 3x + 2$

7.11.2. $v(x) = x - 2$

7.11.3. $c(x) = (x-1)(x-2)^2$

7.11.4. $\mathbf{J} = \begin{pmatrix} \lambda \\ & \lambda \\ & & \lambda \\ & & & \mu \\ & & & & \mu \\ & & & & & \mu & 1 \\ & & & & & & \mu \end{pmatrix}$

7.11.5. Set $\nu_0 = \|\mathbf{I}\|_F = 2$, $\mathbf{U}_0 = \mathbf{I}/2$, and generate the sequence (7.11.2).

$$r_{01} = \langle \mathbf{U}_0 | \mathbf{A} \rangle = 2,$$
$$\nu_1 = \|\mathbf{A} - r_{01}\mathbf{U}_0\|_F = \sqrt{1209}, \quad \mathbf{U}_1 = \frac{\mathbf{A} - r_{01}\mathbf{U}_0}{\nu_1} = \frac{\mathbf{A} - \mathbf{I}}{\sqrt{1209}},$$
$$r_{02} = \langle \mathbf{U}_0 | \mathbf{A}^2 \rangle = 2, \quad r_{12} = \langle \mathbf{U}_1 | \mathbf{A}^2 \rangle = 2\sqrt{1209},$$
$$\nu_2 = \|\mathbf{A}^2 - r_{02}\mathbf{U}_0 - r_{12}\mathbf{U}_1\|_F = 0,$$

so that

$$\mathbf{R} = \begin{pmatrix} 2 & 2 \\ 0 & \sqrt{1209} \end{pmatrix}, \quad \mathbf{c} = \begin{pmatrix} 2 \\ 2\sqrt{1209} \end{pmatrix}, \text{ and } \mathbf{R}^{-1}\mathbf{c} = \begin{pmatrix} -1 \\ 2 \end{pmatrix} = \begin{pmatrix} \alpha_0 \\ \alpha_1 \end{pmatrix}.$$

Consequently, the minimum polynomial is $m(x) = x^2 - 2x + 1 = (x-1)^2$. As a by-product, we see that $\lambda = 1$ is the only eigenvalue of $\mathbf{A}$, and $index(\lambda) = 2$, so the Jordan form for $\mathbf{A}$ must be $\mathbf{J} = \begin{pmatrix} 1 & 1 & 0 & 0 \\ 0 & 1 & 0 & 0 \\ 0 & 0 & 1 & 0 \\ 0 & 0 & 0 & 1 \end{pmatrix}$.

7.11.6. Similar matrices have the same minimum polynomial because similar matrices have the same Jordan form, and hence they have the same eigenvalues with the same indicies.

7.11.10. $\mathbf{x} = (3, -1, -1)^T$

7.11.12. $\mathbf{x} = (-3, 6, 5)^T$

Argument, as usually managed, is the worst sort of conversation,
and in books it is generally the worst sort of reading.
— Jonathan Swift (1667–1745)

Solutions for Chapter 8

Solutions for exercises in section 8. 2

8.2.1. The eigenvalues are $\sigma(\mathbf{A}) = \{12, 6\}$ with $alg\ mult_{\mathbf{A}}(6) = 2$, and it's clear that $12 = \rho(\mathbf{A}) \in \sigma(\mathbf{A})$. The eigenspace $N(\mathbf{A} - 12\mathbf{I})$ is spanned by $\mathbf{e} = (1, 1, 1)^T$, so the Perron vector is $\mathbf{p} = (1/3)(1, 1, 1)^T$. The left-hand eigenspace $N(\mathbf{A}^T - 12\mathbf{I})$ is spanned by $(1, 2, 3)^T$, so the left-hand Perron vector is $\mathbf{q}^T = (1/6)(1, 2, 3)$.

8.2.3. If $\mathbf{p}_1$ and $\mathbf{p}_2$ are two vectors satisfying $\mathbf{Ap} = \rho(\mathbf{A})\mathbf{p}$, $\mathbf{p} > \mathbf{0}$, and $\|\mathbf{p}\|_1 = 1$, then $\dim N(\mathbf{A} - \rho(\mathbf{A})\mathbf{I}) = 1$ implies that $\mathbf{p}_1 = \alpha\mathbf{p}_2$ for some $\alpha < 0$. But $\|\mathbf{p}_1\|_1 = \|\mathbf{p}_2\|_1 = 1$ insures that $\alpha = 1$.

8.2.4. $\sigma(\mathbf{A}) = \{0, 1\}$, so $\rho(\mathbf{A}) = 1$ is the Perron root, and the Perron vector is $\mathbf{p} = (\alpha + \beta)^{-1}(\beta, \alpha)$.

8.2.5. (a) $\rho(\mathbf{A}/r) = 1$ is a simple eigenvalue of $\mathbf{A}/r$, and it's the only eigenvalue on the spectral circle of $\mathbf{A}/r$, so (7.10.33) on p. 630 guarantees that $\lim_{k\to\infty}(\mathbf{A}/r)^k$ exists.

(b) This follows from (7.10.34) on p. 630.

(c) $\mathbf{G}$ is the spectral projector associated with the simple eigenvalue $\lambda = r$, so formula (7.2.12) on p. 518 applies.

8.2.6. If $\mathbf{e}$ is the column of all 1 's, then $\mathbf{Ae} = \rho\mathbf{e}$. Since $\mathbf{e} > \mathbf{0}$, it must be a positive multiple of the Perron vector $\mathbf{p}$, and hence $\mathbf{p} = n^{-1}\mathbf{e}$. Therefore, $\mathbf{Ap} = \rho\mathbf{p}$ implies that $\rho = \rho(\mathbf{A})$. The result for column sums follows by considering $\mathbf{A}^T$.

8.2.7. Since $\rho = \max_i \sum_j a_{ij}$ is the largest row sum of $\mathbf{A}$, there must exist a matrix $\mathbf{E} \geq \mathbf{0}$ such that every row sum of $\mathbf{B} = \mathbf{A} + \mathbf{E}$ is ρ. Use Example 7.10.2 (p. 619) together with Exercise 8.2.7 to obtain $\rho(\mathbf{A}) \leq \rho(\mathbf{B}) = \rho$. The lower bound follows from the Collatz–Wielandt formula. If $\mathbf{e}$ is the column of ones, then $\mathbf{e} \in \mathcal{N}$, so

$$\rho(\mathbf{A}) = \max_{\mathbf{x} \in \mathcal{N}} f(\mathbf{x}) \geq f(\mathbf{e}) = \min_{1 \leq i \leq n} \frac{[\mathbf{Ae}]_i}{e_i} = \min_i \sum_{j=1}^{n} a_{ij}.$$

8.2.8. (a), (b), (c), and (d) are illustrated by using the nilpotent matrix $\mathbf{A} = \begin{pmatrix} 0 & 1 \\ 0 & 0 \end{pmatrix}$.

(e) $\mathbf{A} = \begin{pmatrix} 0 & 1 \\ 1 & 0 \end{pmatrix}$ has eigenvalues ± 1.

8.2.9. If $\xi = g(\mathbf{x})$ for $\mathbf{x} \in \mathcal{P}$, then $\xi\mathbf{x} \geq \mathbf{Ax} > \mathbf{0}$. Let $\mathbf{p}$ and $\mathbf{q}^T$ be the respective the right-hand and left-hand Perron vectors for $\mathbf{A}$ associated with the Perron root r, and use (8.2.3) along with $\mathbf{q}^T\mathbf{x} > 0$ to write

$$\xi\mathbf{x} \geq \mathbf{Ax} > \mathbf{0} \implies \xi\mathbf{q}^T\mathbf{x} \geq \mathbf{q}^T\mathbf{Ax} = r\mathbf{q}^T\mathbf{x} \implies \xi \geq r,$$

so $g(\mathbf{x}) \geq r$ for all $\mathbf{x} \in \mathcal{P}$. Since $g(\mathbf{p}) = r$ and $\mathbf{p} \in \mathcal{P}$, it follows that $r = \min_{\mathbf{x} \in \mathcal{P}} g(\mathbf{x})$.

8.2.10. $\mathbf{A} = \begin{pmatrix} 1 & 2 \\ 2 & 4 \end{pmatrix} \implies \rho(\mathbf{A}) = 5$, but $g(\mathbf{e}_1) = 1 \implies \min_{\mathbf{x} \in \mathcal{N}} g(\mathbf{x}) < \rho(\mathbf{A})$.

Solutions for exercises in section 8. 3

8.3.1. (a) The graph is strongly connected.

(b) $\rho(\mathbf{A}) = 3$, and $\mathbf{p} = (1/6,\, 1/2,\, 1/3)^T$.

(c) $h = 2$ because $\mathbf{A}$ is imprimitive and singular.

8.3.2. If $\mathbf{A}$ is nonsingular then there are either one or two distinct nonzero eigenvalues inside the spectral circle. But this is impossible because $\sigma(\mathbf{A})$ has to be invariant under rotations of $120°$ by the result on p. 677. Similarly, if $\mathbf{A}$ is singular with $alg\ mult_{\mathbf{A}}(0) = 1$, then there is a single nonzero eigenvalue inside the spectral circle, which is impossible.

8.3.3. No! The matrix $\mathbf{A} = \begin{pmatrix} 1 & 1 \\ 0 & 2 \end{pmatrix}$ has $\rho(\mathbf{A}) = 2$ with a corresponding eigenvector $\mathbf{e} = (1,1)^T$, but $\mathbf{A}$ is reducible.

8.3.4. $\mathbf{P}_n$ is nonnegative and irreducible (its graph is strongly connected), and $\mathbf{P}_n$ is imprimitive because $\mathbf{P}_n^n = \mathbf{I}$ insures that every power has zero entries. Furthermore, if $\lambda \in \sigma(\mathbf{P}_n)$, then $\lambda^n \in \sigma(\mathbf{P}_n^n) = \{1\}$, so all eigenvalues of $\mathbf{P}_n$ are roots of unity. Since all eigenvalues on the spectral circle are simple (recall (8.3.13) on p. 676) and uniformly distributed, it must be the case that $\sigma(\mathbf{P}_n) = \{1, \omega, \omega^2, \ldots, \omega^{n-1}\}$.

8.3.5. $\mathbf{A}$ is irreducible because the graph $\mathcal{G}(\mathbf{A})$ is strongly connected—every node is accessible by some sequence of paths from every other node.

8.3.6. $\mathbf{A}$ is imprimitive. This is easily seen by observing that each $\mathbf{A}^{2n}$ for $n > 1$ has the same zero pattern (and each $\mathbf{A}^{2n+1}$ for $n > 0$ has the same zero pattern), so every power of $\mathbf{A}$ has zero entries.

8.3.7. (a) Having row sums less than or equal to 1 means that $\|\mathbf{P}\|_\infty \leq 1$. Because $\rho(\star) \leq \|\star\|$ for every matrix norm (recall (7.1.12) on p. 497), it follows that $\rho(\mathbf{S}) \leq \|\mathbf{S}\|_1 \leq 1$.

(b) If $\mathbf{e}$ denotes the column of all 1's, then the hypothesis insures that $\mathbf{Se} \leq \mathbf{e}$, and $\mathbf{Se} \neq \mathbf{e}$. Since $\mathbf{S}$ is irreducible, the result in Example 8.3.1 (p. 674) implies that it's impossible to have $\rho(\mathbf{S}) = 1$ (otherwise $\mathbf{Se} = \mathbf{e}$), and therefore $\rho(\mathbf{S}) < 1$ by part (a).

8.3.8. If $\mathbf{p}$ is the Perron vector for $\mathbf{A}$, and if $\mathbf{e}$ is the column of 1's, then

$$\mathbf{D}^{-1}\mathbf{ADe} = \mathbf{D}^{-1}\mathbf{Ap} = r\mathbf{D}^{-1}\mathbf{p} = r\mathbf{e}$$

shows that every row sum of $\mathbf{D}^{-1}\mathbf{AD}$ is r, so we can take $\mathbf{P} = r^{-1}\mathbf{D}^{-1}\mathbf{AD}$ because the Perron–Frobenius theorem guarantees that $r > 0$.

8.3.9. Construct the Boolean matrices as described in Example 8.3.5 (p. 680), and show that $\mathbf{B}_9$ has a zero in the $(1,1)$ position, but $\mathbf{B}_{10} > \mathbf{0}$.

8.3.10. According to the discussion on p. 630, $\mathbf{f}(t) \to \mathbf{0}$ if $r < 1$. If $r = 1$, then $\mathbf{f}(t) \to \mathbf{G}\mathbf{f}(0) = \mathbf{p}\big(\mathbf{q}^T\mathbf{f}(0)/\mathbf{q}^T\mathbf{p}\big) > \mathbf{0}$, and if $r > 1$, the results of the Leslie analysis imply that $f_k(t) \to \infty$ for each k.

8.3.11. The only nonzero coefficient in the characteristic equation for $\mathbf{L}$ is c_1, so $\gcd\{2, 3, \ldots, n\} = 1$.

8.3.12. (a) Suppose that $\mathbf{A}$ is essentially positive. Since we can always find a $\beta > 0$ such that $\beta\mathbf{I} + \mathrm{diag}\,(a_{11}, a_{22}, \ldots, a_{nn}) \geq \mathbf{0}$, and since $a_{ij} \geq 0$ for $i \neq j$, it follows that $\mathbf{A} + \beta\mathbf{I}$ is a nonnegative irreducible matrix, so (8.3.5) on p. 672 can be applied to conclude that $(\mathbf{A} + (1 + \beta)\mathbf{I})^{n-1} > \mathbf{0}$, and thus $\mathbf{A} + \alpha\mathbf{I}$ is primitive with $\alpha = \beta + 1$. Conversely, if $\mathbf{A} + \alpha\mathbf{I}$ is primitive, then $\mathbf{A} + \alpha\mathbf{I}$ must be nonnegative and irreducible, and hence $a_{ij} \geq 0$ for every $i \neq j$, and $\mathbf{A}$ must be irreducible (diagonal entries don't affect the reducibility or irreducibility).

(b) If $\mathbf{A}$ is essentially positive, then $\mathbf{A} + \alpha\mathbf{I}$ is primitive for some α (by the first part), so $(\mathbf{A} + \alpha\mathbf{I})^k > \mathbf{0}$ for some k. Consequently, for all $t > 0$,

$$0 < \sum_{k=0}^{\infty} \frac{t^k(\mathbf{A} + \alpha\mathbf{I})^k}{k!} = e^{t(\mathbf{A}+\alpha\mathbf{I})} = e^{t\mathbf{A}}e^{t\alpha\mathbf{I}} = \mathbf{B} \implies 0 < e^{-\alpha t}\mathbf{B} = e^{t\mathbf{A}}.$$

Conversely, if $0 < e^{t\mathbf{A}} = \sum_{k=0}^{\infty} t^k\mathbf{A}^k/k!$ for all $t > 0$, then $a_{ij} \geq 0$ for every $i \neq j$, for if $a_{ij} < 0$ for some $i \neq j$, then there exists a sufficiently small $t > 0$ such that $[\mathbf{I} + t\mathbf{A} + t^2\mathbf{A}^2/2 + \cdots]_{ij} < 0$, which is impossible. Furthermore, $\mathbf{A}$ must be irreducible; otherwise

$$\mathbf{A} \sim \begin{pmatrix} \mathbf{X} & \mathbf{Y} \\ \mathbf{0} & \mathbf{Z} \end{pmatrix} \implies e^{t\mathbf{A}} = \sum_{k=0}^{\infty} t^k\mathbf{A}^k/k! \sim \begin{pmatrix} \star & \star \\ \mathbf{0} & \star \end{pmatrix}, \quad \text{which is impossible.}$$

8.3.13. (a) Being essentially positive implies that there exists some $\alpha \in \Re$ such that $\mathbf{A} + \alpha\mathbf{I}$ is nonnegative and irreducible (by Exercise 8.3.12). If $(r, \mathbf{x})$ is the Perron eigenpair for $\mathbf{A} + \alpha\mathbf{I}$, then for $\xi = r - \alpha$, $(\xi, \mathbf{x})$ is an eigenpair for $\mathbf{A}$.

(b) Every eigenvalue of $\mathbf{A} + \alpha\mathbf{I}$ has the form $z = \lambda + \alpha$, where $\lambda \in \sigma(\mathbf{A})$, so if r is the Perron root of $\mathbf{A} + \alpha\mathbf{I}$, then for $z \neq r$,

$$|z| < r \implies \mathrm{Re}\,(z) < r \implies \mathrm{Re}\,(\lambda + \alpha) < r \implies \mathrm{Re}\,(\lambda) < r - \alpha = \xi.$$

(c) If $\mathbf{A} \leq \mathbf{B}$, then $\mathbf{A} + \alpha\mathbf{I} \leq \mathbf{B} + \alpha\mathbf{I}$, so Wielandt's theorem (p. 675) insures that $r_A = \rho(\mathbf{A} + \alpha\mathbf{I}) \leq \rho(\mathbf{B} + \alpha\mathbf{I}) = r_B$, and hence $\xi_A = r_A - \alpha \leq r_B - \alpha = \xi_B$.

8.3.14. If $\mathbf{A}$ is primitive with $r = \rho(\mathbf{A})$, then, by (8.3.10) on p. 674,

$$\left(\frac{\mathbf{A}}{r}\right)^k \to \mathbf{G} > \mathbf{0} \implies \exists\, k_0 \text{ such that } \left(\frac{\mathbf{A}}{r}\right)^m > \mathbf{0} \quad \forall m \geq k_0$$

$$\implies \frac{a_{ij}^{(m)}}{r^m} > 0 \quad \forall m \geq k_0$$

$$\implies \lim_{m\to\infty} \left(\frac{a_{ij}^{(m)}}{r^m}\right)^{1/m} \to 1 \implies \lim_{m\to\infty} \left[a_{ij}^{(m)}\right]^{1/m} = r.$$

Conversely, we know from the Perron–Frobenius theorem that $r > 0$, so if $\lim_{k \to \infty} \left[a_{ij}^{(k)}\right]^{1/k} = r$, then $\exists\ k_0$ such that $\forall m \geq k_0$, $\left[a_{ij}^{(m)}\right]^{1/m} > 0$, which implies that $\mathbf{A}^m > \mathbf{0}$, and thus $\mathbf{A}$ is primitive by Frobenius's test (p. 678).

Solutions for exercises in section 8. 4

8.4.1. The left-hand Perron vector for $\mathbf{P}$ is $\boldsymbol{\pi}^T = (10/59,\ 4/59,\ 18/59,\ 27/59)$. It's the limiting distribution in the regular sense because $\mathbf{P}$ is primitive (it has a positive diagonal entry—recall Example 8.3.3 (p. 678)).

8.4.2. The left-hand Perron vector is $\boldsymbol{\pi}^T = (1/n)(1, 1, \ldots, 1)$. Thus the limiting distribution is the uniform distribution, and in the long run, each state is occupied an equal proportion of the time. The limiting matrix is $\mathbf{G} = (1/n)\mathbf{e}\mathbf{e}^T$.

8.4.3. If $\mathbf{P}$ is irreducible, then $\rho(\mathbf{P}) = 1$ is a simple eigenvalue for $\mathbf{P}$, so

$$rank\,(\mathbf{I} - \mathbf{P}) = n - \dim N\,(\mathbf{I} - \mathbf{P}) = n - geo\ mult_{\mathbf{P}}\,(1) = n - alg\ mult_{\mathbf{P}}\,(1) = n - 1.$$

8.4.4. Let $\mathbf{A} = \mathbf{I} - \mathbf{P}$, and recall that $rank\,(\mathbf{A}) = n - 1$ (Exercise 8.4.3). Consequently,

$$\mathbf{A}\text{ singular} \implies \mathbf{A}[\mathrm{adj}\,(\mathbf{A})] = \mathbf{0} = [\mathrm{adj}\,(\mathbf{A})]\mathbf{A} \quad \text{(Exercise 6.2.8, p. 484),}$$

and

$$rank\,(\mathbf{A}) = n - 1 \implies rank\,(\mathrm{adj}\,(\mathbf{A})) = 1 \quad \text{(Exercises 6.2.11)}.$$

It follows from $\mathbf{A}[\mathrm{adj}\,(\mathbf{A})] = \mathbf{0}$ and the Perron–Frobenius theorem that each column of $[\mathrm{adj}\,(\mathbf{A})]$ must be a multiple of $\mathbf{e}$ (the column of 1's or, equivalently, the right-hand Perron vector for $\mathbf{P}$), so $[\mathrm{adj}\,(\mathbf{A})] = \mathbf{e}\mathbf{v}^T$ for some vector $\mathbf{v}$. But $[\mathrm{adj}\,(\mathbf{A})]_{ii} = P_i$ forces $\mathbf{v}^T = (P_1, P_2, \ldots, P_n)$. Similarly, $[\mathrm{adj}\,(\mathbf{A})]\mathbf{A} = \mathbf{0}$ insures that each row in $[\mathrm{adj}\,(\mathbf{A})]$ is a multiple of $\boldsymbol{\pi}^T$ (the left-hand Perron vector of $\mathbf{P}$), and hence $\mathbf{v}^T = \alpha\boldsymbol{\pi}^T$ for some α. This scalar α can't be zero; otherwise $[\mathrm{adj}\,(\mathbf{A})] = \mathbf{0}$, which is impossible because $rank\,(\mathrm{adj}\,(\mathbf{A})) = 1$. Therefore, $\mathbf{v}^T\mathbf{e} = \alpha \neq 0$, and $\mathbf{v}^T/(\mathbf{v}^T\mathbf{e}) = \mathbf{v}^T/\alpha = \boldsymbol{\pi}^T$.

8.4.5. If $\mathbf{Q}_{k \times k}$ $(1 \leq k < n)$ is a principal submatrix of $\mathbf{P}$, then there is a permutation matrix $\mathbf{H}$ such that $\mathbf{H}^T\mathbf{P}\mathbf{H} = \begin{pmatrix} \mathbf{Q} & \mathbf{X} \\ \mathbf{Y} & \mathbf{Z} \end{pmatrix} = \widetilde{\mathbf{P}}$. If $\mathbf{B} = \begin{pmatrix} \mathbf{Q} & \mathbf{0} \\ \mathbf{0} & \mathbf{0} \end{pmatrix}$, then $\mathbf{B} \leq \widetilde{\mathbf{P}}$, and we know from Wielandt's theorem (p. 675) that $\rho(\mathbf{B}) \leq \rho\left(\widetilde{\mathbf{P}}\right) = 1$, and if $\rho(\mathbf{B}) = \rho\left(\widetilde{\mathbf{P}}\right) = 1$, then there is a number ϕ and a nonsingular diagonal matrix $\mathbf{D}$ such that $\mathbf{B} = e^{i\phi}\mathbf{D}\widetilde{\mathbf{P}}\mathbf{D}^{-1}$ or, equivalently, $\widetilde{\mathbf{P}} = e^{-i\phi}\mathbf{D}\mathbf{B}\mathbf{D}^{-1}$. But this implies that $\mathbf{X} = \mathbf{0}$, $\mathbf{Y} = \mathbf{0}$, and $\mathbf{Z} = \mathbf{0}$, which is impossible because $\mathbf{P}$ is irreducible. Therefore, $\rho(\mathbf{B}) < 1$, and thus $\rho(\mathbf{Q}) < 1$.

8.4.6. In order for $\mathbf{I} - \mathbf{Q}$ to be an M-matrix, it must be the case that $[\mathbf{I} - \mathbf{Q}]_{ij} \leq 0$ for $i \neq j$, and $\mathbf{I} - \mathbf{Q}$ must be nonsingular with $(\mathbf{I} - \mathbf{Q})^{-1} \geq \mathbf{0}$. It's clear that $[\mathbf{I} - \mathbf{Q}]_{ij} \leq 0$ because $0 \leq q_{ij} \leq 1$. Exercise 8.4.5 says that $\rho(\mathbf{Q}) < 1$, so the Neumann series expansion (p. 618) insures that $\mathbf{I} - \mathbf{Q}$ is nonsingular and $(\mathbf{I} - \mathbf{Q})^{-1} = \sum_{j=1}^{\infty} \mathbf{Q}^j \geq \mathbf{0}$. Thus $\mathbf{I} - \mathbf{Q}$ is an M-matrix.

8.4.7. We know from Exercise 8.4.6 that every principal submatrix of order $1 \leq k < n$ is an M-matrix, and M-matrices have positive determinants by (7.10.28) on p. 626.

8.4.8. You can consider an absorbing chain with eight states

$$\{(1,1,1),(1,1,0),(1,0,1),(0,1,1),(1,0,0),(0,1,0),(0,0,1),(0,0,0)\}$$

similar to what was described in Example 8.4.5, or you can use a four-state chain in which the states are defined to be the *number* of controls that hold at each activation of the system. Using the eight-state chain yields the following mean-time-to-failure vector.

$$
\begin{matrix}
(1,1,1) \\
(1,1,0) \\
(1,0,1) \\
(0,1,1) \\
(1,0,0) \\
(0,1,0) \\
(0,0,1)
\end{matrix}
\begin{pmatrix}
368.4 \\
366.6 \\
366.6 \\
366.6 \\
361.3 \\
361.3 \\
361.3
\end{pmatrix} = (\mathbf{I} - \mathbf{T}_{11})^{-1}\mathbf{e}.
$$

8.4.9. This is a Markov chain with nine states (c, m) in which c is the chamber occupied by the cat, and m is the chamber occupied by the mouse. There are three absorbing states—namely $(1,1), (2,2), (3,3)$. The transition matrix is

$$
\mathbf{P} = \frac{1}{72}
\begin{matrix}
(1,2) \\
(1,3) \\
(2,1) \\
(2,3) \\
(3,1) \\
(3,2) \\
(1,1) \\
(2,2) \\
(3,3)
\end{matrix}
\begin{pmatrix}
18 & 12 & 3 & 6 & 3 & 9 & 6 & 9 & 6 \\
12 & 18 & 3 & 9 & 3 & 6 & 6 & 6 & 9 \\
3 & 3 & 18 & 9 & 12 & 6 & 6 & 9 & 6 \\
4 & 6 & 6 & 18 & 4 & 8 & 2 & 12 & 12 \\
3 & 3 & 12 & 6 & 18 & 9 & 6 & 6 & 9 \\
6 & 4 & 4 & 8 & 6 & 18 & 2 & 12 & 12 \\
0 & 0 & 0 & 0 & 0 & 0 & 72 & 0 & 0 \\
0 & 0 & 0 & 0 & 0 & 0 & 0 & 72 & 0 \\
0 & 0 & 0 & 0 & 0 & 0 & 0 & 0 & 72
\end{pmatrix}
$$

with column headers $(1,2)\ (1,3)\ (2,1)\ (2,3)\ (3,1)\ (3,2)\ (1,1)\ (2,2)\ (3,3)$.

The expected number of steps until absorption and absorption probabilities are

$$
(\mathbf{I} - \mathbf{T}_{11})^{-1}\mathbf{e} =
\begin{matrix}
(1,2) \\
(1,3) \\
(2,1) \\
(2,3) \\
(3,1) \\
(3,2)
\end{matrix}
\begin{pmatrix}
3.24 \\
3.24 \\
3.24 \\
2.97 \\
3.24 \\
2.97
\end{pmatrix}
\quad \text{and} \quad
(\mathbf{I} - \mathbf{T}_{11})^{-1}\mathbf{T}_{12} =
\begin{matrix}
(1,1) & (2,2) & (3,3) \\
\end{matrix}
\begin{pmatrix}
0.226 & 0.41 & 0.364 \\
0.226 & 0.364 & 0.41 \\
0.226 & 0.41 & 0.364 \\
0.142 & 0.429 & 0.429 \\
0.226 & 0.364 & 0.41 \\
0.142 & 0.429 & 0.429
\end{pmatrix}
$$